Monotone Flows and Rapid Convergence for Nonlinear Partial Differential Equations

SERIES IN MATHEMATICAL ANALYSIS AND APPLICATIONS

Series in Mathematical Analysis and Applications (SIMAA) is edited by Ravi P. Agarwal, Florida Institute of Technology, USA and Donal O'Regan, National University of Ireland, Galway, Ireland.

The series is aimed at reporting on new developments in mathematical analysis and applications of a high standard and of current interest. Each volume in the series is devoted to a topic in analysis that has been applied, or is potentially applicable, to the solutions of scientific, engineering and social problems.

Volume 1
Method of Variation of Parameters for Dynamic Systems
V. Lakshmikantham and S.G. Deo

Volume 2
Integral and Integrodifferential Equations: Theory, Methods and Applications
edited by Ravi P. Agarwal and Donal O'Regan

Volume 3
Theorems of Leray-Schauder Type and Applications
Donal O'Regan and Radu Precup

Volume 4
Set Valued Mappings with Applications in Nonlinear Analysis
edited by Ravi P. Agarwal and Donal O'Regan

Volume 5
Oscillation Theory for Second Order Dynamic Equations
Ravi P. Agarwal, Said R. Grace and Donal O'Regan

Volume 6
Theory of Fuzzy Differential Equations and Inclusions
V. Lakshmikantham and R.N. Mohapatra

Volume 7
Monotone Flows and Rapid Convergence for Nonlinear Partial Differential Equations
V. Lakshmikantham and S. Köksal

Monotone Flows and Rapid Convergence for Nonlinear Partial Differential Equations

V. Lakshmikantham

and

S. Köksal

CRC Press
Taylor & Francis Group
Boca Raton London New York

CRC Press is an imprint of the
Taylor & Francis Group, an **informa** business

CRC Press
Taylor & Francis Group
6000 Broken Sound Parkway NW, Suite 300
Boca Raton, FL 33487-2742

First issued in paperback 2019

© 2003 by V. Lakshmikantham and S. Köksal
CRC Press is an imprint of Taylor & Francis Group, an Informa business

No claim to original U.S. Government works

ISBN-13: 978-0-415-30528-0 (hbk)
ISBN-13: 978-0-367-39540-7 (pbk)

British Library Cataloguing in Publication Data
A catalogue record for this book is available from the British Library

Library of Congress Cataloging in Publication Data
A catalog record for this book has been requested

**Visit the Taylor & Francis Web site at
http://www.taylorandfrancis.com**

**and the CRC Press Web site at
http://www.crcpress.com**

Contents

Preface

An interesting and fruitful technique for proving existence results for nonlinear problems is the method of lower and upper solutions. This method coupled with the monotone iterative technique manifests itself as an effective and flexible mechanism that offers theoretical as well as constructive existence results in a closed set, generated by the lower and upper solutions. The lower and upper solutions serve as rough bounds, which can be improved by monotone iterative procedures. Moreover, the iteration schemes can also be employed for the investigation of qualitative properties of solutions. The ideas embedded in these techniques have proved to be of immense value and have played a crucial role in unifying a wide variety of nonlinear problems.

Another fruitful idea of Chaplygin is to obtain approximate solutions of nonlinear problems which are not only monotone but also converge rapidly to the solution. Here strict lower and upper solutions and the assumption of convexity are used for nonlinear initial value problems (IVPs). The method of quasilinearization developed by Bellman and Kalaba, on the other hand, uses the convexity assumption and provides a lower bounding monotone sequence that converges to the assumed unique solution once the initial approximation is chosen in an adroit fashion. If we utilize the technique of lower and upper solutions combined with the method of quasilinearization and employ the idea of Newton and Fourier, it is possible to construct concurrently lower and upper bounding monotone sequences whose elements are the solutions of the corresponding linear problems. Of course, both sequences converge rapidly to the solution. Furthermore, this unification provides a framework to enlarge the class of nonlinear problems considerably to which the method is applicable. For example, it is not necessary to impose the usual convexity assumption on the nonlinear function involved, since one can allow much weaker assumptions. In fact, several possibilities can be investigated with this unified methodology and consequently this technique is known as generalized quasilinearization. Moreover, these ideas are extended, refined and generalized to various other types of nonlinear problems.

In this monograph, we extend the foregoing group of ideas to partial differential equations and provide a unified approach for studying elliptic, parabolic and hyperbolic equations. The monograph is divided, for convenience, into two parts: the first part describes the general methodology systematically utilizing the classical approach and the second part exhibits the development of the same basic ideas via variational techniques. In each case, this methodology is applied to elliptic, parabolic and hyperbolic equations so that one can understand and appreciate the intricacies involved in the various extensions.

Some of the important features of the monograph are as follows: It is the first monograph that

- attempts to describe both the monotone iterative technique and generalized quasilinearization in one unified way,

- incorporates the fundamental ideas of monotone flows and rapid convergence via variational techniques,

- exhibits the general methodology through the classical method and the variational approach,

- introduces the combined methodology which is growing rapidly because of its applicability to various real world problems.

The unified approach that is employed in the book covers several known and new results. We have only indicated, in the remarks, all the possible special cases for some results and did not even mention in other situations to avoid monotonous repetition.

We wish to express our immense thanks to Mrs. Donn Miller-Kermani for her excellent and painstaking typing of the manuscript. We would like to express our appreciation to Ms. Janie Wardle and the copy editor of Taylor and Francis for all the help and cooperation in this project.

The second author would like to thank her parents, Nedret and Tekin, her husband, Steve, and her son, Denis, for their endless support and faith in her.

V. Lakshmikantham and S. Köksal

Part A

Chapter 1

Elliptic Equations

1.1 Introduction

This chapter introduces the theory of lower and upper solutions coupled with the monotone iterative technique and the method of generalized quasilinearization for elliptic boundary value problems (BVPs), and therefore forms the basis for the rest of the book, where the essential ideas are extended to parabolic and hyperbolic initial boundary value problems (IBVPs), by classical as well as variational approaches.

Section 1.2 begins by analyzing the different situations that occur when the nonlinear term in semilinear elliptic BVPs is monotone increasing or decreasing. This analysis leads to the consideration of lower and upper solutions that are coupled in order to reach a common goal of obtaining monotone sequences of the same type. We then discuss that if we desire the constructed monotone sequences to converge rapidly in order to be more useful, we need to profitably utilize the method of quasilinearization, which not only offers monotone sequences but also assures rapid convergence. Of course, we have to pay a price for this, namely, the nonlinear term needs to be convex. After stating some typical known results in the two methodologies we shall provide a unified framework that facilitates a variety of characterizations, extensions and generalizations. Section 1.3 is devoted to proving general results relative to the monotone iterative technique under various types of coupled lower and upper solutions so that existing results are covered, and new ones are generated. Necessary comparison results which are proved are also useful in the development of the method of generalized quasilinearization in Section 1.4. Some of the possible special cases which are included in the unified framework are indicated in each case. Section

1.5 considers the weakly coupled system of semilinear elliptic BVPs when
the nonlinear functions involved are of mixed monotone type. Instead of
dealing with mixed monotone systems directly, we shall first prove a result for monotone increasing systems and then derive the results for mixed
monotone systems as a simple consequence. This will be done by employing
the theory of reflection operators and expanding the given system suitably
to transform it into a monotone increasing system. Section 1.6 is dedicated
to the investigation of existence results in the sector generated by lower and
upper solutions for a weakly coupled system of BVPs in unbounded domains,
while Section 1.7 extends the monotone iterative technique for the systems
considered in Section 1.6.

1.2   Monotone Iterates: A Preview

The well-known method of lower and upper solutions, coupled with the
monotone iterative technique, provides an effective and flexible mechanism
that offers theoretical as well as constructive existence results for nonlinear
problems in a closed set which is generated by the lower and upper solutions. The lower and upper solutions serve as bounds for solutions which
are improved by a monotone iterative process. The ideas embedded in this
technique have proved to be of immense value and have played an important
role in unifying a variety of nonlinear problems.

Let $\Omega \subset R^n$ be a bounded domain with boundary $\partial\Omega$. We consider
the following semilinear elliptic boundary value problem (BVP) in nondivergence form

$$\left[\begin{array}{l} Lu = -\sum_{i,j=1}^{n} a_{ij}(x)u_{x_i x_j} + \sum_{i=1}^{n} b_i(x)u_{x_i} + c(x)u \\ \quad = F(x,u) \text{ in } \Omega, \\ Bu = \phi \text{ on } \partial\Omega, \end{array} \right. \tag{1.2.1}$$

where we assume that $a_{ij}, b_i, c \in C^\alpha[\bar\Omega, R]$, $c(x) \geq 0$ in Ω, $\phi \in C^{1,\alpha}[\bar\Omega, R]$,
$F \in C^\alpha[\bar\Omega \times R, R]$ and the ellipticity condition

$$\sum_{i,j=1}^{n} a_{ij}(x)\xi_i\xi_j \geq \theta \mid \xi \mid^2, \tag{1.2.2}$$

holds in Ω with $\theta > 0$. Moreover, we let $p, q \in C^{1,\alpha}[\partial\Omega, R_+]$ with $p(x) > 0$,
γ be the unit outer normal on $\partial\Omega$, and

$$Bu = p(x)u + q(x)\frac{\partial u}{\partial\gamma}, \ u \in C^1[\bar\Omega, R]. \tag{1.2.3}$$

A vector γ is said to be an outer normal at $x \in \partial\Omega$ if $x - h\gamma \in \Omega$ for small $h > 0$. The outer normal derivative is then given by

$$\frac{du}{d\gamma} = \lim_{h \to 0} \frac{u(x) - u(x - h\gamma)}{h}.$$

We assume that an outer normal exists and the functions in question have outer normal derivatives on $\partial\Omega$. We also assume that $\partial\Omega$ belongs to $C^{2,\alpha}$.

A well-known result in the monotone iterative technique is the following theorem relative to the BVPs (1.2.1).

Theorem 1.2.1 *Assume that*

(A_1) $\alpha_0, \beta_0 \in C^2[\bar{\Omega}, R]$ *with* $\alpha_0(x) \le \beta_0(x)$ *in* Ω *satisfy*

$$\left[\begin{array}{ll} L\alpha_0 \le F(x, \alpha_0) \text{ in } \Omega, & B\alpha_0 \le \phi \text{ on } \partial\Omega, \\ L\beta_0 \ge F(x, \beta_0) \text{ in } \Omega, & B\beta_0 \ge \phi \text{ on } \partial\Omega; \end{array} \right. \quad (1.2.4)$$

(A_2) $F \in C^\alpha[\bar{\Omega} \times R, R]$ *and for some* $M > 0$, *and* $F(x, u) + Mu$ *is nondecreasing in* u *for* $x \in \bar{\Omega}$.

Then there exist monotone sequences $\{\alpha_n(x)\}$, $\{\beta_n(x)\} \in C^{2,\alpha}[\bar{\Omega}, R]$ *such that* $\alpha_n(x) \to \rho(x)$, $\beta_n(x) \to r(x)$ *in* $C^2[\bar{\Omega}, R]$ *and* ρ, r *are the minimal and maximal solutions of* (1.2.1) *respectively.*

If $F(x, u)$ satisfies a one-sided Lipschitz condition, namely,

$$F(x, u_1) - F(x, u_2) \ge -M(u_1 - u_2), \quad (1.2.5)$$

where

$$u_1 \ge u_2, \ M \ge 0, \ x \in \Omega,$$

then $F(x, u) + Mu$ is nondecreasing in u for $x \in \Omega$.

The functions α_0, β_0 satisfying (1.2.4) are known as lower and upper solutions of (1.2.1). If, for any solution u of (1.2.1) existing in the sector $[\alpha_0, \beta_0] = [u \in R : \alpha_0(x) \le u \le \beta_0(x), x \in \bar{\Omega}]$, we have $\rho(x) \le u(x) \le r(x)$ in $\bar{\Omega}$, then (ρ, r) are said to be minimal and maximal solutions of (1.2.1) or equivalently extremal solutions of (1.2.1) relative to the sector $[\alpha_0, \beta_0]$.

The special case when $F(x, u)$ is nondecreasing in u is covered in Theorem 1.2.1 when $M = 0$. However, the other case, when $F(x, u)$ is nonincreasing in u is not included in Theorem 1.2.1 and is of special interest. Under somewhat special conditions, one can prove that when $F(x, u)$ is nonincreasing in u, a single iteration procedure yields an alternative sequence which forms two monotone sequences bounding the solutions of (1.2.1). The iteration scheme in the present case is simply either

$$L\alpha_{n+1} = F(x, \alpha_n) \text{ in } \Omega, \ B\alpha_{n+1} = \phi \text{ on } \partial\Omega \qquad (1.2.6)$$

or

$$L\beta_{n+1} = F(x, \beta_n) \text{ in } \Omega, B\beta_{n+1} = \phi \text{ on } \partial\Omega, \qquad (1.2.7)$$

$n = 0, 1, 2, \dots$. In this case, the following result is valid.

Theorem 1.2.2 *Suppose that $F(x, u)$ is nonincreasing in u for $x \in \Omega$ and $F \in C^\alpha[\bar{\Omega} \times R, R]$. Then*

A_1 *the iterates $\{\alpha_n(x)\}$ satisfy the relation*

$$\alpha_0 \le \alpha_2 \le \alpha_4 \le \dots \le \alpha_{2n} \le \alpha_{2n+1} \le \dots \le \alpha_3 \le \alpha_1 \text{ on } \bar{\Omega},$$

provided $\alpha_0 \le \alpha_2$ on $\bar{\Omega}$. Moreover, the alternating sequences $\{\alpha_{2n}(x)\}$, $\{\alpha_{2n+1}(x)\}$ converge in $C^2[\bar{\Omega}, R]$ to ρ, r, respectively; and

A_2 *the iterates $\{\beta_n(x)\}$ satisfy*

$$\beta_1 \le \beta_3 \le \dots \le \beta_{2n+1} \le \beta_{2n} \le \dots \le \beta_2 \le \beta_0 \text{ on } \bar{\Omega},$$

if we suppose that $\beta_2 \le \beta_0$ on $\bar{\Omega}$. Furthermore, the alternating sequences $\{\beta_{2n+1}(x)\}$, $\{\beta_{2n}(x)\}$ converge in $C^2[\bar{\Omega}, R]$ to r^*, ρ^*, respectively.

Note that we did not assume condition (A_1) of Theorem 1.2.1. In fact, one can show that lower and upper solutions exist satisfying (A_1) in this case. We state this fact as a lemma.

Lemma 1.2.1 *Suppose that $F(x, u)$ is monotone nonincreasing in u for each $x \in \bar{\Omega}$. Then there exist lower and upper solutions α_0, β_0 of (1.2.1) satisfying $\alpha_0(x) \le \beta_0(x)$ in $\bar{\Omega}$.*

Proof   Let $\alpha_0(x) = z(x) - R_0$, $\beta_0(x) = z(x) + R_0$, where $z(x)$ is the solution of

$$Lz(x) = F(x, 0) \text{ in } \Omega, Bz(x) = \phi \text{ on } \ \partial\Omega.$$

Choose $R_0 > 0$ sufficiently large so that

$$\alpha_0(x) \le 0 \le \beta_0(x) \text{ in } \bar{\Omega}.$$

Since $F(x, u)$ is nonincreasing in u, we get

$$L\alpha_0 = Lz = F(x, 0) \leq F(x, \alpha_0) \text{ in } \Omega,$$

$$B\alpha_0 = p\alpha_0 + q\frac{d\alpha_0}{d\gamma} = Bz - pR_0 \leq \phi \text{ on } \partial\Omega,$$

and similarly,

$$L\beta_0 = Lz = F(x, 0) \geq F(x, \beta_0) \text{ in } \Omega,$$

$$B\beta_0 = p\beta_0 + q\frac{d\beta_0}{d\gamma} = Bz + pR_0 \geq \phi \text{ on } \partial\Omega.$$

A natural question that arises is whether it is possible to obtain the monotone sequences $\{\alpha_n(x)\}$, $\{\beta_n(x)\}$, when $F(x, u)$ is nonincreasing in u without the additional assumptions on the iterates, namely $\alpha_0 \leq \alpha_2$ and $\beta_2 \leq \beta_0$ on $\bar{\Omega}$. The answer is positive if we assume the existence of coupled lower and upper solutions. In fact, one can prove the following result in this direction.

Theorem 1.2.3 *Assume that*

(A_1) $\alpha_0, \beta_0 \in C^2[\bar{\Omega}, R]$ *with* $\alpha_0(x) \leq \beta_0(x)$ *in* Ω *and*

$$\begin{aligned} L\alpha_0 &\leq F(x, \beta_0) \text{ in } \Omega, \quad B\alpha_0 \leq \phi \text{ on } \partial\Omega, \\ L\beta_0 &\geq F(x, \alpha_0) \text{ in } \Omega, \quad B\beta_0 \geq \phi \text{ on } \partial\Omega; \end{aligned} \qquad (1.2.8)$$

(A_2) $F \in C^\alpha[\bar{\Omega} \times R, R]$ *and* $F(x, u)$ *is nonincreasing in* u *for* $x \in \Omega$.

Then the conclusion of Theorem 1.2.1 is valid.

The lower and upper solutions defined in (1.2.8) are known as coupled lower and upper solutions of (1.2.1).

The monotone iterates resulting in the foregoing theorems that converge monotonically in $C^2[\bar{\Omega}, R]$ to solutions of (1.2.1) may not converge rapidly enough. Hence the numerical procedures for approximating the solutions which are close to the solution of the original problem may require a large number of computations. If, on the other hand, the monotone iterates constructed converge rapidly enough, for example quadratically, it would be more effective from all aspects. The approach that we describe next takes care of this situation which is popularly known as the method of quasilinearization.

The main idea of the method of quasilinearization is to provide an explicit analytic representation for the solutions of nonlinear differential equations, which yields pointwise lower estimates for the solution whenever the nonlinear function involved is convex. The most important application of this popular method has been to obtain a sequence of lower bounds which are the solutions of linear differential equations that converge quadratically to the unique solution of the given nonlinear problem. When we employ the technique of lower and upper solutions coupled with the method of quasilinearization and utilize the idea of Newton and Fourier, it is possible to construct concurrently lower and upper bounding monotone sequences, which converge quadratically to the solution of the given problem. Moreover, this unification provides a mechanism to enlarge the class of nonlinear problems to which this method is applicable and therefore this unified technique is called the method generalized quasilinearization.

A simple result using this approach may be stated as follows.

Theorem 1.2.4 *Assume that*

(B_1) *condition (A_1) of Theorem 1.2.1 holds;*

(B_2) $F \in C^{\alpha}[\bar{R} \times R, R]$, $F_u(x, u)$, $F_{uu}(x, u)$ *exist and are continuous and* $F_{uu}(x, u) \geq 0$ *on* $\bar{\Omega} \times R$;

(B_3) $0 < N \leq c(x) - F_u(x, \beta_0)$ *in* $\bar{\Omega}$.

Then there exist monotone sequences $\{\alpha_n(x)\}$, $\{\beta_n(x)\} \in C^{2\alpha}[\bar{\Omega}, R]$ *such that* $\alpha_n \to \rho$, $\beta_n \to r$ *in* $C^2[\bar{\Omega}, R]$, $\rho = r = u$ *is the unique solution of* (1.2.1) *satisfying* $\alpha_0(x) \leq u(x) \leq \beta_0(x)$ *in* Ω *and the convergence is quadratic.*

Here we construct the sequences of iterates as follows:

$$\left[\begin{array}{ll} L\alpha_{n+1} = F(x, \alpha_n) + F_u(x, \alpha_n)(\alpha_{n+1} - \alpha_n) \text{ in } \Omega, & B\alpha_{n+1} = \phi \text{ on } \partial\Omega, \\ L\beta_{n+1} = F(x, \beta_n) + F_u(x, \alpha_n)(\beta_{n+1} - \beta_n) \text{ in } \Omega, & B\beta_{n+1} = \phi \text{ on } \partial\Omega, \end{array} \right.$$
$$(1.2.9)$$

$n = 0, 1, 2, \ldots$. We note the special choice of the iterates β_n.

A dual result when $F(x, u)$ is concave is also true, which we state below.

Theorem 1.2.5 *Assume that*

(B_1) *condition (A_1) of Theorem 1.2.1 holds.*

(B_2) $F \in C^{\alpha}[\bar{\Omega} \times R, R]$, $F_u(x, u)$, $F_{uu}(x, u)$ *exist and are continuous and* $F_{uu}(x, u) \leq 0$ *on* $\bar{\Omega} \times R$;

(B_3) $0 < N \le c(x) - F_u(x, \alpha_0)$ *in* $\bar{\Omega}$.

Then the conclusion of Theorem 1.2.4 is valid.

In this case, the iterates are constructed with a special choice for α_n as follows:

$$L\alpha_{n+1} = f(x, \alpha_n) + f_u(x, \beta_n)(\alpha_{n+1} - \alpha_n) \text{ in } \Omega, \ B\alpha_{n+1} = \phi \text{ on } \partial\Omega,$$

$$L\beta_{n+1} = f(x, \beta_n) + f_u(x, \beta_n)(\beta_{n+1} - \beta_n) \text{ in } \Omega, \ B\beta_{n+1} = \phi \text{ on } \partial\Omega$$

for $n = 0, 1, 2, \dots$.

In the following sections, we shall describe the situation when $F(x, u)$ admits a splitting of a difference of two monotone functions or equivalently,

$$F(x, u) = f(x, u) + g(x, u)$$

where $f(x, u)$ is monotone nondecreasing and $g(x, u)$ is monotone nonincreasing in u for $x \in \Omega$. We shall also consider the case when $f(x, u)$ is convex and $g(x, u)$ is concave in u for $x \in \Omega$. We shall see that this simple setting unifies and covers several known results as well as providing some interesting new results relative to the monotone iterative technique and generalized quasilinearization.

1.3 Monotone Iterative Technique

We shall devote this section to proving general results relative to the monotone iterative technique which contain as special cases, several important results of interest. Before we proceed further we need the following comparison theorem and a corollary which are useful in our development of the results.

Theorem 1.3.1 *Let* $\alpha, \beta \in C^2[\bar{\Omega}, R]$ *be lower and upper solutions of* (1.2.1); *namely,* α, β *satisfy*

$$L\alpha \le F(x, \alpha) \text{ in } \Omega, \quad B\alpha \le \varphi \text{ on } \partial\Omega,$$
$$L\beta \ge F(x, \beta) \text{ in } \Omega, \quad B\beta \ge \varphi \text{ on } \partial\Omega.$$

Suppose further that

$$F(x, u_1) - F(x, u_2) \le k(x)(u_1 - u_2), \tag{1.3.1}$$

for $u_1 \ge u_2$, $k \in C^\alpha[\bar{\Omega}, R_+]$, L *and* $c(x) - k(x) > 0$ *in* Ω.
  Then $\alpha(x) \le \beta(x)$ *in* $\bar{\Omega}$.

Proof   Set $m(x) = \alpha(x) - \beta(x)$. If $m(x) \leq 0$ in Ω is not true, then there exist an $\epsilon > 0$ and $x_0 \in \bar{\Omega}$ such that

$$\alpha(x_0) = \beta(x_0) + \epsilon \text{ and } \alpha(x) \leq \beta(x) + \epsilon, \ x \in \bar{\Omega}.$$

If $x_0 \in \partial\Omega$, then $d\alpha(x_0)/d\gamma \geq d\beta(x_0)/d\gamma$ and hence, using the fact that $p(x_0) > 0$, we get

$$\begin{aligned}
B\alpha(x_0) &= p(x_0)\alpha(x_0) + q(x_0)\frac{d\alpha(x_0)}{d\gamma} \\
&\geq p(x_0)[\beta(x_0) + \epsilon] + q(x_0)\frac{d\beta(x_0)}{d\gamma} \\
&> B\beta(x_0),
\end{aligned}$$

which is a contradiction.

If $x_0 \in \Omega$, then $\alpha_{x_i}(x_0) = \beta_{x_i}(x_0)$ and $\sum_{i,j=1}^{n}(\alpha_{x_i x_j}(x_0) - \beta_{x_i x_j}(x_0))\lambda_i\lambda_j \leq 0$. It then follows that using (1.3.1),

$$\begin{aligned}
F(x_0, \alpha(x_0)) &\geq L\alpha(x_0) \\
&\geq L(\beta(x_0) + \epsilon) \\
&\geq F(x_0, \beta(x_0)) + c(x_0)\epsilon \\
&\geq F(x_0, \alpha(x_0)) + [c(x_0) - k(x_0)]\epsilon.
\end{aligned}$$

Since $c(x) - k(x) > 0$, we have a contradiction. Hence the claim is true and the proof is complete.

Corollary 1.3.1 *For any $p \in C^2[\bar{\Omega}, R]$ satisfying*

$$\begin{aligned}
Lp &= -\sum_{i,j=1}^{n} a_{i,j}(x)p_{x_i x_j} + \sum_{i=1}^{n} b_i(x)p_{x_i} + c_0(x)p \qquad (1.3.2) \\
&\leq 0 \text{ in } \Omega, \\
Bp &\leq 0 \text{ on } \partial\Omega, \text{ where } c_0(x) > 0, \qquad (1.3.3)
\end{aligned}$$

we have $p(x) \leq 0$ in $\bar{\Omega}$.

Let us now consider the following elliptic boundary value problems in nondivergence form

$$Lu = f(x,u) + g(x,u) \text{ in } \Omega, \ Bu = \phi \text{ on } \partial\Omega, \qquad (1.3.4)$$

where $f, g \in C^\alpha[\bar{\Omega} \times R, R]$.

Definition 1.3.1 *Relative to the BVP (1.3.4) the functions $\alpha, \beta \in C^2[\bar{\Omega}, R]$ are said to be*

(a) *natural lower and upper solutions if*

$$\left[\begin{array}{ll} L\alpha \leq f(x, \alpha) + g(x, \alpha) \text{ in } \Omega, & B\alpha \leq \phi \text{ on } \partial\Omega, \\ L\beta \geq f(x, \beta) + g(x, \beta) \text{ in } \Omega, & B\beta \geq \phi \text{ on } \partial\Omega; \end{array} \right. \qquad (1.3.5)$$

(b) *coupled lower and upper solutions of type I if*

$$\left[\begin{array}{ll} L\alpha \leq f(x, \alpha) + g(x, \beta) \text{ in } \Omega, & B\alpha \leq \phi \text{ on } \partial\Omega, \\ L\beta \geq f(x, \beta) + g(x, \alpha) \text{ in } \Omega, & B\beta \geq \phi \text{ on } \partial\Omega; \end{array} \right. \qquad (1.3.6)$$

(c) *coupled lower and upper solutions of type II if*

$$\left[\begin{array}{ll} L\alpha \leq f(x, \beta) + g(x, \alpha) \text{ in } \Omega, & B\alpha \leq \phi \text{ on } \partial\Omega, \\ L\beta \geq f(x, \alpha) + g(x, \beta) \text{ in } \Omega, & B\beta \geq \phi \text{ on } \partial\Omega; \end{array} \right. \qquad (1.3.7)$$

(d) *coupled lower and upper solutions of type III if*

$$\left[\begin{array}{ll} L\alpha \leq f(x, \beta) + g(x, \beta) \text{ in } \Omega, & B\alpha \leq \phi \text{ on } \partial\Omega, \\ L\beta \geq f(x, \alpha) + g(x, \alpha) \text{ in } \Omega, & B\beta \geq \phi \text{ on } \partial\Omega. \end{array} \right. \qquad (1.3.8)$$

Whenever $\alpha \leq \beta$ in Ω, we note that the lower and upper solutions defined in (1.3.5) and (1.3.8) also satisfy (1.3.7) and hence it is enough to consider the cases (1.3.6) and (1.3.7), which is precisely what we plan to do.

We are now in a position to prove the first main result.

Theorem 1.3.2 *Assume that*

(A_1) *$\alpha_0, \beta_0 \in C^2[\bar{\Omega}, R]$ are the coupled lower and upper solutions of type I with $\alpha_0(x) \leq \beta_0(x)$ in Ω;*

(A_2) *$f, g \in C^\alpha[\bar{\Omega} \times R, R]$, $f(x, u)$ is nondecreasing in u and $g(x, u)$ is non-increasing in u for $x \in \Omega$.*

Then there exist monotone sequences $\{\alpha_n(x)\}$, $\{\beta_n(x)\} \in C^{2,\alpha}[\bar{\Omega}, R]$ such that $\alpha_n(x) \to \rho(x)$, $\beta_n(x) \to r(x)$ in $C^2[\bar{\Omega}, R]$ and (ρ, r) are the coupled minimal and maximal solutions of (1.3.4) respectively, that is, (ρ, r) satisfy

$$L\rho = f(x, \rho) + g(x, r) \text{ in } \Omega, \ B\rho = \phi \text{ on } \partial\Omega,$$

$$Lr = f(x, r) + g(x, \rho) \text{ in } \Omega, \ Br = \phi \text{ on } \partial\Omega.$$

Proof   Consider the following linear BVPs for each $n = 1, 2, \ldots$,

$$L\alpha_{n+1} = f(x, \alpha_n) + g(x, \beta_n) \text{ in } \Omega, \ B\alpha_{n+1} = \phi \text{ on } \partial\Omega, \qquad (1.3.9)$$

and

$$L\beta_{n+1} = f(x, \beta_n) + g(x, \alpha_n) \text{ in } \Omega, \ B\beta_{n+1} = \phi \text{ on } \partial\Omega. \qquad (1.3.10)$$

In order to conclude the existence of the unique solutions of the BVPs (1.3.9) and (1.3.10) for each $n \geq 1$, we need to show that for any $\eta, \xi \in C^2[\bar{\Omega}, R]$ with $\alpha_0 \leq \eta \leq \xi \leq \beta_0$, $h_1(x) \in C^\alpha[\bar{\Omega}, R]$ and $h_2(x) \in C^\alpha[\bar{\Omega}, R]$, where

$$h_1(x) = f(x, \eta) + g(x, \xi),$$

and

$$h_2(x) = f(x, \xi) + g(x, \eta).$$

We note that if $\eta, \xi \in C^2[\bar{\Omega}, R]$, then $\eta, \xi \in W^{2,\alpha}[\bar{\Omega}, R]$ in view of the boundedness of Ω and $\partial\Omega \in C^{2,\alpha}[\bar{\Omega}, R]$. The Embedding Theorem A.3.4 then shows that $\eta, \xi \in C^{1,\alpha}[\bar{\Omega}, R]$. Consequently, we have

$$
\begin{aligned}
| \, f(x, \eta(x)) - f(y, \eta(y)) \, | \qquad\qquad\qquad\qquad\qquad & \\
\leq \quad & K_1[\| \, x - y \, \|^\alpha + | \, \eta(x) - \eta(y) \, |^\alpha] \\
\leq \quad & K_1[\| \, x - y \, \|^\alpha + | \, \eta \, |^\alpha_{C^1[\bar{\Omega},R]} \| \, x - y \, \|^\alpha] \\
\leq \quad & L_1 \, \| \, x - y \, \|^\alpha, \text{ where } L_1 = K_1[1 + | \, \eta \, |^\alpha_{C^1[\bar{\Omega},R]}].
\end{aligned}
$$

Similarly, we get

$$| \, g(x, \xi(x)) - g(y, \xi(y)) \, | \leq L_2 \, \| \, x - y \, \|^\alpha,$$

where

$$L_2 = K_2[1 + | \, \xi \, |^\alpha_{C^1[\bar{\Omega},R]}].$$

As a result, we find that, because of the definition of $h_1(x)$,

$$
\begin{aligned}
| \, h_1(x) - h_1(y) \, | \ &= \ | \, f(x, \eta(x)) + g(x, \xi(x)) - f(y, \eta(y)) - g(y, \xi(y)) \, | \\
&\leq \ | \, f(x, \eta(x)) - f(y, \eta(y)) \, | + | \, g(x, \xi(x)) - g(y, \xi(y)) \, | \\
&\leq \ C \, \| \, x - y \, \|^\alpha,
\end{aligned}
$$

where

$$C = L_1 + L_2.$$

Hence, $h_1(x) \in C^\alpha[\bar{\Omega}, R]$. In a very similar way, we can show that $h_2(x) \in C^\alpha[\bar{\Omega}, R]$. Consequently, there exist unique solutions $\alpha_n, \beta_n \in C^{2,\alpha}[\bar{\Omega}, R]$ of the BVPs (1.3.9) and (1.3.10) by Theorem A.3.1, provided that $\alpha_k, \beta_k \in [\alpha_0, \beta_0]$ with $\alpha_k \leq \beta_k$, successively, for $k \geq 1$.

Therefore, our aim now is to show that

$$\alpha_0 \leq \alpha_1 \leq \alpha_2 \leq \ldots \leq \alpha_k \leq \beta_k \leq \ldots \beta_2 \leq \beta_1 \leq \beta_0 \text{ in } \Omega. \qquad (1.3.11)$$

We first claim that $\alpha_1 \geq \alpha_0$ in $\bar{\Omega}$. For this, let $p = \alpha_0 - \alpha_1$ so that $Bp \leq 0$ on $\partial\Omega$ and

$$Lp = L\alpha_0 - L\alpha_1 \leq f(x, \alpha_0) + g(x, \beta_0) - f(x, \alpha_0) - g(x, \beta_0) = 0 \text{ in } \Omega.$$

Hence by Corollary 1.3.1, $p(x) \leq 0$ in $\bar{\Omega}$, which implies $\alpha_0 \leq \alpha_1$ in $\bar{\Omega}$. Similarly, we can show that $\beta_1 \leq \beta_0$ in $\bar{\Omega}$. We next prove that $\alpha_1 \leq \beta_1$ in $\bar{\Omega}$. Consider $p = \alpha_1 - \beta_1$ so that $Bp = 0$ on $\partial\Omega$ and

$$Lp = f(x, \alpha_0) + g(x, \beta_0) - f(x, \beta_0) - g(x, \alpha_0) \leq 0 \text{ in } \Omega,$$

using the monotone nature of f, g. Thus we get by Corollary 1.3.1, $p \leq 0$ in $\bar{\Omega}$ which yields $\alpha_1 \leq \beta_1$ in $\bar{\Omega}$. As a result, it follows that

$$\alpha_0 \leq \alpha_1 \leq \beta_1 \leq \beta_0 \text{ in } \bar{\Omega}.$$

Assume that for some $k > 1$, $\alpha_{k-1} \leq \alpha_k \leq \beta_k \leq \beta_{k-1}$ in $\bar{\Omega}$. Then we show that $\alpha_k \leq \alpha_{k+1} \leq \beta_{k+1} \leq \beta_k$ in $\bar{\Omega}$. To do this, let $p = \alpha_{k+1} - \alpha_k$ so that $Bp = 0$ on $\partial\Omega$ and because of the monotone character of f, g we get

$$Lp = f(x, \alpha_k) + g(x, \beta_k) - f(x, \alpha_{k-1}) - g(x, \beta_{k-1}) \leq 0 \text{ in } \Omega.$$

Corollary 1.3.1 then implies $\alpha_k \leq \alpha_{k+1}$ in $\bar{\Omega}$. Similarly, we can show that $\beta_{k+1} \leq \beta_k$ in $\bar{\Omega}$. Now to prove $\alpha_{k+1} \leq \beta_{k+1}$ in $\bar{\Omega}$, consider $\beta = \alpha_{k+1} - \beta_{k+1}$ and note that $Bp = 0$ on $\partial\Omega$. Moreover,

$$Lp = f(x, \alpha_k) + g(x, \beta_k) - f(x, \beta_k) - g(x, \alpha_k) \leq 0 \text{ in } \Omega,$$

using the assumption and monotone nature of f, g. Thus we have by Corollary 1.3.1, $\alpha_{k+1} \leq \beta_k$ in Ω and as a result, it follows that

$$\alpha_k \leq \alpha_{k+1} \leq \beta_{k+1} \leq \beta_k, \text{ for any } k > 1.$$

Hence by induction, we see that (1.3.11) is valid for all $k = 1, 2, \ldots$.

We recall that $\alpha_k, \beta_k \in C^{2,\alpha}[\bar{\Omega}, R]$ for $k = 1, 2, \ldots$. Since $C^{2,\alpha}[\bar{\Omega}, R] \subset W^{2,q}[\bar{\Omega}, R]$ for $q > 1$, by Theorem A.2.3, we have

$$\| \alpha_k \|_{W^{2,q}[\bar{\Omega}, R]} \leq C[\| h_k \|_{L^q[\bar{\Omega}, R]} + \| \phi \|_{W^{1,q}[\bar{\Omega}, R]}], \qquad (1.3.12)$$

where $h_k(x) = f(x, \alpha_{k-1}) + g(x, \beta_{k-1})$.

The continuity of h_k implies that $\{h_k(x)\}$ is uniformly bounded in $C[\bar{\Omega}, R]$. Since $C[\bar{\Omega}, R]$ is dense in $L^q[\bar{\Omega}, R]$, $\{h_k(x)\}$ is also uniformly bounded in $L^q[\bar{\Omega}, R]$. This together with (1.3.12) shows that $\{\alpha_k(x)\}$ is uniformly bounded in $W^{2,q}[\bar{\Omega}, R]$. For $q = \frac{2}{1-\alpha}$, $\alpha_k \in W^{2,q}[\bar{\Omega}, R]$ and hence by the Embedding Theorem A.3.4

$$\| \alpha_k \|_{C^{1,\alpha}[\bar{\Omega}, R]} \leq C \| \alpha_k \|_{W^{2,q}[\bar{\Omega}, R]}, \text{ for } k = 1, 2, \ldots$$

for some constant C independent of the elements of $W^{2,q}$. Thus $\{\alpha_k(x)\}$ is uniformly bounded in $C^{1,\alpha}[\bar{\Omega}, R]$. This implies that $\{h_k(x)\}$ is uniformly bounded in $C^{\alpha}[\bar{\Omega}, R]$. Consequently, by Schauder's estimate given in A.2.2, we find that

$$\| \alpha_k \|_{C^{2,\alpha}[\bar{\Omega}, R]} \leq C[\| h_k \|_{C^{\alpha}[\bar{\Omega}, R]} + \| \phi \|_{C^{1,\alpha}[\bar{\Omega}, R]}] \text{ for all } k,$$

which implies the uniform boundedness of $\{\alpha_k(x)\}$ in $C^{2,\alpha}[\bar{\Omega}, R]$. As a result, we have $\{\alpha_k(x)\}$ is relatively compact in $C^2[\bar{\Omega}, R]$ which yields the existence of a subsequence $\{\alpha_{k_j}\}$ which converges in $C^2[\bar{\Omega}, R]$. Let $\rho^* \in C^2[\bar{\Omega}, R]$ be the limit of $\alpha_{k_j}\{(x)\}$. By the monotone nature of $\{\alpha_k(x)\}$, it converges pointwise to $\rho(x)$ in $\bar{\Omega}$. But the convergence of $\{\alpha_{k_j}(x)\}$ in $C^2[\bar{\Omega}, R]$ implies pointwise convergence and thus $\rho^*(x) = \rho(x)$ in $\bar{\Omega}$. This shows that the entire sequence $\{\alpha_n(x)\}$ converges in $C^2[\bar{\Omega}, R]$ to $\rho(x)$, that is, $\lim_{k \to \infty} \alpha_k(x) = \rho(x)$ in $C^2[\bar{\Omega}, R]$ and $\alpha_0 \leq \rho \leq \beta_0$ in $\bar{\Omega}$.

Similar arguments show that $\lim_{k \to \infty} \beta_k(x) = r(x)$ in $C^2[\bar{\Omega}, R]$ and $\alpha_0 \leq \rho \leq r \leq \beta_0$ in $\bar{\Omega}$. Thus the limits

$$\lim_{k \to \infty} L\alpha_k = L\rho \quad \lim_{k \to \infty} L\beta_n = Lr,$$

$$\lim_{k \to \infty} B\alpha_k = B\rho \quad \lim_{k \to \infty} B\beta_k = Br,$$

$$\lim_{k \to \infty} \{f(x, \alpha_k) + g(x, \beta_k)\} = f(x, \rho) + g(x, r),$$

and

$$\lim_{n \to \infty} \{f(x, \beta_n) + g(x, \alpha_n)\} = f(x, r) + g(x, \rho),$$

exist uniformly in $\bar{\Omega}$. We then see immediately that ρ and r are the solutions of the semilinear BVPs

$$L\rho = f(x, \rho) + g(x, r) \text{ in } \Omega, \ B\rho = \phi \text{ on } \partial\Omega$$

$$Lr = f(x, r) + g(x, \rho) \text{ in } \Omega, \ Br = \phi \text{ on } \partial\Omega.$$

Finally, we claim that ρ and r are the coupled minimal and maximal solutions of (1.3.4), that is, if u is any solution of (1.3.4) such that $\alpha_0(x) \leq u(x) \leq \beta_0(x)$, then

$$\alpha_0(x) \leq \rho(x) \leq u(x) \leq r(x) \leq \beta_0(x) \text{ in } \bar{\Omega}.$$

Suppose that for some k, $\alpha_k \leq u \leq \beta_k$ in $\bar{\Omega}$. Setting $p = \alpha_{n+1} - u$; $Bp = 0$ on $\partial\Omega$, we obtain $p \leq 0$ by Corollary 1.3.1, implying $u \geq \alpha_{k+1}$ in $\bar{\Omega}$. Similarly, $u \leq \beta_{k+1}$ in $\bar{\Omega}$. Hence $\alpha_{k+1} \leq u \leq \beta_{k+1}$ in $\bar{\Omega}$. By induction, $\alpha_k \leq u \leq \beta_k$, in $\bar{\Omega}$ for all $k \geq 1$. Now, taking the limit as $k \to \infty$, we get $\rho \leq u \leq r$ in $\bar{\Omega}$. This completes the proof.

Corollary 1.3.2 *In addition to the assumptions of Theorem 1.3.2, if, for $u_1 \geq u_2$, f and g satisfy*

$$f(x, u_1) - f(x, u_2) \leq N_1(u_1 - u_2),$$

and

$$g(x, u_1) - g(x, u_2) \geq -N_2(u_1 - u_2),$$

where $N_1 > 0$, $N_2 > 0$, then $\rho = u = r$ is the unique solution of (1.3.4) provided $c(x) - (N_1 + N_2) > 0$ in Ω.

Proof   Since we have $\rho \leq r$, it is enough to show that $r \leq \rho$. Setting $p = r - \rho$, we have $Bp = 0$ on $\partial\Omega$, and

$$\begin{aligned} Lp &= f(x, r) + g(x, \rho) - f(x, \rho) - g(x, r) \\ &\leq N_1(r - \rho) + N_2(r - \rho) \text{ in } \Omega. \end{aligned}$$

Then

$$Lp - (N_1 + N_2)p \leq 0 \text{ in } \Omega,$$

which implies, by Corollary 1.3.1, $p \leq 0$ since $c(x) - (N_1 + N_2) > 0$. This proves that $\rho = r = u$ is the unique solution of (1.3.4).

Several remarks are now in order.

Remarks 1.3.1

(i) In Theorem 1.3.2, suppose that $g(x, u) \equiv 0$. Then Theorem 1.2.1 results with $M = 0$.

(ii) If $f(x, u) \equiv 0$ in Theorem 1.3.2, then we get Theorem 1.2.3.

(iii) If $g(x,u) \equiv 0$ and $f(x,u)$ is not nondecreasing in u but $f(x,u) + Mu$, $M > 0$, is nondecreasing in u, then one can consider the BVP

$$\tilde{L}u = Lu + Mu = f(x,u) + Mu = \tilde{f}(x,u) \text{in } \Omega \quad (1.3.13)$$
$$Bu = \phi \text{on } \partial\Omega.$$

It is clear that α_0, β_0 are the lower and upper solutions of (1.3.13), $\tilde{f}(x,u)$ is nondecreasing in u and $c(x) + M > 0$ in Ω. As a result, Theorem 1.3.2 yields the same conclusion as in Theorem 1.2.1.

(iv) If $f(x,u) \equiv 0$ and $g(x,u)$ is not nonincreasing in u but $\tilde{g}(x,u) = g(x,u) - Nu$, $N > 0$, is nonincreasing in u, we consider the BVP

$$Lu = \tilde{f}(x,u) + \tilde{g}(x,u) \text{ in } \Omega, \ Bu = \phi \text{ on } \partial\Omega, \quad (1.3.14)$$

where $\tilde{f}(x,u) = Nu$ and assume that the coupled lower and upper solutions of type I of (1.3.14) exist. Then Theorem 1.3.2 implies the same conclusions for the BVP (1.3.14). In addition, if $c(x)-(N_2+N) > 0$ in Ω with N_2 as in Corollary 1.3.2, then the BVP (1.3.14) has a unique solution.

(v) If $g(x,u)$ is nonincreasing in u and $f(x,u)$ is not nondecreasing in u but $\tilde{f}(x,u) = f(x,u) + Mu$, $M > 0$, is nondecreasing in u, then we consider the BVP

$$\tilde{L}u = Lu + Mu = \tilde{f}(x,u) + g(x,u) \text{ in } \Omega, \ Bu = \phi \text{ on } \partial\Omega. \quad (1.3.15)$$

It is easy to see that α_0, β_0 are coupled lower and upper solutions of type I of (1.3.15). Hence the conclusion of Theorem 1.3.2 applied to (1.3.15) holds for the BVP

$$Lu = f(x,u) + g(x,u) \text{ in } \Omega, \ Bu = \phi \text{ on } \partial\Omega,$$

where $f(x,u)$ is not nondecreasing in u.

(vi) If in BVP (1.3.4), $f(x,u)$ is nondecreasing in u and $g(x,u)$ is not nonincreasing in u but $\tilde{g}(x,u) = g(x,u) - Nu$, $N > 0$, is nonincreasing in u, then consider the BVP

$$\tilde{L}u = Lu - Nu = f(x,u) + \tilde{g}(x,u) \text{ in } \Omega, \ Bu = \phi \text{ on } \partial\Omega, \quad (1.3.16)$$

and assume that the coupled lower and upper solutions of type I of (1.3.16) exist. Then, the assertion of Theorem 1.3.2 is guaranteed for the BVP (1.3.16). In addition, if the uniqueness condition, $c(x) -$ $(N_1 + N_2 + N) > 0$ in Ω with N_1 and N_2 as in Corollary 1.3.2 is satisfied, then Theorem 1.3.2 applied to the BVP (1.3.16) implies the same conclusion for the BVP $Lu = f(x, u) + g(x, u)$ in Ω, $Bu = \phi$ on $\partial\Omega$, where $g(x, u)$ is not nonincreasing in u.

(vii) If both functions $f(x, u)$, $g(x, u)$ do not satisfy the required monotone character in Theorem 1.3.2 but $\tilde{f}(x, u) = f(x, u) + Mu$, $\tilde{g}(x, u) = g(x, u) - Nu$, $M, N > 0$, are nondecreasing and nonincreasing in u, respectively, we consider the BVP

$$\tilde{L}u = Lu + Mu - Nu = \tilde{f}(x, u) + \tilde{g}(x, u) \text{ in } \Omega, \ Bu = \phi \text{ on } \partial\Omega,$$
(1.3.17)

and assume that the coupled lower and upper solutions of type I of (1.3.17) exist. Clearly, the conditions of Theorem 1.3.2 are fulfilled by $\tilde{f}$, $\tilde{g}$, and α_0, β_0. Hence this shows that the conclusion remains the same for the BVP (1.3.17). In addition, if $c(x) - (N_2 + N_2 + M + N) > 0$ in Ω with N_1 and N_2 as in Corollary 1.3.2. Then the conclusion of Theorem 1.3.2 holds for the BVP

$$Lu = f(x, u) + g(x, u) \quad \text{in} \quad \Omega, \quad Bu = \phi \quad \text{on} \quad \partial\Omega$$

where $f(x, u)$ and $g(x, u)$ are not monotone functions.

Let us next consider utilizing the coupled lower and upper solutions of type II. We prove the following result.

Theorem 1.3.3 *Assume that (A_2) of Theorem 1.3.2 holds. Then for any solution $u(x)$ of (1.3.4) with $\alpha_0 \le u \le \beta_0$ in $\bar{\Omega}$, we have the iterates $\{\alpha_n(x)\}$, $\{\beta_n(x)\}$ satisfying*

$$\alpha_0 \le \alpha_2 \le \ldots \le \alpha_{2n} \le u \le \alpha_{2n+1} \le \ldots \le \alpha_3 \le \alpha_1, \text{ in } \bar{\Omega}, \qquad (1.3.18)$$

$$\beta_1 \le \beta_3 \le \ldots \le \beta_{2n+1} \le u \le \beta_{2n} \le \ldots \le \beta_2 \le \beta_0, \text{ in } \bar{\Omega}, \qquad (1.3.19)$$

provided $\alpha_0 \le \alpha_2$ and $\beta_2 \le \beta_0$ in $\bar{\Omega}$, where the iteration schemes are given by

$$L\alpha_{n+1} = f(x, \beta_n) + g(x, \alpha_n) \text{ in } \Omega, \ B\alpha_{n+1} = \phi \text{ on } \partial\Omega, \qquad (1.3.20)$$

$$L\beta_{n+1} = f(x, \alpha_n) + g(x, \beta_n) \text{ in } \Omega, \ B\beta_{n+1} = \phi \text{ on } \partial\Omega. \qquad (1.3.21)$$

Moreover, the monotone sequences $\{\alpha_{2n}\}$, $\{\alpha_{2n+1}\}$, $\{\beta_{2n}\}$, $\{\beta_{2n+1}\} \in$
$C^{2,\alpha}[\bar{\Omega}, R]$ *converge to* ρ, r, ρ^*, r^* *in* $C^2[\bar{\Omega}, R]$ *respectively and they satisfy*

$$
\left[
\begin{array}{ll}
Lr = f(x,\rho^*) + g(x,\rho) \text{ in } \Omega, & Br = \phi \text{ on } \partial\Omega, \\
L\rho = f(x,r^*) + g(x,r) \text{ in } \Omega, & B\rho = \phi \text{ on } \partial\Omega, \\
Lr^* = f(x,\rho) + g(x,\rho^*) \text{ in } \Omega, & Br^* = \phi \text{ on } \partial\Omega, \\
L\rho^* = f(x,r) + g(x,r^*) \text{ in } \Omega, & B\rho^* = \phi \text{ on } \partial\Omega.
\end{array}
\right.
\tag{1.3.22}
$$

Also, $\rho \leq u \leq r$ *and* $r^* \leq u \leq \rho^*$ *in* $\bar{\Omega}$.

Proof   We shall first show that the coupled lower and upper solutions of type II of (1.3.4) exist. Since $f(x,0) + g(x,0) \in C^\alpha[\bar{\Omega}, R]$ for the BVP

$$ Lz = f(x,0) + g(x,0) \text{ in } \Omega, \ Bz = \phi \text{ on } \partial\Omega, $$

define $\alpha_0 = z - R_0$, $\beta_0 = z + R_0$ by choosing $R_0 > 0$ sufficiently large so that we have $\alpha_0(x) \leq 0 \leq \beta_0(x)$ in $\bar{\Omega}$. Then using the monotone character of $f(x,u)$ and $g(x,u)$, we get

$$ L\alpha_0 = Lz - c(x)R_0 = f(x,0) + g(x,0) - c(x)R_0 \leq f(x,\beta_0) + g(x,\alpha_0) \text{ in } \Omega, $$

$$ B\alpha_0 = Bz - pR_0 \leq \phi \text{ on } \partial\Omega. $$

Similarly, we obtain $L\beta_0 \geq f(x,\alpha_0) + g(x,\beta_0)$ in Ω, $B\beta_0 \geq \phi$ on $\partial\Omega$. Thus, α_0, β_0 are the coupled lower and upper solutions of type II for the BVP (1.3.4).

Following the arguments used in Theorem 1.3.2, we see that the linear BVPs (1.3.20), (1.3.21) have unique solutions α_{n+1}, $\beta_{n+1} \in C^{2,\alpha}[\bar{\Omega}, R]$ for each $n = 1, 2, \ldots$. Our aim is therefore to prove the relations (1.3.18) and (1.3.19). Let u be any solution of (1.3.4) such that $\alpha_0 \leq u \leq \beta_0$ in $\bar{\Omega}$. We shall show that

$$ \alpha_0 \leq \alpha_2 \leq u \leq \alpha_3 \leq \alpha_1, \quad \beta_1 \leq \beta_3 \leq u \leq \beta_2 \leq \beta_0 \text{ in } \bar{\Omega}. \tag{1.3.23} $$

Setting $p = u - \alpha_1$, we get $Bp = 0$ on $\partial\Omega$ and

$$ Lp = f(x,u) + g(x,u) - f(x,\beta_0) - g(x,\alpha_0) \leq 0 \text{ in } \Omega, $$

using the monotone nature of f,g and the fact $\alpha_0 \leq u \leq \beta_0$ in $\bar{\Omega}$. This implies by Corollary 1.3.1, that $p \leq 0$ in $\bar{\Omega}$ and this implies $u \leq \alpha_1$ in $\bar{\Omega}$. Similarly, we can show that $u \leq \beta_1$ in $\bar{\Omega}$. Next let $p = \alpha_2 - u$ so that $Bp = 0$ on $\partial\Omega$ and $Lp = f(x,\beta_1) + g(x,\alpha_1) - f(x,u) - g(x,u) \leq 0$ in Ω. Hence by Corollary 1.3.1, it follows that $\alpha_2 \leq u$ in $\bar{\Omega}$. A similar argument yields $u \leq \beta_2$ in $\bar{\Omega}$. Considering $p = \alpha_3 - \alpha_1$, we see that $Bp = 0$ on $\partial\Omega$, and, as before,

$$Lp = f(x, \beta_2) + g(x, \alpha_2) - f(x, \beta_0) - g(x, \alpha_0) \leq 0 \text{ in } \Omega,$$

which implies by Corollary 1.3.1, $\alpha_3 \leq \alpha_1$ in $\bar{\Omega}$. In the same way, we get $\beta_1 \leq \beta_3$ in $\bar{\Omega}$. Also using similar reasoning, we obtain $u \leq \alpha_3$, $\beta_3 \leq u$ in $\bar{\Omega}$, proving (1.3.23).

Now assuming for some $n > 2$, the inequalities

$$\alpha_{2n-4} \leq \alpha_{2n-2} \leq u \leq \alpha_{2n-1} \leq \alpha_{2n-3},$$

$$\beta_{2n-3} \leq \beta_{2n-1} \leq u \leq \beta_{2n-2} \leq \beta_{2n-4} \text{ in } \bar{\Omega},$$

hold, then it can be shown, by employing similar arguments, that

$$\alpha_{2n-2} \leq \alpha_{2n} \leq u \leq \alpha_{2n-1}, \quad \beta_{2n-1} \leq \beta_{2n+1} \leq u \leq \beta_{2n} \leq \beta_{2n-2} \text{ in } \bar{\Omega}.$$

Thus, by induction, (1.3.18) and (1.3.19) are valid for all $n = 0, 1, 2, \ldots$.

Since $\alpha_n, \beta_n \in C^{2,\alpha}[\Omega, R]$ for all n, employing a similar reasoning as in Theorem 1.3.2, we conclude that the limits

$$\lim_{n \to \infty} \alpha_{2n} = \rho, \quad \lim_{n \to \infty} \alpha_{2n+1} = r, \quad \lim_{n \to \infty} \beta_{2n+1} = r^* \text{ and } \lim_{n \to \infty} \beta_{2n} = \rho^*,$$

exist in $C^2[\bar{\Omega}, R]$. Thus we have the limits

$$\begin{aligned}
\lim_{n \to \infty} L\alpha_{2n} &= L\rho, \\
\lim_{n \to \infty} L\alpha_{2n+1} &= Lr, \\
\lim_{n \to \infty} L\beta_{2n+1} &= Lr^*, \\
\lim_{n \to \infty} L\beta_{2n} &= L\rho^* \\
\lim_{n \to \infty} B\alpha_{2n} &= B\rho, \\
\lim_{n \to \infty} B\alpha_{2n+1} &= Br, \\
\lim_{n \to \infty} B\beta_{2n+1} &= Br^*, \\
\lim_{n \to \infty} B\beta_{2n} &= B\rho^*, \\
\lim_{n \to \infty} [f(x, \beta_{2n+1}) + g(x, \alpha_{2n+1})] &= f(x, r^*) + g(x, r), \\
\lim_{n \to \infty} [f(x, \beta_{2n}) + g(x, \alpha_{2n})] &= f(x, \rho^*) + g(x, \rho), \\
\lim_{n \to \infty} [f(x, \alpha_{2n}) + g(x, \beta_{2n})] &= f(x, \rho) + g(x, \rho^*), \\
\lim_{n \to \infty} [f(x, \alpha_{2n+1}) + g(x, \beta_{2n+1})] &= f(x, r) + g(x, r^*).
\end{aligned}$$

We then find that ρ, r, ρ^*, r^* satisfy

$$
\begin{aligned}
L\rho &= f(x, r^*) + g(x, r) \text{ in } \Omega, \ B\rho = \phi \text{ on } \partial\Omega, \\
Lr &= f(x, \rho^*) + g(x, \rho) \text{ in } \Omega, \ Br = \phi \text{ on } \partial\Omega, \\
L\rho^* &= f(x, r) + g(x, r^*) \text{ in } \Omega, \ B\rho^* = \phi \text{ on } \partial\Omega, \\
Lr^* &= f(x, \rho) + g(x, \rho^*) \text{ in } \Omega, \ Br^* = \phi \text{ on } \partial\Omega,
\end{aligned}
$$

and from (1.3.18) and (1.3.19), it follows that

$$
\rho \leq u \leq r \text{ and } r^* \leq u \leq \rho^* \text{ in } \bar{\Omega}.
$$

The proof is therefore complete.

Corollary 1.3.3 *Under the assumption of Theorem 1.3.3 if $f(x, u)$, $g(x, u)$ satisfy the relations*

$$
f(x, u_1) - f(x, u_2) \leq N_1(u_1 - u_2), \ u_1 \geq u_2, \ N_1 > 0, \ x \in \Omega,
$$

$$
g(x, u_1) - g(x, u_2) \geq -N_2(u_1 - u_2), \ u_1 \geq u_2, \ N_2 > 0, \ x \in \Omega,
$$

and $c(x) - (N_1 + N_2) > 0$, then $u = \rho = r = \rho^* = r^*$ is the unique solution of (1.3.4).

Proof   Let $v_1 = r - \rho$, $v_2 = \rho^* - r^*$ so that we have $v_1 \geq 0$, $v_2 \geq 0$ on $\bar{\Omega}$. It then follows that $Bv_1 = 0$, $Bv_2 = 0$ on $\partial\Omega$ and

$$
Lv_1 \leq N_1(\rho^* - r^*) + N_2(r - \rho),
$$

$$
Lv_2 \leq N_1(r - \rho) + N_2(\rho^* - r^*)
$$

and therefore $L(v_1 + v_2) \leq (N_1 + N_2)(v_1 + v_2)$ in Ω, $B(v_1 + v_2) = 0$ on $\partial\Omega$. Corollary 1.3.1 now yields $v_1 + v_2 \leq 0$ in $\bar{\Omega}$, which implies $\rho = r$, $\rho^* = r^*$ in $\bar{\Omega}$. We claim that $\rho = r^*$, $r = \rho^*$ in $\bar{\Omega}$. If not, assuming the contrary and proceeding as above, we are led to a contradiction. Hence $u = \rho = \rho^* = r = r^*$ is the unique solution of (1.3.4) and the proof is complete.

Theorem 1.3.3 contains several special cases which we give below.

Remarks 1.3.2

(i) In Theorem 1.3.3, suppose that $f(x, u) \equiv 0$. Then we get Theorem 1.3.2.

(ii) If $g(x, u) \equiv 0$ in Theorem 1.3.3, then we obtain a new result dual to Theorem 1.2.1 with $M = 0$.

(iii) If $g(x, u) \equiv 0$ and $f(x, u)$ is not nondecreasing in u but $\tilde{f}(x, u) = f(x, u) + Mu$, $M > 0$, is nondecreasing in u, then we consider the BVP

$$\tilde{L}u = Lu + Mu = \tilde{f}(x, u) \text{ in } \Omega, \ Bu = \phi \text{ on } \partial\Omega. \tag{1.3.24}$$

If there exist α_0, β_0 as in Theorem 1.3.3 such that

$$\tilde{L}\alpha_0 \leq \tilde{f}(x, \beta_0) \text{ in } \Omega, \ B\alpha_0 \leq \phi \text{ on } \partial\Omega,$$

$$\tilde{L}\beta_0 \geq \tilde{f}(x, \alpha_0) \text{ in } \Omega, \ B\beta_0 \geq \phi \text{ on } \partial\Omega,$$

with $\alpha_0 \leq 0 \leq \beta_0$ on $\bar{\Omega}$, then Theorem 1.3.3 yields a new result dual to Theorem 1.2.3.

(iv) If $f(x, u) \equiv 0$ and $g(x, u)$ is not nonincreasing in u but $\tilde{g}(x, u) = g(x, u) - Nu$, $N > 0$ is nonincreasing in u and $c(x) - N > 0$, then we consider the BVP

$$\tilde{L}u = Lu - Nu = \tilde{g}(x, u) \text{ in } \Omega, \ Bu = \phi \text{ on } \partial\Omega. \tag{1.3.25}$$

Evidently, α_0, β_0 exist satisfying

$$\tilde{L}\alpha_0 \leq \tilde{g}(x, \alpha_0) \text{ in } \Omega, \ B\alpha_0 \leq \phi \text{ on } \partial\Omega,$$

$$\tilde{L}\beta_0 \geq \tilde{g}(x, \beta_0) \text{ in } \Omega, \ B\beta_0 \geq \phi \text{ on } \partial\Omega.$$

Hence Theorem 1.3.3 provides a new result which is an extension of Theorem 1.2.2.

(v) If $g(x, u)$ is nonincreasing in u and $f(x, u)$ is not nondecreasing in u but $\tilde{f}(x, u) = f(x, u) + Mu$, $M > 0$, is nondecreasing in u for $x \in \Omega$, then we consider the BVP

$$\tilde{L}u = Lu + Mu = \tilde{f}(x, u) + g(x, u) \text{ in } \Omega, \ Bu = \phi \text{ on } \partial\Omega. \tag{1.3.26}$$

If the coupled lower and upper solutions α_0, β_0 of type II exist for the BVP (1.3.26) such that $\alpha_0 \leq \beta_0$ on $\bar{\Omega}$, then the conclusion of Theorem 1.3.3 holds for the BVP

$$Lu = f(x, u) + g(x, u) \text{ in } \Omega, \ Bu = \phi \text{ on } \partial\Omega,$$

where $f(x, u)$ is not nondecreasing in u, when the uniqueness assumptions of Corollary 1.3.3 are satisfied.

(vi) If, in BVP (1.3.4), $f(x, u)$ is nondecreasing and $g(x, u)$ is not nonincreasing in u but $\tilde{g}(x, u) = g(x, u) - Nu$, $N > 0$, is nonincreasing in u for $x \in \Omega$ and $c(x) - N > 0$, then consider the BVP

$$\tilde{L}u = Lu - Nu = f(x, u) + \tilde{g}(x, u) \text{ in } \Omega, \ Bu = \phi \text{ on } \partial\Omega. \quad (1.3.27)$$

As before, α_0, β_0 exist as coupled lower and upper solutions of type II for the BVP (1.3.27) and therefore the conclusion of Theorem 1.3.3 holds for the BVP

$$Lu = f(x, u) + g(x, u) \text{ in } \Omega, \ Bu = \phi \text{ on } \partial\Omega,$$

when $g(x, u)$ is not nondecreasing in u, provided uniqueness conditions hold as in Corollary 1.3.3.

(vii) If, in Theorem 1.3.3 both $f(x, u)$, $g(x, u)$ are not monotone but $\tilde{f}(x, u) = f(x, u) + Mu$, $M > 0$, is nondecreasing, $\tilde{g}(x, u) = g(x, u) - Nu$, $N > 0$ is nonincreasing in u for $x \in \Omega$, we consider the BVP

$$\tilde{L}u = L + (M - N)u = \tilde{f}(x, u) + \tilde{g}(x, u) \text{ in } \Omega, \ Bu = \phi \text{ on } \partial\Omega. \quad (1.3.28)$$

As before, if α_0, β_0 exist as coupled lower and upper solutions of type II for the BVP (1.3.28), then the conclusion of Theorem 1.3.4 holds for the BVP (1.3.28)

$$Lu = f(x, u) + g(x, u) \text{ in } \Omega, \ Bu = \phi \text{ on } \partial\Omega,$$

when f, g are not monotone, provided the uniqueness conditions of Corollary 1.3.3 are satisfied.

1.4   Generalized Quasilinearization

As we have seen, the method of upper and lower solutions together with the monotone iterative technique offers monotone sequences which converge to the extremal solutions of the original nonlinear problem. As observed earlier, when we employ the technique of lower and upper solutions coupled with the method of quasilinearization and utilize the idea of Newton and Fourier, it is possible to construct concurrently lower and upper bounding sequences, whose elements are the solutions of the given problem. Furthermore, this unification provides a mechanism to enlarge the class of nonlinear problems to which this method is applicable. For example, it is not necessary to impose the convexity assumption on the function involved and this leads to several other possibilities. Moreover, these ideas can be refined, generalized, and extended to various types of nonlinear problems, and consequently, this technique is known as generalized quasilinearization.

In this section, we shall consider the extension of the method of generalized quasilinearization to semilinear elliptic boundary value problems and prove very general results that include several important special cases.

We consider the semilinear elliptic BVP (1.3.4), namely,

$$\left[\begin{array}{ll} Lu = f(x,u) + g(x,u) & \text{in } \Omega, \\ Bu = \varphi & \text{on } \partial\Omega, \end{array}\right. \tag{1.4.1}$$

where $f, g \in C^\alpha[\bar\Omega \times R, R]$ and prove the following result.

Theorem 1.4.1 *Assume that*

(A_1) $\alpha_0, \beta_0 \in C^2[\bar\Omega, R]$ *with* $\alpha_0(x) \leq \beta_0(x)$ *in* Ω *satisfy*

$$L\alpha_0 \leq f(x,\alpha_0) + g(x,\alpha_0) \text{ in } \Omega, \ B\alpha_0 \leq \varphi \text{ on } \partial\Omega,$$

$$L\beta_0 \geq f(x,\beta_0) + g(x,\beta_0) \text{ in } \Omega, \ B\beta \geq \varphi \text{ on } \partial\Omega;$$

(A_2) f, g *satisfy, in addition, that* f_u, g_u, f_{uu}, g_{uu} *exist, are continuous, and* $f_{uu}(x,u) \geq 0$, $g_{uu}(x,u) \leq 0$ *on* $\bar\Omega \times R$;

(A_3) $0 < N \leq c(x) - [f_u(x,\beta_0) + g_u(x,\alpha_0)]$ *in* Ω.

Then there exist monotone sequences $\{\alpha_n(x)\}$, $\{\beta_n(x)\} \in C^{2,\alpha}[\bar\Omega, R]$ *such that* $\alpha_n \to \rho$, $\beta_n \to r$ *in* $C^2[\bar\Omega, R]$, $\rho = r = u$ *is the unique solution of* (1.4.1) *satisfying* $\alpha_0 \leq u \leq \beta_0$ *in* Ω, *and the convergence is quadratic.*

Proof   We consider the following linear BVPs for each $k = 1, 2, \ldots$

$$L\alpha_{k+1} = F(x, \alpha_{k+1}; \alpha_k, \beta_k) \text{ in } \Omega, \ B\alpha_{k+1} = \varphi \text{ on } \partial\Omega, \tag{1.4.2}$$

$$L\beta_{k+1} = G(x, \beta_{k+1}; \alpha_k, \beta_k) \text{ in } \Omega, \ B\beta_{k+1} = \varphi \text{ on } \partial\Omega, \tag{1.4.3}$$

where

$$\begin{aligned} F(x, u; , \alpha_k, \beta_k) &= f(x,\alpha_k) + g(x,\alpha_k) \\ &+ f_u(x,\alpha_k)(u - \alpha_k) + g_u(x,\beta_k)(u - \alpha_k), \end{aligned} \tag{1.4.4}$$

and

$$\begin{aligned} G(x, u; \alpha_k, \beta_k) &= f(x,\beta_k) + g(x,\beta_k) \\ &+ f_u(x,\alpha_k)(u - \beta_k) + g_u(x,\beta_k)(u - \beta_k). \end{aligned} \tag{1.4.5}$$

The assumptions $f_{uu}(x, u) \geq 0$, $g_{uu}(x, u) \leq 0$ yield the inequalities

$$f(x, u) \geq f(x, v) + f_u(x, v)(u - v), \ u \geq v \qquad (1.4.6)$$

$$g(x, u) \geq g(x, v) + g_u(x, v)(u - v), \ u \geq v. \qquad (1.4.7)$$

Since $f_u(x, u)$ is nondecreasing and $g_u(x, u)$ is nonincreasing in u for each $x \in \bar{\Omega}$, we find that for any $\eta, \mu \in C^2[\bar{\Omega}, R]$ with $\alpha_0 \leq \eta, \mu \leq \beta_0$, $c(x) - f_u(x, \eta) - g_u(x, \mu) \geq c(x) - f_u(x, \beta_0) - g_u(x, \alpha_0) \geq N > 0$ in Ω. In order to conclude the existence of unique solutions of the BVPs (1.4.2) and (1.4.3) for each $k \geq 1$, we need to show that for any $\eta, \mu \in C^2[\bar{\Omega}, R]$ with $\alpha_0 \leq \eta, \mu \leq \beta_0$, $h(x) \in C^\alpha[\bar{\Omega}, R]$, where

$$h(x) = f(x, \eta) + g(x, \eta) - f_u(x, \eta)\eta - g_u(x, \mu)\eta.$$

For this purpose, we note that if $\eta, \mu \in C^2[\bar{\Omega}, R]$, then $\eta, \mu \in W^{2,q}[\bar{\Omega}, R]$ in view of the boundedness of Ω and $\partial\Omega \in C^{2,\alpha}[\bar{\Omega}, R]$. The embedding theorem (Theorem A.2.5) then shows that $\eta, \mu \in C^{1,\alpha}[\bar{\Omega}, R]$.

We have

$$
\begin{aligned}
&\mid f(x, \eta(x)) - f(y, \eta(y)) \mid \\
&\leq \quad K[\parallel x - y \parallel^\alpha + \mid \eta(x) - \eta(y) \mid^\alpha] \\
&\leq \quad K[\parallel x - y \parallel^\alpha + \mid \eta \mid_{C^1[\bar{\Omega}, R]}^\alpha \parallel x - y \parallel^\alpha] \\
&\leq \quad L_0 \parallel x - y \parallel^\alpha,
\end{aligned}
$$

where

$$L_0 = K[1 + \mid \eta \mid_{C^1[\bar{\Omega}, R]}^\alpha].$$

Also,

$$
\begin{aligned}
&\mid f_u(x, \eta(x))\eta(x) - f_u(y, \eta(y))\eta(y) \mid \\
&\leq \quad \mid f_u(x, \eta(x))\eta(x) - f_u(x, \eta(x))\eta(y) \mid \\
&+ \quad \mid f_u(x, \eta(x)) - f_u(y, \eta(y)) \mid \mid \eta(y) \mid \\
&\leq \quad \mid f_u(x, \eta(x)) \mid \mid \eta(x) - \eta(y) \mid + \mid \eta(y) \mid \mid f_{uu}(x, \xi) \mid \mid \eta(x) - \eta(y) \mid \\
&\leq \quad K_1 \mid \eta \mid_{C^1[\bar{\Omega}, R]}^\alpha \parallel x - y \parallel^\alpha + K_2 K_3 \mid \eta \mid_{C^1[\bar{\Omega}, R]}^\alpha \parallel x - y \parallel^\alpha,
\end{aligned}
$$

where for $\alpha_0 \leq u \leq \beta_0$,

$$\mid f_u(x, u) \mid \leq K_1, \ \mid \eta(u) \mid \leq K_2, \ \mid f_{uu}(x, u) \mid \leq K_3,$$

and ξ is between $\eta(x)$ and $\eta(y)$.

Thus $|\,f_u(x,\eta(x))\eta(x) - f_u(y,\eta(y))\eta(y)\,| \leq L_1 \parallel x - y \parallel^\alpha$, where $L_1 = [K_1 + K_2K_3]\,|\,\eta\,|^\alpha_{C^1[\bar\Omega,R]}$. We can obtain similar estimates for $g(x,u)$. As a result, we find that, because of the definition of $h(x)$,

$$
\begin{aligned}
|\,h(x) - h(y)\,| &\leq |f(x,\eta(x)) - f(y,\eta(y))\,| + |\,g(x,\eta(x)) - g(y,\eta(y))\,| \\
&\quad + |\,f_u(x,\eta(x))\eta(x) - f_u(y,\eta(y))\eta(y)\,| \\
&\quad + |\,g_u(x,\mu(x))\eta(x) - g_u(y,\mu(y))\eta(y)\,| \\
&\leq C \parallel x - y \parallel^\alpha,
\end{aligned}
$$

where $C = L_0 + L_1 + M_0 + M_1$, M_0, M_1, being corresponding constants relative to the function g. Hence $h \in C^\alpha[\bar\Omega, R]$. If for each $k \geq 1$, $\alpha_k, \beta_k \in [\alpha_0, \beta_0]$ with $\alpha_k \leq \beta_k$, then there exist unique solutions $\alpha_k, \beta_k \in C^{2,\alpha}[\bar\Omega, R]$ of the BVPs

$$
\begin{bmatrix}
L_{c_0}\alpha_{k+1} &= f(x,\alpha_k) + g(x,\alpha_k) \\
&\quad - f_u(x,\alpha_k)\alpha_k - g_u(x,\beta_k)\alpha_k & \text{in } \Omega, \\
B\alpha_{k+1} &= \varphi & \text{on } \partial\Omega
\end{bmatrix}
\tag{1.4.8}
$$

and

$$
\begin{bmatrix}
L_{c_0}\beta_{k+1} &= f(x,\beta_k) + g(x,\beta_k) \\
&\quad - f_u(x,\alpha_k)\beta_k - g_u(x,\beta_k)\beta_k & \text{in } \Omega, \\
B\beta_{k+1} &= \varphi & \text{on } \partial\Omega,
\end{bmatrix}
\tag{1.4.9}
$$

where $L_{c_0}u$ is as defined in (1.3.2) with $c_0(x) = c(x) - f_u(x,\alpha_k) - g_u(x,\beta_k) > 0$. Our aim is therefore to show that

$$
\alpha_0 \leq \alpha_1 \leq \alpha_2 \leq \ldots \leq \alpha_k \leq \beta_k \leq \ldots \leq \beta_2 \leq \beta_1 \leq \beta_0 \text{ in } \bar\Omega. \tag{1.4.10}
$$

We shall first prove that

$$
\alpha_0 \leq \alpha_1 \leq \beta_1 \leq \beta_0 \text{ in } \bar\Omega. \tag{1.4.11}
$$

Since α_0 is a lower solution of (1.4.1), we see that

$$
L\alpha_0 \leq f(x,\alpha_0) + g(x,\alpha_0) = F(x,\alpha_0;\alpha_0,\beta_0) \text{ in } \Omega
$$

$$
B\alpha_0 \leq \varphi \text{ on } \partial\Omega,
$$

and

$$
L\alpha_1 = F(x,\alpha_1;\alpha_0,\beta_0) \text{ in } \Omega, B\alpha_1 = \varphi \text{ on } \partial\Omega.
$$

Also F satisfies (1.4.2) with $L(x) = f_u(x,\alpha_0) + g_u(x,\beta_0)$ and by (A_3), $c(x) - f_u(x,\alpha_0) - g_u(x,\beta_0) \geq N > 0$. Theorem 1.3.1 therefore yields $\alpha_0 \leq \alpha_1$ in $\bar\Omega$.

Similar reasoning gives $\beta_1 \leq \beta_0$ in $\bar{\Omega}$. We next show that $\alpha_1 \leq \beta_0$ and $\alpha_0 \leq \beta_1$ in $\bar{\Omega}$. To show that $\alpha_1 \leq \beta_0$, we get, using (1.4.6) and (1.4.7),

$$
\begin{aligned}
L\alpha_1 &= F(x, \alpha_1; \alpha_0, \beta_0) \\
&= f(x, \alpha_0) + g(x, \alpha_0) + f_u(x, \alpha_0)(\alpha_1 - \alpha_0) \\
&+ g_u(x, \beta_0)(\alpha_1 - \alpha_0) \\
&\leq f(x, \beta_0) - f_u(x, \alpha_0)(\beta_0 - \alpha_0) + g(x, \beta_0) \\
&- g_u(x, \beta_0)(\beta_0 - \alpha_0) \\
&+ f_u(x, \alpha_0)(\alpha_1 - \alpha_0) + g_u(x, \beta_0)(\alpha_1 - \alpha_0) \\
&= f(x, \beta_0) + g(x, \beta_0) + f_u(x, \alpha_0)[\alpha_1 - \alpha_0 + \alpha_0 - \beta_0] \\
&+ g_u(x, \beta_0)[\alpha_0 - \beta_0 + \alpha_1 - \alpha_0] \\
&= f(x, \beta_0) + g(x, \beta_0) + f_u(x, \alpha_0)(\alpha_1 - \beta_0) \\
&+ g_u(x, \beta_0)(\alpha_1 - \beta_0) \\
&= G(x, \alpha_1; \alpha_0, \beta_0) \text{ in } \Omega.
\end{aligned}
$$

But $L\beta_1 \geq G(x, \beta_0; \alpha_0, \beta_0)$ in Ω and hence Theorem 1.3.1 implies that $\alpha_1 \leq \beta_0$ in $\bar{\Omega}$. Similarly, one can prove that $\alpha_0 \leq \beta_1$ in $\bar{\Omega}$. To show that $\alpha_1 \leq \beta_1$ in $\bar{\Omega}$, we utilize (1.4.6) and (1.4.7) and the fact that $g_u(x, u)$ is nonincreasing in u, to get

$$
\begin{aligned}
L\alpha_1 &= F(x, \alpha_1; \alpha_0, \beta_0) & (1.4.12) \\
&\leq f(x, \alpha_1) + g(x, \alpha_1) + g_u(x, \alpha_1)(\alpha_0 - \alpha_1) \\
&\quad + g_u(x, \beta_0)(\alpha_1 - \alpha_0) \\
&\leq f(x, \alpha_1) + g(x, \alpha_1) + [g_u(x, \beta_0) - g_u(x, \alpha_1)](\alpha_1 - \alpha_0) \\
&\leq F(x, \alpha_1; \alpha_1, \beta_1).
\end{aligned}
$$

Similarly, since $f_u(x, u)$ is nondecreasing in u, we arrive at

$$
\begin{aligned}
L\beta_1 &= G(x, \beta_1; \alpha_0, \beta_0) & (1.4.13) \\
&\geq f(x, \beta_1) + f_u(x, \beta_1)(\beta_0 - \beta_1) + f_u(x, \alpha_0)(\beta_1 - \beta_0) + g(x, \beta_1) \\
&= f(x, \beta_1) + g(x, \beta_1) + [f_u(x, \alpha_0) - f_u(x, \beta_1)](\beta_1 - \beta_0) & (1.4.14) \\
&\geq f(x, \beta_1) + g(x, \beta_1) \\
&= F(x, \beta_1; \alpha_1, \beta_1).
\end{aligned}
$$

It then follows by Theorem 1.3.1 that $\alpha_1 \leq \beta_1$ in $\bar{\Omega}$, proving (1.4.11).

We shall next consider that if $\alpha_{k-1} \leq \alpha_k \leq \beta_k \leq \beta_{k-1}$ in $\bar{\Omega}$, for some $k > 1$, then it follows that

$$
\alpha_k \leq \alpha_{k+1} \leq \beta_{k+1} \leq \beta_k \text{ in } \bar{\Omega}. \tag{1.4.15}
$$

Since α_k satisfies

$$L\alpha_k = F(x, \alpha_k; \alpha_{k-1}, \beta_{k-1}) \text{ in } \Omega, \ B\alpha_k = \varphi \text{ on } \partial\Omega,$$

we get, using (1.4.6) and (1.4.7) and the assumption $\alpha_{k-1} \leq \alpha_k \leq \beta_k \leq \beta_{k-1}$ in $\bar{\Omega}$,

$$L\alpha_k \leq F(x, \alpha_k; \alpha_k, \beta_k) \text{ in } \Omega, \ B\alpha_k = \varphi \text{ on } \partial\Omega. \tag{1.4.16}$$

Similarly, we can obtain

$$L\beta_k \geq G(x, \beta_k; \alpha_k, \beta_k) \text{ in } \Omega, \ B\beta_k = \varphi \text{ on } \partial\Omega. \tag{1.4.17}$$

Moreover, $\alpha_{k+1}, \beta_{k+1}$ are the solutions of (1.4.2) and (1.4.3). Hence by Theorem 1.3.1, we get $\alpha_k \leq \alpha_{k+1}, \ \beta_{k+1} \leq \beta_k$ in $\bar{\Omega}$. Next we show that $\alpha_{k+1} \leq \beta_k$ in $\bar{\Omega}$. We find that

$$
\begin{aligned}
L\alpha_{k+1} &= F(x, \alpha_{k+1}; \alpha_k, \beta_k) \\
&\leq f(x, \beta_k) - f_u(x, \alpha_k)(\beta_k - \alpha_k) + g(x, \beta_k) \\
&\quad -g_u(x, \beta_k)(\beta_k - \alpha_k) \\
&\quad +f_u(x, \alpha_k)(\alpha_{k+1} - \alpha_k) + g_u(x, \beta_k)(\alpha_{k+1} - \alpha_k).
\end{aligned}
$$

Hence

$$
\begin{aligned}
L\alpha_{k+1} &\leq f(x, \beta_k) + g(x, \beta_k) \tag{1.4.18} \\
&\quad +f_u(x, \alpha_k)[-\beta_k + \alpha_k + \alpha_{k+1} - \alpha_k] \\
&\quad +g_u(x, \beta_k)[\alpha_k - \beta_k + \alpha_{k+1} - \alpha_k] \\
&= G(x, \alpha_{k+1}; \alpha_k, \beta_k) \text{ in } \Omega.
\end{aligned}
$$

Theorem 1.3.1 yields, because of (1.4.17) and (1.4.18), $\alpha_{k+1} \leq \beta_k$ in $\bar{\Omega}$. Using similar arguments, we see that

$$
\begin{aligned}
L\beta_{k+1} &\geq f(x, \alpha_k) + f_u(x, \alpha_k)[\beta_k - \alpha_k] + g(x, \alpha_k) \tag{1.4.19} \\
&\quad +f_u(x, \beta_k)(\beta_k - \alpha_k) + f_u(x, \alpha_k)(\beta_{k+1} - \beta_k) \\
&\quad +g_u(x, \beta_k)(\beta_{k+1} - \beta_k) \\
&= f(x, \alpha_k) + g(x, \alpha_k) + f_u(x, \alpha_k)(\beta_{k+1} - \alpha_k) \\
&\quad +g_u(x, \beta_k)(\beta_{k+1} - \alpha_k) \\
&= F(x, \beta_{k+1}; \alpha_k, \beta_k) \text{ in } \Omega.
\end{aligned}
$$

Consequently, (1.4.19) together with $L\alpha_k = F(x, \alpha_k; \alpha_k, \beta_k)$ in Ω, gives by Theorem 1.3.1, $\alpha_k \leq \beta_{k+1}$ in $\bar{\Omega}$. Finally, we show that $\alpha_{k+1} \leq \beta_{k+1}$ in $\bar{\Omega}$.

This follows from (1.4.2) and (1.4.19) and Theorem 1.3.1 proving (1.4.15). Thus by induction (1.4.10) is valid for all k.

We recall that α_k, $\beta_k \in C^{2,\alpha}[\bar{\Omega}, R]$, for $k = 1, 2, \ldots$. Since $C^{2,\alpha}[\bar{\Omega}, R] \subset W^{2,q}[\bar{\Omega}, R]$ for $q > 1$, by Theorem A.2.3, we have

$$\| \alpha_k \|_{W^{2,q}[\bar{\Omega},R]} \leq C[\| h_k \|_{L^q[\bar{\Omega},R]} + \| \varphi \|_{W^{1,q}[\bar{\Omega},R]}], \qquad (1.4.20)$$

where

$$\begin{aligned} h_k(x) \;=\; & f(x, \alpha_k(x)) + g(x, \alpha_k(x)) \\ & -[f_u(x, \alpha_k(x))\alpha_k(x) + g_u(x, \beta_k(x))\alpha_k(x)]. \end{aligned}$$

The continuity of f, g, f_u, g_u and the definition of $h_k(x)$ implies that $\{h_k(x)\}$ is uniformly bounded in $C[\bar{\Omega}, R]$. Since $C[\bar{\Omega}, R]$ is dense in $L^q[\bar{\Omega}, R]$, $\{h_k(x)\}$ is also uniformly bounded in $W^{2,q}[\bar{\Omega}, R]$. For $q = \frac{2}{1-\alpha}$, $\alpha_k \in W^{2,q}[\bar{\Omega}, R]$ and hence by the Embedding Theorem A.3.4,

$$\| \alpha_k \|_{C^{1,\alpha}[\bar{\Omega},R]} \leq C \| \alpha_k \|_{W^{2,q}[\bar{\Omega},R]}, \text{ for } k = 1, 2, \ldots, \qquad (1.4.21)$$

for some constant C independent of the elements of $W^{2,q}$. Thus $\{\alpha_k(x)\}$ is uniformly bounded in $C^{1,\alpha}[\bar{\Omega}, R]$. This implies from the earlier reasoning that $\{h_k(x)\}$ is uniformly bounded in $C^{1,\alpha}[\bar{\Omega}, R]$. Consequently by Schauder's estimate in Theorem A.2.5, we find that

$$\| \alpha_k \|_{C^{2,\alpha}[\bar{\Omega},R]} \leq C \| h_k \|_{C^{\alpha}[\bar{\Omega},R]} + \| \varphi \|_{C^{1,\alpha}[\bar{\Omega},R]} \text{ for all } k,$$

which implies the uniform boundedness of $\{\alpha_k(x)\}$ in $C^{2,\alpha}[\bar{\Omega}, R]$. As a result, we have $\{\alpha_k(x)\}$ is relatively compact in $C^2[\bar{\Omega}, R]$, which yields the existence of a subsequence $\{\alpha_{k_j}(x)\}$ that converges in $C^2[\bar{\Omega}, R]$. Let $\alpha^* \in C^2[\bar{\Omega}, R]$ be the limit of this subsequence $\{\alpha_{k_j}(x)\}$. By the monotone nature of it, $\{\alpha_k(x)\}$ converges pointwise to $\rho(x)$ in $\bar{\Omega}$. But the convergence of $\{\alpha_{k_j}(x)\}$ in $C^2[\bar{\Omega}, R]$ implies pointwise convergence and thus $\alpha^*(x) = \rho(x)$ in $\bar{\Omega}$. This shows that the entire sequence $\{\alpha_k(x)\}$ converges in $C^2[\bar{\Omega}, R]$ to $\rho(x)$, that is, $\lim_{k\to\infty} \alpha_k(x) = \rho(x)$ in $C^2[\bar{\Omega}, R]$ and $\alpha_0 \leq \rho \leq \beta_0$ in $\bar{\Omega}$. Similar arguments prove that $\lim_{k\to\infty} \beta_k(x) = r(x)$ in $C^2[\Omega, R]$ and $\alpha_0 \leq \rho \leq r \leq \beta_0$ in $\bar{\Omega}$. Thus the limits

$$\lim_{k\to\infty} L\alpha_k \;=\; L\rho,$$

$$\lim_{k\to\infty} L\beta_k \;=\; L_{c_0}r,$$

$$\lim_{k\to\infty} B\alpha_k \;=\; B\rho, \quad \lim_{k\to\infty} B\beta_k = Br,$$

$$\lim_{k \to \infty} \{f(x, \alpha_k) + g(x, \alpha_k) - [f_u(x, \alpha_k) + g_u(x, \beta_k)]\alpha_k\}$$
$$= f(x, \rho) + g(x, \rho) - [f_u(x, \rho) + g_u(x, r)]\rho,$$

and

$$\lim_{k \to \infty} \{f(x, \beta_k) + g(x, \beta_k) - [f_u(x, \alpha_k) + g_u(x, \beta_k)]\beta_k\}$$
$$= f(x, r) + g(x, r) - [f_u(x, \rho) + g_u(x, r)]r,$$

exist uniformly in $\bar{\Omega}$. Noting the definition of L, we see immediately that ρ, r are solutions of (1.4.1).

Since $\rho \leq r$ in $\bar{\Omega}$, taking $\alpha = r$, $\beta = \rho$, and using Theorem 1.3.1 we obtain $r \leq \rho$ in $\bar{\Omega}$, proving $\rho = r = u$ is the unique solution of (1.4.1).

To prove the quadratic convergence of $\{\alpha_k\}$, $\{\beta_k\}$ to the unique solution u, respectively, we consider $P_{k+1} = u - \alpha_{k+1}$, $Q = \beta_{k+1} - u$ so that $BP_{k+1} = 0$ and $BQ_{k+1} = 0$.

Then we have

$$
\begin{aligned}
LP_{k+1} &= f(x, u) + g(x, u) \\
&\quad -[f(x, \alpha_k) + g(x, \alpha_k) + f_u(x, \alpha_k)(\alpha_{k+1} - \alpha_k) \\
&\quad +g_u(x, \beta_k)(\alpha_{k+1} - \alpha_k)] \\
&\leq [f_u(x, u) - f_u(x, \alpha_k)]P_k - [g_u(x, \beta_k) - g_u(x, \alpha_k)]P_k \\
&\quad +[f_u(x, \alpha_k) + g_u(x, \beta_k)]P_{k+1} \\
&= f_{uu}(x, \xi)P_k^2 - g_{uu}(x, \sigma)(\beta_k - \alpha_k)P_k \\
&\quad +[f_u(x, \alpha_k) + g_u(x, \beta_k)]P_{k+1},
\end{aligned}
$$

where $\alpha_k \leq \xi \leq u$, $\alpha_k \leq \sigma \leq \beta_k$. But

$$
\begin{aligned}
-g_{uu}(x, \sigma)(\beta_k - \alpha_k)P_k &\leq N_2(Q_k + P_k)P_k \\
&\leq N_2[P_k^2 + P_kQ_k] \\
&\leq \frac{3}{2}N_2P_k^2 + \frac{N_2}{2}Q_k^2,
\end{aligned}
$$

where $\mid g_{uu}(x, u) \mid \leq N_2$ for $x \in \bar{\Omega}$ and $\alpha_0 \leq u \leq \beta_0$. Hence, with $N_0 = \max(N_1 + \frac{3}{2}N_2, \frac{N_2}{2})$ where $\mid f_{uu}(x, u) \mid \leq N$, for $x \in \bar{\Omega}$ and $\alpha_0 \leq u \leq \beta_0$, we get

$$LP_{k+1} \leq [f_u(x, \alpha_k) + g_u(x, \beta_k)]P_{k+1} + N_0(P_k^2 + Q_k^2)$$

or equivalently,

$$LP_{k+1} \leq N_0(P_k^2 + Q_k^2),$$

where $0 < N \leq c_0(x) = [c(x) - f_u(x, \alpha_k) - g_u(x, \beta_k)]$. Taking $\alpha = P_{k+1}$ and $\beta = (N_0/N)[\max_\Omega \mid P_k(x) \mid^2 + \max_{\bar{\Omega}} \mid Q_k(x) \mid^2]$, we see that $B\beta \geq$

BP_{k+1}. Hence by Theorem 1.3.1, we get $P_{k+1}(x) \le \beta$ in $\bar{\Omega}$, which implies the estimate

$$\max_{\bar{\Omega}} \mid P_{k+1}(x) \mid \le \frac{N_0}{N} \left[\max_{\bar{\Omega}} \mid P_k(x) \mid^2 + \max_{\bar{\Omega}} \mid Q_k(x) \mid^2 \right].$$

One can get a similar estimate for Q_{k+1}. The proof is therefore complete.

The following remarks are now in order.

Remarks 1.4.1

(i) If $g(x, u) \equiv 0$ in (1.4.1) we get a result with $f(x, u)$ being convex.

(ii) If $f(x, u) \equiv 0$, we obtain the dual result where $g(x, u)$ is concave.

(iii) Consider the case $g(x, u) \equiv 0$ and $f(x, u)$ is not convex. Suppose that $f(x, u) + G(x, u)$ is convex with $G(x, u)$ convex. Then the BVP (1.4.1) can be written as

$$Lu = \tilde{f}(x, u) + \tilde{g}(x, u), \ Bu = \varphi,$$

where $\tilde{f}(x, u) = f(x, u) + G(x, u)$ and $\tilde{g}(x, u) = -G(x, u)$ so that the conditions of Theorem 1.4.1 are fulfilled and we get the same conclusion when $f(x, u)$ is not convex.

(iv) A dual situation of (3) arises when $f(x, u) \equiv 0$ and $g(x, u)$ is not concave. Suppose that $g(x, u) + F(x, u)$ is concave with $F(x, u)$ concave. Then (1.4.1) is valid with $\tilde{f}(x, u) = -F(x, u)$ and $\tilde{g}(x, u) = g(x, u) + F(x, u)$. Then also the conditions of Theorem 1.4.1 are satisfied and we get the same conclusion.

Thus we see that Theorem 1.4.1 includes several interesting results. We note that if in (3) above, $f(x, u)$ is not convex, we can always find a function $G(x, u)$ which is convex such that $f(x, u) + G(x, u)$ is convex. For example, we can choose $G(x, u) = Mu^2$, $M > 0$, where $-M = \min f_{uu}(x, u)$ for $x \in \bar{\Omega}$ and $\alpha_0 \le u \le \beta_0$ on $\bar{\Omega}$. Thus it is clear that $f(x, u)$ need not be convex to obtain the conclusion. A similar comment holds in case (4) as well.

1.5   Weakly Coupled Mixed Monotone Systems

We consider, in this section, a weakly coupled system of semilinear elliptic BVPs, in the vectorial form

$$\left[\begin{array}{ll} Lu = f(x, u) & \text{in } \Omega, \\ Bu = \phi & \text{on } \partial\Omega, \end{array} \right. \tag{1.5.1}$$

where for each $k = 1, 2, \ldots, N$,

$$L_k u^k = - \sum_{i,j=1}^{n} a_{ij}^k(x) u_{x_i x_j}^k + \sum_{i=1}^{n} b_i^k(x) u_{x_i}^k + c^k(x) u^k,$$

and

$$B_k u^k = p_k(x) u^k + q_k(x) \frac{du^k}{d\gamma}, \quad u \in c^1[\Omega, R^N].$$

We assume that $a_{ij}^k, b_i^k, c^k \in C^\alpha[\bar{\Omega}, R^N]$, $c^k(x) \geq 0$ in Ω, $\varphi^k \in C^{1,\alpha}[\bar{\Omega}, R^N]$ and $f_k \in C^\alpha[\bar{\Omega} \times R^N, R^N]$. Moreover, we let $p_k, q_k \in C^{1,\alpha}[\partial\Omega, R_+^N]$, $p_k(x) > 0$ and γ be the unit outer normal on $\partial\Omega$ and assume the ellipticity condition for each k, namely,

$$\sum_{i,j}^{n} a_{ij}^k(x) \xi_i^k \xi_j^k \geq \theta \mid \xi^k \mid^2, \quad \theta > 0.$$

Also, the vectorial inequalities between the vectors are to be understood componentwise, that is, $u \leq v$ implies $u_i \leq v_i$, $i = 1, 2, \ldots, N$.

Our aim is to extend the monotone iterative technique to mixed monotone systems of BVPs as a generalization of Theorem 1.3.2 of Section 1.3. We shall, however, discuss only monotone increasing systems first and then investigate mixed monotone systems.

Theorem 1.5.1 *Assume that*

(A_1) $\alpha_0, \beta_0 \in C^2[\bar{\Omega}, R^N]$ *are lower and upper solutions of* (1.5.1) *such that* $\alpha_0(x) \leq \beta_0(x)$ *in* Ω;

(A_2) $f \in C^\alpha[\bar{\Omega} \times R^N, R^N]$ *and* $f(x, u)$ *is monotone nondecreasing in* u *for* $x \in \Omega$.

Then there exist monotone sequences $\{\alpha_n(x)\}$, $\{\beta_n(x)\} \in C^{2,\alpha}[\bar{\Omega}, R^N]$ *such that* $\alpha_n(x) \to \rho(x)$, $\beta_n(x) \to r(x)$ *in* $C^2[\bar{\Omega}, R^N]$ *and* (ρ, r) *are coupled minimal and maximal solutions of* (1.5.1), *respectively. If, in addition,* $f(x, u)$ *satisfies*

$$\mid f(x, u) - f(x, v) \mid \leq N \mid u - v \mid, \quad u \geq v, \ u, v \in [\alpha_0, \beta_0], \tag{1.5.2}$$

and $c(x) - N > 0$, *then* $\rho = r = u$ *is the unique solution of* (1.5.1) *such that*

$$\alpha_0(x) \leq u(x) \leq \beta_0 \quad \text{in } \bar{\Omega}. \tag{1.5.3}$$

Remark 1.5.1 If $f(x, u)$ in (A_2) is only quasimonotone nondecreasing in u for $x \in \Omega$, that is,

$$u \le v \text{ and } u_i = v_i \text{ for some } i, \ 1 \le i \le N, \text{ then } f_i(x, u) \le f_i(x, v), \quad (1.5.4)$$

then one normally supposes a further condition

$$f_k(x, u) - f_k(x, v) \ge -M_k(u_k - v_k), \ u \ge v, \ u, v \in [\alpha_0, \beta_0] \qquad (1.5.5)$$

with $M_k \ge 0$ for each k. It then follows that we have

$$\left[\begin{array}{ll} \tilde{L}u = F(x, u) & \text{in } \Omega, \\ Bu = \varphi & \text{on } \partial\Omega, \end{array} \right. \qquad (1.5.6)$$

where $F_k(x, u) = f_k(x, u) + M_k u_k$ and $\tilde{L}_k u^k = L_k u^k + M_k u^k$. Because of condition (1.5.5), $F(x, u)$ is monotone nondecreasing in u for $x \in \Omega$ and consequently, it follows that BVP (1.5.6) satisfies the assumptions of Theorem 1.5.1.

Proof (Theorem 1.5.1) We need to consider in this case, the following linear BVPs for each k,

$$L_k \alpha_{n+1}^k = f_k(x, \alpha_n) \text{ in } \Omega, \ B_k \alpha_{n+1}^k = \varphi^k \text{ on } \partial\Omega, \qquad (1.5.7)$$

$$L_k \beta_{n+1}^k = f_k(x, \beta_k) \text{ in } \Omega, \ B_k \beta_{n+1}^k = \varphi^k \text{ on } \partial\Omega. \qquad (1.5.8)$$

Since for each k, (1.5.7) and (1.5.8) are scalar linear BVPs, the proof can be carried out following the proof of Theorem 1.3.2 and Corollary 1.3.2 with $g(x, u) \equiv 0$ by making appropriate modifications in a routine way, and hence we do not provide the proof.

Let us now consider the mixed monotone weakly coupled system of BVPs, given by

$$\left[\begin{array}{ll} Lu = f(x, u) + g(x, u) & \text{in } \Omega, \\ Bu = \varphi & \text{on } \partial\Omega, \end{array} \right. \qquad (1.5.9)$$

where $f(x, u)$ is nondecreasing in u and $g(x, u)$ is nonincreasing in u for $x \in \Omega$. In order to prove a result on the monotone iterative technique corresponding to Theorem 1.3.2, we need to employ the notion of coupled lower and upper solutions of type I and modify the proof of Theorem 1.3.2 suitably. Instead of proceeding with that approach, we shall utilize the idea of a reflection operator and expand the given system suitably to transform it into the system of the type (1.5.1) so that we can apply Theorem 1.5.1 to yield the desired result analogous to Theorem 1.3.2. This is what we will do next. We prove the following result.

Theorem 1.5.2 *Assume that*

(A_1) $\alpha_0, \beta_0 \in C^2[\bar{\Omega}, R^N]$ *are coupled lower and upper solutions of* (1.5.9) *which are of type I as given in Definition 1.3.1 such that* $\alpha_0(x) \leq \beta_0(x)$ *in* Ω;

(A_2) $f, g \in C^\alpha[\bar{\Omega} \times R^N, R^N]$, $f(x, u)$ *is nondecreasing in* u *and* $g(x, u)$ *is nonincreasing in* u *for* $x \in \Omega$;

(A_3) f, g *satisfy*

$$\left[\begin{array}{ll} f(x, u) - f(x, v) & \leq N_1(u - v) \\ g(x, u) - g(x, v) & \geq -N_2(u - v), \end{array} \right. \tag{1.5.10}$$

for $u \geq v$, $u, v \in [\alpha_0, \beta_0]$, $N_1, N_2 \geq 0$ *and* $c(x) - (N_1 + N_2) > 0$ *on* Ω.

Then there exist monotone sequences $\{\alpha_n(x)\}$, $\{\beta_n(x)\} \in C^{2,\alpha}[\bar{\Omega}, R^N]$ *such that* $\alpha_n(x) \to \rho(x)$, $\beta_n(x) \to r(x)$ *in* $C^2[\bar{\Omega}, R^N]$ *and* $\rho = r = u$ *is the unique solution of* (1.5.9) *satisfying* $\alpha_0(x) \leq u(x) \leq \beta_0(x)$ *in* $\bar{\Omega}$.

Proof We define $G \in C^\alpha[\bar{\Omega} \times R^{2N}, R^{2N}]$ by

$$G(x, z) = G(x, u, v) = \left[\begin{array}{l} f(x, u) + g(x, v), \\ f(x, v) + g(x, u) \end{array} \right]. \tag{1.5.11}$$

Introduce an ordering in $K = R^N_+ \times \{-R^N_+\}$ by $z_1 \leq z_2$ if and only if $u_1 \leq u_2$, $v_1 \geq v_2$ where $z_1 = (u_1, v_1)$ and $z_2 = (u_2, v_2)$. Define the reflection operator $P : R^{2N} \to R^{2N}$ by

$$P(u, v) = (v, u) \tag{1.5.12}$$

and let $\xi_0 = (\alpha_0, \beta_0)$ so that $\eta_0 = P\xi_0 = (\beta_0, \alpha_0)$ and $\xi_0 \leq \eta_0$. We then have

$$L\xi_0 \leq G(x, \xi_0) \text{ in } \Omega, \ B\xi_0 = \tilde{\phi} = (\phi, \phi) \text{ on } \partial\Omega,$$

$$L\eta_0 \geq G(x, \eta_0) \text{ in } \Omega, \ B\eta_0 = \tilde{\phi} = (\phi, \phi) \text{ on } \partial\Omega.$$

Then the BVP

$$Lz = G(x, z) \text{ in } \Omega, \ Bz = \tilde{\phi} \text{ on } \partial\Omega, \tag{1.5.13}$$

satisfies the assumptions of Theorem 1.5.1, since

$$G(x, z_1) - G(x, z_2) \leq (N_1 + N_2)(z_1 - z_2), \ z_1 \geq z_2,$$

where $z_1 = (u_1, v_1)$ and $z_2 = (u_2, v_2)$. Then defining $z_0 = \tilde{\phi}$ so that $\xi_0 \leq z_0 \leq \eta_0$ on $\partial\Omega$, we get from Theorem 1.5.1 that there exist monotone sequences

$\{\xi_n(x)\}$, $\{\eta_n(x)\} \in C^2[\bar{\Omega}, R^{2N}]$ which converge uniformly and monotonically to the unique solution $z(x)$ of (1.5.13) such that $\xi_0(x) \leq z(x) \leq \eta_0(x)$ on $\bar{\Omega}$. This implies $\alpha_0(x) \leq u(x) \leq \beta_0(x)$ on $\bar{\Omega}$, where $u(x)$ is the unique solution of (1.5.9) in view of the definition of the reflection operator P. The proof is therefore complete.

We note that if $f(x, u)$ in (1.5.9) is only assumed to be quasimonotone in u and satisfies (1.5.5), then the BVP we need to consider translates to

$$\left[\begin{array}{ll} \tilde{L}u = F(x, u) + g(x, u) & \text{in } \Omega, \\ Bu = \phi & \text{on } \Omega, \end{array} \right.$$

which satisfies the assumptions of Theorem 1.5.2 and consequently, we get the same conclusion of Theorem 1.5.2 for the BVP (1.5.9).

1.6   Elliptic Systems in Unbounded Domains

We shall investigate, in this section, existence results in a closed set generated by lower and upper solutions for the weakly coupled elliptic boundary value problem in the vectorial form

$$\left[\begin{array}{ll} Lu = f(x, u) & \text{in } \Omega, \\ Bu = \phi & \text{on } \partial\Omega, \end{array} \right. \tag{1.6.1}$$

in various unbounded domains Ω in R^n, including the exterior of a bounded domain. Specifically, we consider the BVP (1.6.1) in the exterior of a bounded domain, the Dirichlet BVP

$$\left[\begin{array}{ll} Lu = f(x, u) & \text{on } \partial\Omega, \\ u = \phi & \text{on } \partial\Omega, \end{array} \right. \tag{1.6.2}$$

in a general bounded domain, and the system

$$Lu = f(x, u) \text{ in } R^n. \tag{1.6.3}$$

Here for each $k = 1, 2, \ldots, N$,

$$L_k u^k = -\sum_{i,j}^{n} a_{ij}^k(x) u_{x_i x_j}^k + \sum_{i=1}^{n} b_i^k(x) u_{x_i}^k, \tag{1.6.4}$$

$$B_k u^k = p_k(x) u^k + q_k(x) \frac{du^k}{d\gamma}, \tag{1.6.5}$$

L_k is uniformly elliptic on $\bar{\Omega} = \Omega \cup \partial\Omega$, $a_{ij}^k \in C_{\text{loc}}^{2,\alpha}(\bar{\Omega})$, $b_i^k \in C_{\text{loc}}^{1,\alpha}(\bar{\Omega})$, $p^k, q^k \in C_{\text{loc}}^{1,\alpha}(\partial\Omega)$, $p^k > 0$ and $q_k \geq 0$. Also, $f_k \in C^\alpha[\bar{\Omega}, R^N]$ and $\phi_k \in C_{[\bar{\Omega}, R^N]}^{1,\alpha}$.

We shall prove an existence theorem for each of the three BVPs (1.6.1), (1.6.2), and (1.6.3). We shall note, as usual, that the inequalities between the vectors $u, v \in R^N$ are understood component-wise.

Let Ω be a general unbounded domain in R^n and let $\bar{\Omega} \in \Omega \cup \partial\Omega$.

Definition 1.6.1 *We shall say that $u(x)$ is a solution of (1.6.1) if $u \in C_{\text{loc}}^{2,\alpha}[\bar{\Omega}, R^N]$ and satisfies (1.6.1) for every bounded subdomain of Ω.*

Since the BVPs (1.6.2) and (1.6.3) are special cases of (1.6.1), the same definition of a solution holds for these problems. For proving the existence of a solution, we require the following Müller-type lower and upper solutions Müller-type lower and upper solutions.

Definition 1.6.2 *The functions $\alpha_0, \beta_0 \in C_{\text{loc}}^2[\bar{\Omega}, R^N]$ are said to be Müller-type lower and upper solutions of (1.6.1) if $\alpha_0(x) \leq \beta_0(x)$ in $\bar{\Omega}$,*

$$L_k \alpha_0^k \leq f_k(x, \sigma) \text{ in } \Omega, \quad B_k \alpha_0^k \leq \phi \text{ on } \partial\Omega,$$

for $\alpha_0(x) \leq \sigma \leq \beta_0(x)$ with $\sigma_k = \alpha_{0k}$, and

$$L_k \beta_0^k \geq f_k(x, \sigma) \text{ in } \Omega, \quad B_k \beta_0^k \geq \phi \text{ on } \partial\Omega,$$

for $\alpha_0(x) \leq \sigma \leq \beta_0(x)$ with $\sigma_k = \beta_{0k}$ for $k = 1, 2, \ldots, N$.

We set, as usual, $[\alpha_0, \beta_0] = [v \in C_{\text{loc}}^2[\bar{\Omega}, R^N] : \alpha_0 \leq v \leq \beta_0 \text{ in } \bar{\Omega}]$. We shall also suppose that there exists an increasing sequence of smooth bounded domains $\{\Omega_m\}$ in R^n with boundaries Γ_m such that

$$\Omega_m \subset \Omega_{m+1} \subset \Omega, \quad \bigcup_{m=1}^{\infty} \Omega_m = \Omega \text{ and } \Gamma_m = \partial\Omega_m \cup \partial\Omega_m^0, \tag{1.6.6}$$

where $\partial\Omega_m$ is that portion of $\partial\Omega$ contained in $\bar{\Omega}_m$ and $\partial\Omega_m^0$ is the boundary of Ω_m lying in Ω. The smoothness of Ω_m is in the sense that it belongs to $C^{2,\alpha}[\bar{\Omega}_m, R^N]$. When Ω is the exterior of a bounded domain, it is enough to take $\Omega_m = \Omega \cap B_m$, where $\{B_m\}$ is an increasing sequence of balls in R^n containing $\partial\Omega$. In this case, $\partial\Omega_m = \partial\Omega$ and $\partial\Omega_m^0 = S_m$, where S_m is the surface of the ball B_m.

We shall first investigate the exterior problem (1.6.1) where Ω is the exterior of a bounded domain.

Consider the BVP in the vectorial form

$$
\left[
\begin{array}{ll}
Lu = f(x,u) & \text{in } \Omega_m, \\
Bu = \phi & \text{on } \partial\Omega
\end{array}
\right.
\tag{1.6.7}
$$

in a fixed bounded domain Ω_m. We first prove the following lemma.

Lemma 1.6.1 *For each fixed Ω_m, the BVP (1.6.7) has a sequence of so-lutions $\{u^{m+n}\}$ in $C^{2,\alpha}[\bar{\Omega}_m, R^N]$ such that $\alpha_0 \le u^{m+n} \le \beta_0$ in $\bar{\Omega}_m$ and $\mid u^{m+n} \mid_{C^{2,\alpha}_{[\bar{\Omega}_m,R^N]}} \le c$, $n = 1, 2, \ldots$, where c is a constant independent of n.*

Proof   We consider the BVP (1.6.7) with $\Omega_m = \Omega \cap B_m$ and the additional boundary condition

$$
u(x) = u^0(x) \text{ on } S_m,
\tag{1.6.8}
$$

where $u^0(x)$ is equal to either $\alpha_0(x)$ or $\beta_0(x)$. Let us fix $u^0 = \beta_0$, since the other case yields a similar result. Since by Definition 1.6.2, the restriction α_0, β_0 to $\bar{\Omega}_m$ are Müller-type lower and upper solutions of (1.6.7), (1.6.8), the standard existence theorem for an elliptic system in a bounded domain as-sures that the problem (1.6.7) with (1.6.8) has a solution $u^m \in C^{2,\alpha}[\bar{\Omega}_m, R^N]$ satisfying $x_0 \le u^m \le \beta_0$ in $\bar{\Omega}_m$. Moreover, if u^{m+u} is a solution of (1.6.7) with (1.6.8), where Ω_m is replaced by a larger domain Ω_{m+n}, $n = 1, 2, \ldots$, then the restriction of u^{m+n} to $\bar{\Omega}_m$ satisfies (1.6.7) in $\bar{\Omega}_m$. Hence, for each fixed $\bar{\Omega}_m$, the problem (1.6.7) has a sequence of solutions $\{u^{m+n}\}$ such that

$$
\alpha_0(x) \le u^{m+n}(x) \le \beta_0(x) \text{ in } \bar{\Omega}_m, \ n = 1, 2, \ldots .
\tag{1.6.9}
$$

This implies that for each $k = m + n$, $u^k(x)$ fulfills the relation

$$
\left[
\begin{array}{l}
Lu^k = f(x, u^k) \text{ in } \Omega_n, \quad Bu^k = \phi \text{ on } \partial\Omega, \\
\quad\quad \alpha_0 \le u^k \le \beta_0 \quad \text{on } S_m.
\end{array}
\right.
\tag{1.6.10}
$$

Define

$$
h^k(x) = \left[
\begin{array}{ll}
\phi(x), & x \in \partial\Omega, \\
u^k(x), & x \in S_m.
\end{array}
\right.
\tag{1.6.11}
$$

Since $u^k \in W^{1,p}[\Omega_m, R^N]$ and is uniformly bounded in $\bar{\Omega}_m$, for every k, the estimate

$$
\mid u^k \mid_{W^{1,p}_{[\Omega_m,R^N]}} \le C[\mid f(\cdot, u^k) \mid_{L^q[\Omega_m,R^N]} + \mid h^k \mid_{L^p[\Gamma_m,R^N]} + \mid h^k \mid_{L^q[\Gamma^0_m,R^N]}
$$

$$
\tag{1.6.12}
$$

and the hypothesis of f shows that u^k is uniformly bounded in $W^{1,p}_{[\Omega_m,R^N]}$, where c is some positive constant and $p > 1$, $q = \frac{p}{p-1}$. By the Agmon–Douglas–Nirenberg estimate

$$| u^k |_{W^{2,p}_{[\Omega_m,R^N]}} \leq c[| f(\cdot, u^k) |_{L^p[\Omega_m,R^N]} + | h^k |_{1-\frac{1}{p}}], \qquad (1.6.13)$$

there exists a constant c independent of k, such that $| u^k |_{W^{2,p}_{[\Omega_m,R^N]}} \leq c$, where $| h^k |_{1-\frac{1}{p}} = \inf [| w |_{W^{1,p}_{[\Omega_m,R^N]}} : w \in C^1[\Omega_m, R^N]$ and $w = h^k$ on $\Gamma_m]$. Choose $p > N$ such that $\alpha = 1 - \frac{N}{p} > 0$. By the embedding theorem $\{u^k\}$ is uniformly bounded in $C^{1,\alpha}[\bar{\Omega}_m, R^N]$ by some constant c. It then follows from Schauder's estimate

$$| u^k |_{C^{2,\alpha}_{[\bar{\Omega}_m,R^N]}} \leq C[| f(\cdot, u^k) |_{C^2_{[\bar{\Omega}_m,R^N]}} + | h^k |_{C^{1,\alpha}_{[\Gamma_m,R^N]}}], \qquad (1.6.14)$$

that $u^k \in C^{2,\alpha}_{[\bar{\Omega},R^N]}$ and $| u^k |_{C^{2,\alpha}_{[\bar{\Omega}_m,R^N]}} \leq c$ for some constant c, independent of k. This leads to the desired estimate of the lemma.

Based on Lemma 1.6.1, we can prove the following existence result.

Theorem 1.6.1 *Let α_0, β_0 be lower and upper solutions of Müller-type as given in Definition 1.6.2. Then there exists a solution $u(x)$ of (1.6.1) such that $\alpha_0(x) \leq u(x) \leq \beta_0(x)$ in $\bar{\Omega}$.*

Proof In view of Lemma 1.6.1 and (1.6.8), the Ascoli–Arzela theorem implies that the sequence $\{u^{m+n}\}$ contains a subsequence denoted by $\{u^{(m+n)}\}$, which converges in $C^2[\bar{\Omega}_m, R^N]$ to a function $u^{(m)}$ as $n \to \infty$. It is clear that $u^{(m)}$ is a solution of (1.6.7). Starting with $m = 1$, the subsequence $\{u^{(1,n)}\}$ converges to a solution $u^{(1)}$ of (1.6.7) in $\bar{\Omega}_1$ and $\alpha_0 \leq u^{(1)} \leq \beta_0$ in $\bar{\Omega}_1$. Similarly, the subsequence $\{u^{(2,n)}\}$ converges to $u^{(2)}$ of (1.6.7) in $\bar{\Omega}_2$ and $\alpha_0 \leq u^{(2)} \leq \beta_0$ in $\bar{\Omega}_2$. Proceeding in this way, we get that the subsequence $\{u^{(m,n)}\}$ converges to a solution $u^{(m)}$ of (1.6.7) in $\bar{\Omega}_m$ and $\alpha_0 \leq u^{(m)} \leq \beta_0$ in $\bar{\Omega}_m$. By the diagonalization process, we obtain a subsequence $\{u^{(m,n)}\}$ which converges in $C^2[\bar{\Omega}_m, R^N]$ for every m. Define $\bar{u} = \lim_{m\to\infty} u^{(m,m)}$. Then by Lemma 1.6.1, $\alpha_0 \leq \bar{u} \leq \beta_0$ in $\bar{\Omega}$. Moreover, $\bar{u}$ is a solution of (1.6.1) in every domain $\bar{\Omega}_m$. If $\bar{\Omega}^*$ is a bounded subdomain in $\bar{\Omega}$, then $\bar{\Omega}^*$ is contained in $\bar{\Omega}_m$ for some m, and the foregoing conclusion assures that $\bar{u}$ satisfies (1.6.1) in $\bar{\Omega}^*$. This shows that $\bar{u}$ is a solution of (1.6.1) completing the proof.

We note that if we take in (1.6.8), $u^0(x) = \alpha_0(x)$ instead of $\beta_0(x)$, the foregoing argument shows that there exists a subsequence $\{u^{m,m}\}$ which converges to a solution $\underaccent{\sim}{u}$ of (1.6.1).

We shall next discuss the BVP (1.6.2) in a general unbounded domain Ω, including the half-space $\Omega = R^n_+$. Since this problem may be considered as a special case of (1.6.1) with $p(x) \equiv 0$ and $q(x) = 1$, the definition (1.6.2) remains the same. In addition, we require that

$$\alpha_0(x) = \beta_0(x) = d(x), \quad x \in \partial\Omega. \tag{1.6.15}$$

Let Ω_m be any one of the domains in (1.6.6) and consider the BVP

$$\left[\begin{array}{ll} Lu = f(x, u) & \text{in } \Omega_m, \\ u = \phi & \text{on } \partial\Omega_m, \end{array}\right. \tag{1.6.16}$$

together with the additional boundary condition

$$u(x) = u_0(x) \text{ on } \partial\Omega^0_m, \tag{1.6.17}$$

where $u_0(x)$ is either $\alpha_0(x)$ or $\beta_0(x)$. Because of (1.6.15), the boundary condition h on $\Gamma_m = \partial\Omega_m \cap \partial\Omega^0_m$ given by

$$h(x) = \left[\begin{array}{ll} \phi(x), & x \in \partial\Omega_m, \\ u_0(x), & x \in \partial\Omega^0_m, \end{array}\right. \tag{1.6.18}$$

is in $C^{2,\alpha}[\bar{\Omega}_m, R^N]$. By the standard existence theorem in bounded domains, the BVP given in (1.6.16) and (1.6.17) has a solution $u^m \in C^{2,\alpha}[\bar{\Omega}_m, R^N]$ and $\alpha_0 \le u^m \le \beta_0$ in $\bar{\Omega}_m$. As before, for each $n = 1, 2, \ldots$, the solution u^{m+n} of (1.6.16) and (1.6.17) in the larger domain Ω_{m+n} satisfies (1.6.16) and $\alpha_0 \le u^{m+n} \le \beta_0$ in $\bar{\Omega}_{m+n}$. Hence for each domain Ω_m, the problem (1.6.16) has a sequence of solutions $\{u^{m+n}\}$ $n = 1, 2, \ldots$, such that $\alpha_0 \le u^{m+n} \le \beta_0$ in $\bar{\Omega}_{m+n}$. Employing the argument as in the proof of Lemma 1.6.1, this sequence is uniformly bounded in $C^{2,\alpha}[\bar{\Omega}_m, R^N]$. It then follows from the proof of Theorem 1.6.1 that $\{u^{m+n}\}$ contains a subsequence $\{u^{(m,m)}\}$ that converges to a solution u of (1.6.2) as $m \to \infty$. This reasoning gives the following analogous result for problem (1.6.2).

Theorem 1.6.2 *Let α_0, β_0 be Müller-type lower and upper solutions of (1.6.2) and suppose that the condition (1.6.15) holds. Then there exists a solution u of (1.6.2) with $\alpha_0 \le u \le \beta_0$ in $\bar{\Omega}$.*

When $\Omega = R^n$, the problem (1.6.1) reduces to (1.6.3). Here the lower and upper solutions α_0, β_0 are required to satisfy only the differential inequalities with boundary inequalities in Definition 1.6.2. By choosing Ω_m as an increasing sequence of balls in R^n and ignoring the boundary condition in (1.6.7) the same reasoning as in the proof of Lemma 1.6.1 shows that the equation

$$Lu = f(x, u) \text{ in } \Omega_m,$$

has a sequence of solutions $\{u^{m+n}\}$ which are uniformly bounded in $C^{2,\alpha}[\bar{\Omega}_m, R^N]$ and satisfy (1.6.9). It follows by a diagonal selection process as in the proof of Theorem 1.6.1 that this sequence contains a subsequence which converges to a solution of (1.6.3). This observation leads to the following conclusion.

Theorem 1.6.3 *Let α_0, β_0 be Müller-type lower and upper solutions of (1.6.3) Then there exists a solution u of (1.6.3) such that $\alpha_0 \leq u \leq \beta_0$ in R^n.*

1.7 Monotone Iterative Technique (MIT) for Systems in Unbounded Domains

To extend the monotone iterative technique for the BVPs (1.6.1), (1.6.2), and (1.6.3), we need $f(x, u)$ to be quasimonotone nondecreasing in u for $x \in \Omega$ and also, for each $i = 1, 2, \ldots, N$, the condition

$$f_i(x, u) - f(x, v) \geq -M_i(u_i - v_i), \ u \geq v, \ u, v \in [\alpha_0, \beta_0], M_i > 0. \quad (1.7.1)$$

The condition (1.7.1) together with the quasimonotonicity assumption implies that $F_i(x, u) = f_i(x, u) + M_i u_i$ is monotone nondecreasing in u for $x \in \Omega$. Also, it is clear that, because of the quasimonotonicity of $f(x, u)$ the definition 1.6.2 leads to

$$L\alpha_0 \leq f(x, \alpha_0) \text{ in } \Omega, \ B\alpha_0 \leq \phi \text{ on } \partial\Omega, \quad (1.7.2)$$

$$L\beta_0 \geq f(x, \beta_0) \text{ in } \Omega, \ B\beta_0 \geq \phi \text{ on } \partial\Omega, \quad (1.7.3)$$

respectively. Then the problem (1.6.1) may be rewritten as

$$\left[\begin{array}{ll} L_i u_i + M_i u_i = F_i(x, u) & \text{in } \Omega, \\ Bu = \phi & \text{on } \partial\Omega, \ i = 1, 2, \ldots, N. \end{array} \right. \quad (1.7.4)$$

Consider the sequence of linear BVPs

$$\left[\begin{array}{lll} L_i u_i^m + M_i u_i^m & = F_i(x, u^{m-1}) \text{ in } \Omega_m, \\ Bu^m & = \phi \text{ on } \partial\Omega, \\ u^m & = u_0 \text{ on } S_m, \end{array} \right. \quad (1.7.5)$$

where $m = 1, 2, \ldots$ and U^m is the extension of u^m given by

$$U^m(x) = \begin{bmatrix} u^m(x), & x \in \bar{\Omega}_m, \\ u_0(x), & x \in \bar{\Omega}/\bar{\Omega}_m. \end{bmatrix} \qquad (1.7.6)$$

We note that $U^0(x) = u_0(x)$ for all $x \in \bar{\Omega}$. As before, we choose $u_0 = \alpha_0$ or β_0. Denote the solution of (1.7.5) and its extension by $\bar{u}^m$ and $\bar{U}^m$ when $u_0 = \beta_0$ and by $\underline{u}^m$ and $\underline{U}^m$, when $u_0 = \alpha_0$, respectively. It is easy to see by an inductive argument that the sequences $\{\bar{u}^m\}$, $\{\underline{u}^m\}$ are in $C^{2,\alpha}[\bar{\Omega}_m, R^N]$ for every m. The following lemma provides a monotone property of the sequences $\{\bar{U}^m\}$ and $\{\underline{U}^m\}$.

Lemma 1.7.1 *Let $f(x, u)$ possess the quasimonotone property and satisfy the condition (1.7.1). Then the sequences $\{\bar{U}^m\}$, $\{\underline{U}^m\}$ defined by (1.7.6) possess the monotone property,*

$$\alpha_0 \leq \underline{U}^m \leq \underline{U}^{m+1} \leq \bar{U}^{m+1} \leq \bar{U}^m \leq \beta_0 \text{ in } \bar{\Omega}, \ m = 1, 2, \ldots. \qquad (1.7.7)$$

Proof   Since for each $i = 1, 2, \ldots N$, (1.7.5) is a scalar BVP, the proof follows similar arguments as in the case of the scalar BVP in Section 1.3. Let us provide a sketch as follows. By (1.7.2), (1.7.5) and $\bar{U}^0 = \beta_0$, the functions $w_i^0 = \bar{u}_i^0 - \bar{u}_i^1 = \beta_{0i} - \bar{u}_i^1$, $i = 1, 2, \ldots, N$, satisfy the relations

$$L_i w_i + M_i w_i = L_i \beta_{0i} + M_i \beta_{0i} - F_i(x, \bar{U}^0) = L_i \beta_{0i} - f_i(x, \beta_0) \geq 0 \text{ in } \Omega_1$$

$$Bw = B(\beta_0 - \phi) \geq 0 \text{ on } \partial\Omega, \ w_i^1 = \bar{u}_i^1 \bar{u}_i^0 - \geq \beta_{0i} - u_i^0 \geq 0 \text{ on } S_i.$$

By Corollary 1.3.1, $w_1^0 \geq 0$ in Ω_1, which yields $\bar{u}_i^1 \leq \bar{u}^0$ in Ω_1. The fact that $\bar{U}^0 = \bar{u}^0$ in $\bar{\Omega}/\Omega_1$ shows that $\bar{U}^1 \leq \bar{U}^0$ in Ω_1. Similar reasoning using (1.7.3) gives $\underline{U}^1 \geq \bar{U}^0$ in Ω_1. Let $w_i^1 = \bar{u}_i^{-1} - \underline{u}_i^1$. Then by (1.7.5) and the monotone character of $F(x, u)$, we arrive at

$$L_i w_i^1 + M_i w_i^1 = F_i(x, \bar{U}^0) - F_i(x, \bar{U}^0) \geq 0 \text{ in } \Omega_1,$$

$$Bw^1 \geq 0 \text{ on } \partial\Omega \text{ and } w_i^1 = \bar{u}_i^0 - \underline{u}_i^0 \geq 0 \text{ on } S_1.$$

This implies by Corollary 1.3.1, that $\bar{u}_i^1 \geq \underline{u}_i^1$ in Ω_1. Since $\bar{U}_i^0 - \underline{U}^0 = 0$ on $\bar{\Omega}/\bar{\Omega}_1$, the inequality $\bar{U}^1 \geq \underline{U}^1$ holds on $\bar{\Omega}$. The relation (1.7.7) follows by induction arguments as in Section 1.3.

In view of (1.7.7), the pointwise limits

$$\lim \bar{U}^m(x) = \bar{u}(x), \ \lim \underline{U}^m(x) = \underline{u}(x) \text{ as } m \to \infty, \qquad (1.7.8)$$

exist and satisfy

$$\alpha_0 \leq \underline{u} \leq \bar{u} \leq \beta_0 \text{ in } \bar{\Omega}. \qquad (1.7.9)$$

We shall show that $(\underline{u}, \bar{u})$ are minimal and maximal solutions of (1.6.1) in $[\alpha_0, \beta_0]$.

Theorem 1.7.1 *Let α_0, β_0 be lower and upper solutions of the exterior BVP (1.6.1). Suppose that $f(x,u)$ is quasimonotone nondecreasing in u and (1.7.1) is satisfied. Then there exist monotone sequences $\{\underline{U}^m\}, \{\bar{U}^m\}$ which converge monotonically to minimal and maximal solutions of (1.6.1) and $\alpha_0 \leq \underline{u} \leq \bar{u} \leq \beta_0$ in $\bar{\Omega}$.*

Proof   Given any bounded domain Ω^* in Ω, there exists Ω_m such that $\bar{\Omega}^* \subset \bar{\Omega}_m$. Let us show that $\bar{u}$ is a solution of (1.6.1) in $\bar{\Omega}_m$. Consider the restriction of $\bar{U}^k$ on $\bar{\Omega}_m$ where $k > m$ and is arbitrary. Then $\bar{U}^k = \bar{u}^k$ in $\bar{\Omega}_m$ and

$$L_i\bar{U}_i^k + M_i\bar{U}_i^k = F_i(x, \bar{U}^{k-1}) \text{ in } \Omega_m, \; B\bar{U} = \phi \text{ on } \partial\Omega, \; i = 1, 2, \ldots, N. \tag{1.7.10}$$

Since $\alpha_0 \leq \bar{U}^k \leq \beta_0$ on S_m and $F_i(x, \bar{U}^{k-1})$ is uniformly bounded on $\bar{\Omega}_m$, the estimates (1.6.12), (1.6.13), with f replaced by F, assure that $\{\bar{U}^k\}$ is uniformly bounded in $W^{2,p}(\Omega_m)$ for every $p > 1$. An application of the embedding theorem and the Schauder estimate (1.6.14) implies that the sequence $\{\bar{U}^k\}$ contains a subsequence which converges in $C^{2,\alpha}[\bar{\Omega}_m, R^N]$ to a function $\bar{U}$ as $k \to \infty$. Since $\{\bar{U}^k\}$ converges to $\bar{u}$ as $k \to \infty$, we must have $\bar{u} = \bar{U}$ and the entire sequence $\{\bar{U}^k\}$ converges in $C^2[\bar{\Omega}_m, R^N]$ to $\bar{u}$. Letting $k \to \infty$ in (1.7.10) shows that $\bar{u}$ is a solution of (1.6.1) in $\bar{\Omega}_m$, and, in particular, $\bar{u}$ satisfies (1.6.1) in $\bar{\Omega}^*$. This proves the conclusion for $\bar{u}$. The proof of $\underline{u}$ is the same.

To show that $(\underline{u}, \bar{u})$ are extremal solutions of (1.6.1) in $[\alpha_0, \beta_0]$, we observe that if u^* is a solution of (1.6.1) in $[\alpha_0, \beta_0]$, then the pairs (α_0, u^*) (u^*, β_0) are also ordered lower and upper solutions. It is clear from the iteration process (1.7.5) that the sequence of iterations with $u_0 = u^*$ consists of the same element $\{u^*\}$. By considering β_0 and u^* as a pair of upper and lower solutions in Lemma 1.7.1, we obtain $u^* \leq \bar{U}^m$ in $\bar{\Omega}$ for every m. This leads to $u^* \leq \bar{u}$ in $\bar{\Omega}$. A similar argument gives $\underline{u} \leq u^*$ in $\bar{\Omega}$. Hence $(\underline{u}, \bar{u})$ are the extremal solutions of (1.6.1) completing the proof.

Consider next the Dirichlet BVP (1.6.2) in a general bounded domain Ω. In view of the quasimonotonicity of $f(x,u)$, the requirement of upper and lower solutions is reduced to

$$\left[\begin{array}{ll} L_i\beta_{0i} \geq f_i(x, \beta_0), & L_i\alpha_{0i} \leq f_i(x, \alpha_0) \text{ in } \Omega, \\ \alpha_0(x) \leq \phi(x) \leq \beta_0(x) \text{ on } \partial\Omega, & i = 1, 2, \ldots, N. \end{array} \right. \tag{1.7.11}$$

Using u^0 as α_0 or β_0, we can construct a sequence $\{u^m\}$ from the linear BVPs

$$\left[\begin{array}{l} L_i u_i^m + M_i u_i^m = F_i(x, U^{m-1}) \text{ in } \Omega_m, \\ u_i^m(x) = \phi(x) \text{ on } \partial\Omega_m, \\ u_i^m(x) = u_{0i}(x) \text{ on } \partial\Omega_m^0, \end{array} \right. \qquad (1.7.12)$$

where U^m is given by (1.7.6). Because of (1.6.15), a unique solution u^m to (1.7.12) exists and is in $C^{2,\alpha}[\bar{\Omega}_m, R^N]$ for each $m = 1, 2, \ldots$. We again denote its extension to $\bar{\Omega}$ by $\bar{U}^m$ when $u_0 = \beta_0$ and by $\underline{U}^m$ when $u_0 = \alpha_0$. This extension leads to the following analogous result for BVP (1.6.2).

Theorem 1.7.2 *Let α_0, β_0 be ordered lower and upper solutions of (1.6.2) and satisfy (1.6.15). Suppose that $f(x, u)$ is quasimonotone nondecreasing in u and (1.7.1) holds. Then the conclusion of Theorem 1.7.1 holds for BVP (1.6.2).*

Proof   By the same reasoning as in the proof of Theorem 1.7.1, it is clear that the sequences $\{\bar{U}^m\}$, $\{\underline{U}^m\}$ possess the monotone property (1.7.7) and converge to some limits $\bar{u}$ and $\underline{u}$ respectively as $m \to \infty$. A similar argument as in the proof of Theorem 1.7.1 shows that $(\underline{u}, \bar{u})$ are extremal solutions of (1.6.2). We omit the details.

When $\Omega = R^N$, the definition of lower and upper solutions for the BVP (1.6.3) becomes

$$L_i \beta_{0i} \geq f_i(x, \beta_0), \quad L_i \alpha_{0i} \leq f_i(x, \alpha_0) \text{ in } R^n. \qquad (1.7.13)$$

Let $\{u^m\}$ be the sequence of iterates given by (1.7.5) with $\Omega_m = B_m$ and without the boundary condition on $\partial\Omega$. Denote the extension of $\{u^m\}$ to $\bar{\Omega}$ by $\{\bar{U}^m\}$ when $u_0 = \beta_0$, and by $\{\underline{U}^m\}$ when $u_0 = \alpha_0$. It is easily seen by following a similar argument as in the proof of Theorem 1.7.1 and Lemma 1.7.1, without the boundary condition on $\partial\Omega$, that the following result holds.

Theorem 1.7.3 *Let α_0, β_0 be ordered lower and upper solutions of (1.6.3). Suppose further that $f(x, u)$ is quasimonotone nondecreasing in u and (1.7.1) holds. Then the conclusion of Theorem 1.7.1 is valid for the problem (1.6.3).*

1.8   Notes and Comments

Although the method of lower and upper solutions and the construction of monotone sequences for proving the existence of extremal solutions of nonlinear problems has been known from the early part of the last century, see

Müller [70], Nagumo [72], Ako [2], Kalaba [37], Keller [38], Keller and Cohen [39], it is in the latter part that the method now known as the monotone iterative technique, was developed systematically by using lower and upper solutions and constructing monotone sequences that converge to the minimal and maximal solutions of the given nonlinear problem; refer to Amman [3], Sattinger [78]. The ideas embedded in the method of lower and upper solutions combined with the monotone iterative technique have been extended, generalized and refined for a variety of nonlinear problems for ordinary and partial differential equations in Ladde et al. [47]. Moreover, by means of the generalized iteration principle, this technique has also been extended to deal with discontinuous differential equations very recently by Carl and Heikkilä [15].

Similarly, obtaining rapid convergence of the constructed monotone sequences to the solution, starts with the fruitful idea of Chaplygin [21], where strict lower and upper solutions and the assumption of convexity on the nonlinear term are utilized. The method of quasilinearization developed by Bellman [5], Bellman and Kalaba [6], on the other hand, uses only the convexity assumption and by choosing the initial approximation carefully, obtains a lower bounding monotone sequence which converges rapidly to the solution, once its existence is assumed. Recently, employing the method of lower and upper solutions together with the method of quasilinearization and utilizing the idea of Newton and Fourier, it is realized that one can construct concurrently lower and upper monotone sequences that converge to the unique solution under less restrictive assumptions. For developments in this direction, which is known as generalized quasilinearization, see Lakshmikantham and Vatsala [60].

Section 1.2 provides a preview of the foregoing framework in detail. The two different situations covered by the monotone iterative technique in Section 1.3 are based on the work of Köksal and Lakshmikantham [44]. The unified mechanism developed in each situation proves a single result from which several known as well as new results can be derived as corollaries. The setting for the method of generalized quasilinearization is precisely the same, that is, one can derive several special cases of interest from one single result, which is the content of Section 1.4 and is due to Lakshmikantham and Vatsala [62]. The contents of Section 1.5 on coupled mixed monotone systems are new in the given set-up and are modeled on the work of Lakshmikantham and Vatsala [61] for monotone flows. See Gouze and Hadeler [35] for ideas in monotone flows. All the results of Sections 1.6 and 1.7 are adapted from the work of Pao [74], [75] which are related to existence and monotone iterative techniques for systems of elliptic BVPs in unbounded domains.

For allied results, see Bebernes and Schmitt [4], Wildenauer [81], Zygourakis and Aris [85].

Chapter 2

Parabolic Equations

2.1 Introduction

Chapter 2 is devoted to the study of second-order parabolic initial boundary value problems (IBVPs) relative to the extension of the monotone iterative technique and the method of generalized quasilinearization to such parabolic problems. Section 2.2 states the IBVP and provides the required comparison results. In Section 2.3, we go directly to develop the monotone iterative technique in the general set-up and point out the various special cases that can be derived from the general results considered. The method of generalized quasilinearization forms the content of Section 2.4 together with the details of interesting particular cases. Section 2.5 is dedicated to weakly coupled mixed monotone parabolic systems and corresponding elliptic systems of BVPs. Utilizing the theory of monotone flows and the ideas of the reflection operator, the convergence of solutions of mixed monotone systems to the solutions of the corresponding elliptic systems is investigated. In Section 2.6, we consider the comparison results of weakly coupled parabolic systems of IBVPs in a very general framework so as to bring out ideas involved in such a set-up. The method of vector Lyapunov functions is developed for weakly coupled parabolic systems of IBVPs in Section 2.7, which leads to stability theory of solutions of such systems. An example is worked out in detail, which demonstrates how the diffusion, convection and reaction terms contribute to stability in various ways and play an important role.

2.2   Comparison Theorems

We plan to extend, in the following sections, the monotone iterative technique and the method of generalized quasilinearization to second-order parabolic initial and boundary value problems (IBVPs). For this purpose, we require suitable comparison results and therefore we shall develop the needed results in this section together with the preliminary notions.

Let Ω be a bounded domain in R^n with boundary $\partial\Omega$ and closure $\bar{\Omega}$. We let $Q_T = (0,T) \times \Omega$, $\Gamma_T = (0,T) \times \partial\Omega$, $\bar{Q}_T = [0,T] \times \bar{\Omega}$ and $\bar{\Gamma}_T = [0,T] \times \partial\Omega$ for $T > 0$. Let a_{ij}, b_i and $c \in C^{\frac{\alpha}{2},\alpha}[\bar{Q}_T, R]$. We let $\mathcal{L}$ be the second-order differential operator defined by

$$\mathcal{L}u = u_t + Lu \tag{2.2.1}$$

where

$$Lu = -\sum_{i,j=1}^{n} a_{ij}(t,x)u_{x_i x_j} + \sum_{i=1}^{n} b_i(t,x)u_{x_i} + c(t,x)u.$$

$\mathcal{L}$ is said to be uniformly parabolic if there exists positive constants θ_1, θ_2 such that for $(t,x) \in \bar{Q}_T$,

$$\theta_1 \mid \xi \mid^2 \le \sum_{i,j}^{n} a_{i,j}(t,x)\xi_i\xi_j \le \theta_2 \mid \xi \mid^2, \ \xi \in R^n. \tag{2.2.2}$$

Let $p,q \in C^{\frac{1+\alpha}{2},1+\alpha}[\Gamma_T, R_+]$ which do not vanish simultaneously, i.e., $p(t,x) + q(t,x) > 0$, and $\gamma(t,x)$ be the unit outward normal on $\partial\Omega$ for $t \in [0,T]$. For $u \in C^{0,1}[\bar{Q}_T, R]$ we define the boundary operator

$$Bu = p(t,x)u + q(t,x)\frac{du}{d\gamma}, \tag{2.2.3}$$

where $\frac{du}{d\gamma}$ denotes the normal derivative of u.

We consider the second-order parabolic IBVP

$$\begin{bmatrix} \mathcal{L}u & = f(t,x,u) \text{ in } Q_T, \\ Bu & = \phi(t,x) \text{ on } \Gamma_T, \\ u(0,x) & = \phi_0(x) \text{ in } \bar{\Omega}, \end{bmatrix} \tag{2.2.4}$$

where $\phi \in C^{\frac{1+\alpha}{2},1+\alpha}[\Gamma_T, R]$, $\phi_0 \in C^{2+\alpha}[\bar{\Omega}, R]$ and $f \in C^{\frac{\alpha}{2},\alpha}[[0,T] \times \bar{\Omega} \times R, R]$.

We begin by proving the following comparison result.

Theorem 2.2.1 *Assume that*

(A_1) $v, w \in C^{1,2}[\bar{Q}_T, R]$ *and satisfy*

$$\mathcal{L}v \le f(t,x,v), \ \mathcal{L}w \ge f(t,x,w) \ in \ Q_T,$$

$$Bv \le Bw \ on \ \Gamma_T, \ v(0,x) \le w(0,x) \ x \in \bar{\Omega}.$$

(A_2) $f(t,x,u_1) - f(t,x,u_2) \le N(u_1 - u_2)$, $u_1 \ge u_2$, $N > 0$, $(t,x) \in Q_T$.

Then $v(t,x) \le w(t,x)$ *on* $\bar{Q}_T$.

Proof   Let $\tilde{w} = w + \epsilon e^{2NT}$ for $\epsilon > 0$. Then we have

$$
\left[
\begin{aligned}
\mathcal{L}\tilde{w} = \mathcal{L}w + 2\epsilon N e^{2Nt} \ &\ge f(t,x,w) + 2\epsilon N e^{2Nt} \\
&\ge f(t,x,\tilde{w}) - N\epsilon e^{2Nt} + 2\epsilon N e^{2Nt} \\
&> f(t,x,\tilde{w}) \ \text{on} \ Q_T, \\
B\tilde{w} = Bw + p(t,x)\epsilon e^{2Nt} \ &> Bv, \ \text{and} \ v(0,x) < \tilde{w}(0,x) \ \text{on} \ \bar{\Omega}.
\end{aligned}
\right.
\qquad (2.2.5)
$$

We shall show that $v(t,x) < \tilde{w}(t,x)$ on $\bar{Q}_T$. If not, setting $m(t,x) = v(t,x) - \tilde{w}(t,x)$, we find that there exists a $t_1 > 0$, $x_1 \in \bar{\Omega}$, such that

$$m(t_1, x) < 0 \ \text{on} \ [0,t_1) \times \bar{\Omega} \ \text{and} \ m(t_1, x_1) = 0.$$

Clearly, $m(t_1, x)$ has its maximum at x_1, which is equal to zero. Also $(t_1, x_1) \notin \Gamma_T$. Then we would have $\lim_{h \to 0} \frac{m(t_1,x_1) - m(t_1, x_1 - h\gamma)}{h} \ge 0$, which implies that $\frac{dv}{d\gamma} - \frac{d\tilde{w}}{d\gamma} = \frac{dm}{d\gamma} \ge 0$ at (t_1, x_1). Hence, we have $Bv = pv + q\frac{dv}{d\gamma} \ge p\tilde{w} + q\frac{d\tilde{w}}{d\gamma} = B\tilde{w}$, which contradicts (2.2.4). Hence $(t_1, x_1) \in Q_T$. Consequently, it follows that $m_{x_i}(t_1, x_1) = 0$ and the quadratic form $\sum_{i,j}^{n} a_{ij}(t_1, x_1)$ $\lambda_i \lambda_j \le 0$ for $\lambda \in R^N$. This leads to $v(t_1, x_1) = \tilde{w}(t_1, x_1)$, $v_{x_i}(t_1, x_1) = \tilde{w}_{x_i}(t_1, x_1)$ and $\sum_{i,j=1}^{n}[v(t_1, x_1) - \tilde{w}(t_1, x_1)]_{x_i x_j} \lambda_i \lambda_j \le 0$. Also, $m(t_1 - h, x_1) - m(t_1, x_1) < 0$ for small $h > 0$, which yields $m_t(t_1, x_1) \ge 0$. Therefore, using the appropriate inequalities in (A_1) and (2.2.4), we arrive at, for $(t_1, x_1) \in Q_T$,

$$
\begin{aligned}
0 \ &\le \ m_t \\
&= \ v_t - \tilde{w}_t \\
&< \ \sum_{i,j=1}^{n} a_{ij}(t_1, x_1)(v_{x_i x_j} - \tilde{w}_{x_i x_j}) \\
&\quad - \sum_{i=1}^{n} b_i(t_1, x_1)(v_{x_i} - \tilde{w}_{x_i}) - c(t_1, x_1)(v - \tilde{w}) \\
&< \ f(t_1, x_1, v) - f(t_1, x_1, \tilde{w}) \\
&\le \ 0
\end{aligned}
$$

which is a contradiction. Consequently, we have proved that $v(t,x) < \tilde{w}(t,x)$ on $\bar{Q}_T$. As $\epsilon \to 0$, we obtain the desired result $v(t,x) \le w(t,x)$ on $\bar{Q}_T$ and the proof is complete.

The following corollary is useful in itself.

Corollary 2.2.1 *Suppose that* $m \in C^{1,2}[\bar{Q}_T, R]$ *and satisfies*

$$\mathcal{L}m \le 0 \text{ on } Q_T, \ Bm \le 0 \text{ on } \Gamma_T, \text{ and } m(0,x) \le 0, \ x \in \bar{\Omega}.$$

Then $m(t,x) \le 0$ *on* $\bar{Q}_T$.

The following comparison result estimates the function satisfying the parabolic inequality by the maximal solution of an ordinary differential equation.

Theorem 2.2.2 *Let* $m \in C^{1,2}[\bar{Q}_T, R_+]$ *such that*

$$\mathcal{L}m \le f(t,x,m) \text{ on } Q_T.$$

Suppose further that

$$f(t,x,z) \le g(t,z), \ x \in \bar{\Omega}, \tag{2.2.6}$$

where $g \in C[[0,T] \times R_+, R]$. *Let* $r(t)$ *be the maximal solution of*

$$z' = g_0(t,z) = g(t,z) - c_0 z, z(0) = z_0 \ge 0, \tag{2.2.7}$$

existing on $[0,T]$ *where* $c_0 \le c(t,x)$ *on* $\bar{Q}_T$, *such that*

$$Bm \le Br \text{ on } \Gamma_T, \ m(0,x) \le z_0, \ x \in \bar{\Omega} \text{ and } p(t,x) > 0 \text{ on } \Gamma_T.$$

Then $m(t,x) \le r(t)$ *on* $\bar{Q}_T$.

Proof The function $r(t)$ clearly satisfies, in view of (2.2.6),

$$\mathcal{L}r(t) = g(t,r(t)) \ge f(t,x,r(t)) \text{ on } Q_T.$$

Let $z(t) = z(t,\epsilon)$ be any solution of

$$z' = g_0(t,z) + \epsilon, \ z(0) = z_0 + \epsilon, \ \epsilon > 0, \tag{2.2.8}$$

which exists on $[0,T]$ if $\epsilon > 0$ is sufficiently small. We recall that $\lim_{\epsilon \to 0} z(t,\epsilon) = r(t)$ on $[0,T]$ and $r(t) < z(t,\epsilon)$, $t \in [0,T]$. Thus we find

$$\mathcal{L}z(t,\epsilon) = g(t,z(t,\epsilon)) + \epsilon \ge f(t,x,z(t,\epsilon)) + \epsilon > f(t,x,z(t,\epsilon)) \text{ on } Q_T,$$

$Bm \leq Br(t) < Bz(t, \epsilon)$ on Γ_T since $p(t, x) > 0$ and $m(0, x) < z_0 + \epsilon = z(0, \epsilon)$, $x \in \bar{\Omega}$.

Hence by following the proof of Theorem 2.2.1 for strict inequalities, we arrive at
$$m(t, x) < z(t, \epsilon) \text{ on } \bar{Q}_T,$$

which, in view of the definition of the maximal solution of (2.2.7), yields $m(t, x) \leq \lim_{\epsilon \to 0} z(t, \epsilon) = r(t)$ on $\bar{Q}_T$, completing the proof.

2.3 Monotone Iterative Technique

We investigate, in this section, the monotone iterative technique for the IBVP
$$\left[\begin{array}{rl} \mathcal{L}u & = f(t, x, u) + g(t, x, u) \text{ on } Q_T, \\ Bu & = \phi(t, x) \text{ on } \Gamma_T, \\ u(0, x) & = \phi_0(x) \text{ on } \bar{\Omega}, \end{array} \right. \tag{2.3.1}$$

where $f, g \in C^{\frac{\alpha}{2}, \alpha}[[0, T] \times \bar{\Omega} \times R, R]$, the other assumptions being the same as described for the IBVP (2.2.4). We recall that the method of lower and upper solutions coupled with the monotone iterative technique offers an effective mechanism to provide constructive existence results for nonlinear problems in general, the lower and upper solutions serving as bounds for solutions which can be improved by the monotone iterative process. Also, as we have seen, a unified approach to the problem yields very general results which cover several cases of importance. Since we already have the necessary comparison results, we can go directly to prove the following theorem on the monotone iterative technique.

Theorem 2.3.1 *Assume that*

(A_1) $v_0, w_0 \in C^{1,2}[\bar{Q}_T, R]$ *are coupled lower and upper solutions of* (2.3.1), *that is,*
$$\mathcal{L}v_0 \leq f(t, x, v_0) + g(t, x, w_0) \text{ in } Q_T,$$
$$Bv_0 \leq \phi \text{ on } \Gamma_T, \; v_0(0, x) \leq \phi_0, \; x \in \bar{\Omega},$$
$$\mathcal{L}w_0 \geq f(t, x, w_0) + g(t, x, v_0) \text{ in } Q_T,$$
$$Bw_0 \geq \phi \text{ on } \Gamma_T, \; w_0(0, x) \geq \phi_0, \; x \in \bar{\Omega},$$

such that $v_0(t, x) \leq w_0(t, x)$ *on* $\bar{Q}_T$;

(A_2) $f, g \in C^{\frac{\alpha}{2}, \alpha}[[0, T] \times \bar{\Omega} \times R, R]$, $f(t, x, u)$ *is nondecreasing in* u *and* $g(t, x, u)$ *is nonincreasing in* u *for* $(t, x) \in Q_T$;

(A_3) *the IBVP* (2.3.1) *satisfies the compatibility condition of order* $[\frac{\alpha+1}{2}]$.

Then there exist monotone sequences $\{v_n\}$, $\{w_n\}$ *which converge in* $C^{1,2}[\bar{Q}_T, R]$ *to* (ρ, r) *respectively and* (ρ, r) *are coupled minimal and maximal solutions of* (2.3.1), *that is,* (ρ, r) *satisfy*

$$\mathcal{L}\rho = f(t, x, \rho) + g(t, x, r) \text{ in } Q_T, \ B\rho = \phi \text{ on } \Gamma_T, \ \rho(0, x) = \phi, x \in \bar{\Omega},$$

and

$$\mathcal{L}r = f(t, x, r) + g(t, x, \rho) \text{ in } Q_T, \ Br = \phi \text{ on } \Gamma_T, \ r(0, x) = \phi_0, \ x \in \bar{\Omega}.$$

Proof   Consider the IBVP

$$\left[\begin{array}{ll} \mathcal{L}u &= f(t, x, \eta) + g(t, x, \mu) \text{ in } Q_T, \\ Bu &= \phi \text{ on } \Gamma_T, \ u(0, x) = \phi_0 \text{ on } \bar{\Omega}, \end{array}\right. \tag{2.3.2}$$

for any $\eta, \mu \in C^{1,2}[\bar{Q}_T, R]$ such that $v \leq \eta$, $\mu \leq w$ on $\bar{Q}_T$. Setting $h_1(t, x) = f(t, x, \eta(t, x))$, we see that

$$\begin{aligned} \mid h_1(t, x) - h_1(\tau, x) \mid &= \mid f(t, x, \eta(t, x)) - f(\tau, x, \eta(\tau, x)) \mid \\ &\leq c_0[\mid t - \tau \mid^{\frac{\alpha}{2}} + \mid \eta(t, x) - \eta(\tau, x) \mid^\alpha] \\ &\leq c_0[\mid t - \tau \mid^{\frac{\alpha}{2}} + c_1 \mid t - \tau \mid^\alpha] \\ &\leq c_0 \mid t - \tau \mid^{\frac{\alpha}{2}} [1 + c_1 T^{\frac{\alpha}{2}}] \\ &\leq C \mid t - \tau \mid^{\frac{\alpha}{2}} \end{aligned}$$

where $C = c_0[1 + c_1 T^{\frac{\alpha}{2}}]$ and $\|\eta_t\| \leq c_1$. This shows that $h_1(t, x)$ is Hölder continuous in t with exponent $\frac{\alpha}{2}$. The proof of Hölder continuity of $h_1(t, x)$ in x with exponent α is analogous to the proof given in Theorem 1.3.2. A similar estimate holds for $h_2(t, x) = g(t, x, \mu(t, x))$ and hence there exists a unique solution $u \in C^{\frac{1+\alpha}{2}, 2+\alpha}[\bar{Q}_T, R]$ for the linear IBVP (2.3.2) by Theorem A.3.1.

Now we consider the following linear IBVPs for each $k = 1, 2, \ldots$

$$\left[\begin{array}{ll} \mathcal{L}v_{k+1} &= f(t, x, v_k) + g(t, x, w_k) \text{ in } Q_T, \\ Bv_{k+1} &= \phi, \text{ on } \Gamma_T, \ v_{k+1}(0, x) = \phi_0 \text{ on } \bar{\Omega}, \end{array}\right. \tag{2.3.3}$$

and

$$\left[\begin{array}{ll} \mathcal{L}w_{k+1} &= f(t, x, w_k) + g(t, x, v_k) \text{ in } Q_T, \\ Bw_{k+1} &= \phi \text{ on } \Gamma_T, \ w_{k+1}(0, x) = \phi_0 \text{ on } \bar{\Omega}. \end{array}\right. \tag{2.3.4}$$

For $k = 0$, we have $v_1, w_1 \in C^{\frac{1+\alpha}{2}, 2+\alpha}[\bar{Q}_T, R]$ as the unique solutions of (2.3.3), (2.3.4) respectively, in view of Theorem A.3.1. Set $p = v_0 - v_1$ so that we have

$$\mathcal{L}p \le f(t, x, v_0) + g(t, x, w_0) - f(t, x, v_0) - g(t, x, w_0) = 0 \text{ in } Q_T,$$

$$Bp \le 0 \text{ on } \Gamma_T, \ p(0, x) \le 0 \text{ on } \bar{\Omega}.$$

An application of Corollary 2.2.1 yields that $v_0 \le v_1$ on $\bar{Q}_T$ and a similar argument shows $w_1 \le w_0$ on $\bar{Q}_T$. We next prove that $v_1 \le w_1$ on $\bar{Q}_T$. Consider $p = v_1 - w_1$ so that $Bp = 0$ on Γ_T, $p(0, x) = 0$ on $\bar{\Omega}$ and

$$\mathcal{L}p = f(t, x, v_0) + g(t, x, w_0) - f(t, x, w_0) - g(t, x, v_0) \le 0 \text{ in } Q_T,$$

using the monotone nature of f, g. Thus we get by Corollary 2.2.1 that $v_1 \le w_1$ on $\bar{Q}_T$. As a result, it follows that

$$v_0 \le v_1 \le w_1 \le w_0 \text{ on } \bar{Q}_T.$$

Assume that for some $k > 1$,

$$v_{k-1} \le v_k \le w_k \le w_{k-1} \text{ on } \bar{Q}_T. \tag{2.3.5}$$

Then we wish to show that

$$v_k \le v_{k+1} \le w_{k+1} \le w_k \text{ on } \bar{Q}_T. \tag{2.3.6}$$

To do this, let $p = v_k - v_{k+1}$ so that $Bp = 0$ on Γ_T, $p(0, x) = 0$ on $\bar{\Omega}$ and

$$\mathcal{L}p = f(t, x, v_{k-1}) + g(t, x, w_{k-1}) - f(t, x, v_k) - g(t, x, w_k) \le 0$$

in Q_T, because of the monotonicity of f, g and (2.3.5). Hence Corollary 2.2.1 gives $v_k \le v_{k+1}$ on $\bar{Q}_T$. Similarly $w_{k+1} \le w_k$ on $\bar{Q}_T$. Letting $p = v_{k+1} - w_{k+1}$, we find that $Bp = 0$ on Γ_T, $p(0, x) = 0$ on $\bar{\Omega}$ and

$$\mathcal{L}p = f(t, x, v_k) + g(t, x, w_k) - f(t, x, w_k) - f(t, x, v_k) \le 0 \text{ in } \bar{Q}_T,$$

in view of (2.3.5) and the monotone nature of f, g. Thus we get by Corollary 2.2.1, $v_{k+1} \le w_{k+1}$ on $\bar{Q}_T$, proving (2.3.6). Hence by induction, it follows that, for all $k = 1, 2, \ldots$,

$$v_0 \le v_1 \le \ldots \le v_k \le w_k \le \ldots \le w_1 \le w_0 \text{ on } \bar{Q}_T. \tag{2.3.7}$$

Let us consider the sequence $\{v_k\}$. Note that $C^{\frac{1+\alpha}{2},2+\alpha}[\bar{Q}_T, R] \subset W_q^{1,2}[\bar{Q}_T, R]$ for $q \geq \frac{n+2}{1-\alpha}$. By Theorem A.3.3, v_k satisfies

$$\| v_k \|_{W_q^{1,2}[\bar{Q}_T,R]} \leq C[\| h_k \|_{L^q[\bar{Q}_T,R]} + \| \phi \|_{W^{\frac{1}{2}-\frac{1}{2q},1-\frac{1}{q}}_{[\bar{\Gamma}_T,R]}} + \| \phi_0 \|_{W^{2-\frac{2}{q}}_{[\bar{\Omega},R]}}$$

$$(2.3.8)$$

where $h_k(t,x) = f(t,x,v_k) + g(t,x,w_k)$, $k = 1, 2, \ldots$.

Since $v_k, w_k \in [v_0, w_0]$, the continuity of h_k implies that the sequence $\{h_k\}$ is uniformly bounded in $C[\bar{Q}_T, R]$. Because of the fact that $C[\bar{Q}_T, R]$ is dense in $L^q[\bar{Q}_T, R]$, it follows that $\{h_k\}$ is a bounded sequence in $L^q[\bar{Q}_T, R]$. This, together with (2.3.8), shows that $\{v_k\}$ is uniformly bounded in $W_q^{1,2}[\bar{Q}_T, R]$. The Embedding Theorem A.3.4 therefore gives

$$\| v_k \|_{C^{\frac{(1+\alpha)}{2},1+\alpha}[\bar{Q}_T,R]} \leq C \| v_k \|_{W_q^{1,2}[\bar{Q}_T,R]}, \qquad (2.3.9)$$

for $k = 1, 2, \ldots$, where c is independent of $W_q^{1,2}[\bar{Q}_T, R]$.

From (2.3.9), we conclude that every uniformly bounded sequence in $W_q^{1,2}[\bar{Q}_T, R]$ is also uniformly bounded in $C^{\frac{1+\alpha}{2},1+\alpha}[\bar{Q}_T, R]$. By Theorem A.3.4, $\{h_k\}$ is a bounded sequence in $C^{\frac{\alpha}{2},\alpha}[\bar{Q}_T, R]$ and therefore by Schauder's estimate given in Theorem A.3.2, we get

$$\| v_k \|_{C^{1+\frac{\alpha}{2},2+\alpha}[\bar{Q}_T,R]} \leq C[\| h_k \|_{C^{\frac{\alpha}{2},\alpha}[\bar{Q}_T,R]} + \| \phi \|_{C^{\frac{(1+\alpha)}{2},1+\alpha}[\bar{\Gamma}_T,R]}$$

$$+ \| \phi_0 \|_{C^{2+\alpha}[\bar{\Omega},R]},$$

for all $k = 1, 2, \ldots$, which implies uniform boundedness of $\{v_k\}$ in $C^{1+\frac{\alpha}{2},2+\alpha}[\bar{Q}_T, R]$. Hence by the compact embedding of $C^{1+\frac{\alpha}{2},1+\alpha}[\bar{Q}_T, R]$ into $C^{1,2}[\bar{Q}_T, R]$, the sequence $\{v_k\}$ is relatively compact in $C^{1,2}[\bar{Q}_T, R]$ and consequently, there exists a subsequence of $\{v_k\}$ which converges in $C^{1,2}[\bar{Q}_T, R]$. Let $v^* \in C^{1,2}[\bar{Q}_T, R]$ be a limit of this subsequence. But (2.3.7) shows that $\{v_k\}$ converges pointwise to $\rho(t,x)$ on $\bar{Q}_T$ and the convergence of a subsequence of $\{v_k\}$ in $C^{1,2}[\bar{Q}_T, R]$ implies pointwise convergence. Therefore, $\rho(t,x) = v^*(t,x)$ on $\bar{Q}_T$. This shows that the entire sequence $\{v_k\}$ converges in $C^{1,2}[\bar{Q}_T, R]$ to $\rho(t,x)$, that is $\lim_{k\to\infty} v_k(t,x) = \rho(t,x)$ in $C^{1,2}[\bar{Q}_T, R]$ and $v_0(t,x) \leq \rho(t,x) \leq w_0(t,x)$ on $\bar{Q}_T$. Similarly, we can prove that the sequence $\{w_k\}$ converges to $r(t,x)$ in $C^{1,2}[\bar{Q}_T, R]$ and $v_0 \leq r \leq w_0$ on $\bar{Q}_T$, which implies by (2.2.7) that $v_0 \leq \rho \leq r \leq w_0$ on $\bar{Q}_T$. Thus the limits

$$\lim_{k\to\infty} \mathcal{L}v_k = \mathcal{L}\rho, \ \lim_{k\to 0} \mathcal{L}w_k = \mathcal{L}r,$$

$$\lim_{k\to\infty} Bv_k = B\rho, \ \lim_{k\to\infty} Bw_k = Br,$$

$$\lim_{k\to\infty} \left[f(t,x,v_k) + g(t,x,w_k) \right] = f(t,x,\rho) + g(t,x,r),$$

and

$$\lim_{k\to\infty} \left[f(t,x,w_k) + g(t,x,v_k) \right] = f(t,x,r) + g(t,x,\rho),$$

exist uniformly in $\bar{Q}_T$ and satisfy

$$\mathcal{L}\rho = f(t,x,\rho) + g(t,x,r) \text{ in } Q_T,$$

$$B\rho = \phi \text{ on } \Gamma_T, \ \rho(0,x) = \phi_0 \text{ on } \bar{\Omega}$$

and

$$\mathcal{L}r = f(t,x,r) + g(t,x,\rho) \text{ in } Q_T,$$

$$Br = \phi \text{ on } \Gamma_T, \ r(0,x) = \phi_0 \text{ on } \bar{\Omega},$$

proving that (ρ, r) are coupled solutions of (2.3.2). We need to show that (ρ, r) are coupled minimal and maximal solutions of (2.3.2).

Let us suppose that u is any solution of (2.3.1) with $v_0 \leq u \leq w_0$ on $\bar{Q}_T$. Assume that for some $k > 1$, we have

$$v_k \leq u \leq w_k \text{ on, } \bar{Q}_T. \tag{2.3.10}$$

Set $p = v_{k+1} - u$ so that $Bp = 0$ on Γ_T, $p(0,x) = 0$ on $\bar{\Omega}$ and

$$\mathcal{L}p = f(t,x,v_k) + g(t,x,w_k) - f(t,x,u) - g(t,x,u) \leq 0 \text{ in } Q_T,$$

because of (2.3.10) and the monotonicity of f, g. Corollary 2.2.1 now yields that $v_{k+1} \leq u$ on $\bar{Q}_T$. Similarly, one can show that $u \leq w_{k+1}$ on $\bar{Q}_T$, so that we have $v_{k+1} \leq u \leq w_{k+1}$ on $\bar{Q}_T$. It then follows by induction that $v_k \leq u \leq w_k$ on $\bar{Q}_T$ for all $k = 1, 2, \ldots$ and this implies that $v_0 \leq \rho \leq r \leq w_0$ on $\bar{Q}_T$ proving that (ρ, r) are coupled extremal solutions of (2.3.1).

In order to prove uniqueness, we have the following lemma.

Lemma 2.3.1 *Suppose that, in addition to the assumptions of Theorem 2.3.1, f, g satisfy for $u_1, u_2 \in [v_0, w_0]$, $u_1 \geq u_2$,*

$$f(t,x,u_1) - f(t,x,u_2) \leq N_1(u_1 - u_2),$$

$$g(t,x,u_1) - g(t,x,u_2) \geq -N_2(u_1 - u_2) \text{ for } (t,x) \in Q_T,$$

where $N_1, N_2 > 0$. Then $\rho = r = u$ is the unique solution of (2.3.1) such that $v_0 \leq u \leq w_0$ on $\bar{Q}_T$.

Proof   Since $\rho \leq r$ on $\bar{Q}_T$, we need only show that $r \leq \rho$. Setting $p = r - \rho$, we observe that $Bp = 0$ on Γ_T, $p(0,x) = 0$ on $\bar{\Omega}$ and

$$
\begin{aligned}
\mathcal{L}p &= f(t,x,r) + g(t,x,\rho) - f(t,x,\rho) - g(t,x,r) \\
&\leq (N_1 + N_2)p \text{ on } Q_T.
\end{aligned}
$$

Theorem 2.2.1 then shows that $r \leq \rho$ on $\bar{Q}_T$ and the proof of uniqueness is complete.

As we have seen earlier, Theorem 2.3.1 includes several interesting special cases as given below.

Remarks 2.3.1

(i) If in Theorem 2.3.1, $g(t,x,u) \equiv 0$, we then get the result where (ρ, r) are minimal and maximal solutions of (2.3.1).

(ii) If $f(t,x,u) \equiv 0$ in Theorem 2.3.1, then we have a result corresponding to the nonincreasing case.

(iii) If $g(t,x,u) \equiv 0$ and $f(t,x,u)$ is not nondecreasing in u but $\tilde{f}(t,x,u) = f(t,x,u) + Mu$ is nondecreasing in u where $M > 0$, then we consider the IBVP

$$
\mathcal{L}u + Mu = \tilde{f}(t,x,u) \text{ in } Q_T, \tag{2.3.11}
$$

$$
Bu = \phi \text{ on } \Gamma_T, \; u(0,x) = \phi_0 \text{ on } \bar{\Omega}.
$$

One easily verifies that v_0, w_0 are lower and upper solutions of (2.3.11) and Theorem 2.3.1 applied to (2.3.11) yields the same conclusions as in (i) for the IBVP (2.3.1).

(iv) If in Theorem 2.3.1, $g(t,x,u)$ is nonincreasing in u and $f(t,x,u)$ is not nondecreasing in u but $\tilde{f}(t,x,u) = f(t,x,u) + Mu$, $M > 0$, is nondecreasing in u, then we consider the IBVP

$$
\left[
\begin{array}{l}
\mathcal{L}u + Mu = \tilde{f}(t,x,u) + g(t,x,u) \text{ in } Q_T, \\
\quad Bu = \phi \text{ on } \Gamma_T, \; u(0,x) = \phi_0 \text{ on } \bar{\Omega},
\end{array}
\right. \tag{2.3.12}
$$

and note that v_0, w_0 are coupled lower and upper solutions of (2.3.12). Hence the conclusion of Theorem 2.3.1 applied to IBVP (2.3.12) holds for the IBVP (2.3.1).

(v) If $f(t, x, u) \equiv 0$ and $g(t, x, u)$ is not nonincreasing in u but $\tilde{g}(t, x, u) = g(t, x, u) - N_2 u$, $N_2 > 0$, is nonincreasing in u, in Theorem 2.3.1, then we consider the IBVP

$$\mathcal{L}u - N_2 u = \tilde{g}(t, x, u) \text{ in } Q_T, \tag{2.3.13}$$

$$Bu = \phi \text{ on } \Gamma_T, \ u(0, x) = \phi_0 \text{ on } \bar{\Omega}.$$

Assuming that v_0, w_0 are coupled lower and upper solutions of (2.3.13), Theorem 2.3.1 applied to (2.3.13) provides the same conclusion as in (ii) for IBVP (2.3.13).

(vi) Suppose that in Theorem 2.3.1, $f(t, x, u)$ is nondecreasing in u and $g(t, x, u)$ is not nonincreasing in u but $\tilde{g}(t, x, u) = g(t, x, u) - N_2 u$, $N_2 > 0$, is nonincreasing in u. We then consider the IBVP

$$\mathcal{L}u - N_2 u = f(t, x, u) + \tilde{g}(t, x, u) \text{ in } Q_T, \tag{2.3.14}$$

$$Bu = \phi \text{ on } \Gamma_T, u(0, x) = \phi_0 \text{ on } \bar{\Omega}.$$

If v_0, w_0 are coupled lower and upper solutions of (2.3.14), then the conclusion of Theorem 2.3.1 applied to IBVP (2.3.14) holds for the IBVP (2.3.14).

(vii) Assume that in Theorem 2.3.1, f and g are both not monotone functions in u but $\tilde{f}(t, x, u) = f(t, x, u) + Mu$, $M > 0$, is nondecreasing in u and $\tilde{g}(t, x, u) - N_2 u$, $N_2 > 0$, is nonincreasing in u. Then we consider the IBVP

$$\mathcal{L}u + (M - N_2)u = \tilde{f}(t, x, u) + \tilde{g}(t, x, u) \text{ in } Q_T, \tag{2.3.15}$$

$$Bu = \phi \text{ on } \Gamma_T, u(0, x) = \phi_0 \text{ on } \Omega.$$

If v_0, w_0 are the coupled lower and upper solutions of (2.3.15), then the conclusion of Theorem 2.3.1 applied to IBVP (2.3.15) holds for IBVP (2.3.15).

2.4 Generalized Quasilinearization

We shall continue to consider, in this section, the IBVP (2.3.1), under the same assumptions as in Section 2.3, and develop the method of generalized quasilinearization, which provides not only monotone sequences that converge to the unique solution of (2.3.1) but also shows that the convergence is quadratic. We shall prove the following very general result which includes several special cases of interest.

Theorem 2.4.1 *Assume that*

(A_1) $v_0, w_0 \in C^{1,2}[\bar{Q}_T, R]$, $\mathcal{L}v_0 \leq f(t, x, v_0) + g(t, x, v_0)$ in Q_T, $Bv_0 \leq \phi$ on Γ_T, $v_0(0, x) \leq \phi_0$ on $\bar{\Omega}$ and $\mathcal{L}w_0 \geq f(t, x, w_0) + g(t, x, w_0)$ in Q_T, $Bw_0 \geq \phi$ on Γ_T, $w_0(0, x) \geq \phi_0$ on $\bar{\Omega}$ such that $v_0(t, x) \leq w_0(t, x)$ on $\bar{Q}_T$.

(A_2) $f, g \in C^{\frac{\alpha}{2}, \alpha}[[0, T] \times \bar{\Omega} \times R, R]$, f_u, g_u, f_{uu}, g_{uu} exist and are continuous, $f_u, g_u \in C^{\frac{\alpha}{2}, \alpha}[[0, T] \times \bar{\Omega} \times R, R]$, $f_{uu}(t, x, u) \geq 0$, $g_{uu}(t, x, u) \leq 0$ for $(t, x) \in \bar{Q}_T$ and $u \in [v_0, w_0]$.

(A_3) *Compatibility condition* (A_3) *of Theorem 2.3.1 holds.*

Then there exist monotone sequences $\{v_k\}$, $\{w_k\} \in C^{\frac{(1+\alpha)}{2}, 2+\alpha}[\bar{Q}_T, R]$ *which converge quadratically in* $C^{1,2}[\bar{Q}_T, R]$ *to the unique solution* u *of* (2.3.1) *satisfying* $v_0 \leq u \leq w_0$ *on* $\bar{Q}_T$.

Proof   Since $f_{uu} \geq 0$, $g_{uu} \leq 0$ by assumption (A_2), we obtain the inequalities

$$f(t, x, u) \geq f(t, x, v) + f_u(t, x, v)(u - v), \qquad (2.4.1)$$

$$g(t, x, u) \geq g(t, x, v) + g_u(t, x, u)(u - v), \qquad (2.4.2)$$

for $u \geq v$ and $u, v \in [v_0, w_0]$ and $(t, x) \in \bar{Q}_T$. Consider the linear IVPBs

$$\mathcal{L}u = F(t, x, u; \eta, \mu) \text{ in } Q_T, Bu = \phi \text{ on } \Gamma_T, u(0, x) = \phi_0 \text{ on } \bar{\Omega}, \qquad (2.4.3)$$

$$\mathcal{L}u = G(t, x, u; \eta, \mu) \text{ in } Q_T, Bu = \phi \text{ on } \Gamma_T, u(0, x) = \phi_0 \text{ on } \bar{\Omega}, \qquad (2.4.4)$$

for $\eta, \mu \in [v_0, w_0]$, $\eta, \mu \in C^{1,2}[\bar{Q}_T, R]$, where

$$
\begin{aligned}
F(t, x, u; \eta, \mu) &= f(t, x, \eta) + f_u(t, x, \eta)(u - \eta) && (2.4.5) \\
&\quad + g(t, x, \eta) + g_u(t, x, \mu)(u - \eta), \\
G(t, x, u; \eta, \mu) &= f(t, x, \mu) + f_u(t, x, \eta)(u - \mu) && (2.4.6) \\
&\quad + g(t, x, \mu) + g_u(t, x, \mu)(u - \mu).
\end{aligned}
$$

Set $F(t, x, u; \eta, \mu) = d(t, x)u + h(t, x)$, where

$$d(t, x) = f_u(t, x, \eta) + g_u(t, x, \mu), \qquad (2.4.7)$$

$$h(t, x) = f(t, x, \eta) + g(t, x, \eta) - f_u(t, x, \eta)\eta - g_u(t, x, \mu)\eta. \qquad (2.4.8)$$

To guarantee the unique solution of IBVP (2.4.3), we need to show that $d(t, x)$ and $h(t, x)$ belong to $C^{\frac{\alpha}{2}, \alpha}[\bar{Q}_T, R]$. We will only show the details for $f_u(t, x, \eta(t, x))\eta(t, x)$ to avoid repetition. Now

$$
\begin{aligned}
& \mid f_u(t, x, \eta(t, x))\eta(t, x) - f_u(t, y, \eta(t, y))\eta(t, y) \mid \\
& \leq \ \mid f_u(t, x, \eta(t, x))\eta(t, x) - f_u(t, y, \eta(t, y))\eta(t, x) \mid \\
& \quad + \mid f_u(t, y, \eta(t, y))\eta(t, x) - f_u(t, y, \eta(t, y))\eta(t, y) \mid \\
& \leq \ \mid \eta(t, x) \mid c_0(\parallel x - y \parallel^\alpha + \mid \eta(t, x) - \eta(t, y) \mid^\alpha) \\
& \quad + \mid f_u(t, y, \eta(t, y)) \parallel \eta(t, x) - \eta(t, y) \mid^\alpha \\
& \leq \ k_1 c_0(\parallel x - y \parallel^\alpha + c_1^\alpha \parallel x - y \parallel^\alpha) + k_2 c_1^\alpha \parallel x - y \parallel^\alpha \\
& = \ C \parallel x - y \parallel^\alpha
\end{aligned}
$$

where $\mid \eta \mid \leq k_1$, $\parallel \eta_x \parallel \leq c_1$, $\mid f_u \mid \leq k_2$ and $C = k_1 c_0 + (k_1 c_0 + k_2)c_1^\alpha$. This implies that $f_u(t, x, \eta(t, x))\eta(t, x)$ is Hölder continuous in x with exponent α. Hölder continuity in t with exponent $\frac{\alpha}{2}$ follows as in Theorem 2.3.1. Since Hölder continuity of the rest of the terms in $d(t, x)$ and $h(t, x)$ can be proven in a similar way, it is clear that $d(t, x)$, $h(t, x)$ belong to $C^{\frac{\alpha}{2}, \alpha}[\bar{Q}_T, R]$. Consequently, the linear IBVP (2.4.3) possesses the unique solution $u \in C^{\frac{(1+\alpha)}{2}, 2+\alpha}[\bar{Q}_T, R]$ for each η, μ by Theorem A.3.1. A similar conclusion is valid for IBVP (2.4.4).

Now we consider the linear IBVPs for $k = 1, 2, 3, \ldots$,

$$Lv_{k+1} = F(t, x, v_{k+1}; v_k, w_k) \text{ in } Q_T, \tag{2.4.9}$$

$$Bv_{k+1} = \phi \text{ on } \Gamma_T, \ v_{k+1}(0, x) = \phi_0 \text{ on } \bar{\Omega},$$

$$Lw_{k+1} = G(t, x, w_{k+1}; v_k, w_k) \text{ in } Q_T, \tag{2.4.10}$$

$$Bw_{k+1} = \phi \text{ on } \Gamma_T, \ w_{k+1}(0, x) = \phi_0 \text{ on } \bar{\Omega}.$$

For $k = 0$, we therefore have $v_1, w_1 \in C^{\frac{(1+\alpha)}{2}, 2+\alpha}[\bar{Q}_T, R]$ as the unique solutions of (2.4.9), (2.4.10) respectively. Set $p = v_0 - v_1$ so that $Bp \leq 0$ on Γ_T, $p(0, x) \leq 0$ on $\bar{\Omega}$ and

$$
\begin{aligned}
\mathcal{L}p & \leq \ f(t, x, v_0) + g(t, x, v_0) - F(t, x, v_1; v_0, w_0) \\
& \leq \ [f_u(t, x, v_0) + g_u(t, x, w_0)]p \text{ in } Q_T.
\end{aligned}
$$

Corollary 2.2.1 shows that $v_0 \leq v_1$ on $\bar{Q}_T$. Similarly, we can show that $w_1 \leq w_0$ on $\bar{Q}_T$. Next we set $p = v_1 - w_1$ so that $Bp = 0$ on Γ_T, $p(0, x) = 0$

on $\bar{\Omega}$ and using (2.4.1), (2.4.2), we find

$$
\begin{aligned}
\mathcal{L}p &= F(t,x,v_1;v_0,w_0) - G(t,x,w_1;v_0,w_0) \\
&= f(t,x,v_0) + f_u(t,x,v_0)(v_1 - v_0) \\
&\quad + g(t,x,v_0) + g_u(t,x,w_0)(v_1 - v_0) \\
&\quad - [f(t,x,w_0) + f_u(t,x,v_0)(w_1 - w_0) \\
&\quad + g(t,x,w_0) + g_u(t,x,w_0)(w_1 - w_0)] \\
&\leq -f_u(t,x,v_0)(w_0 - v_0) + f_u(t,x,v_0)(v_1 - v_0) \\
&\quad - f_u(t,x,v_0)(w_1 - w_0) \\
&\quad - g_u(t,x,w_0)(w_0 - v_0) + g_u(t,x,w_0)(v_1 - v_0) \\
&\quad - g_u(t,x,w_0)(w_1 - w_0), \\
&\leq [f_u(t,x,v_0) + g_u(t,x,w_0)]p \quad \text{in } Q_T.
\end{aligned}
$$

Hence Corollary 2.2.1 yields $v_1 \leq w_1$ on $\bar{Q}_T$ and as a result we have

$$
v_0 \leq v_1 \leq w_1 \leq w_0 \text{ on } \bar{Q}_T. \tag{2.4.11}
$$

Assume that for some $k > 1$,

$$
v_{k-1} \leq v_k \leq w_k \leq w_{k-1} \text{ on } \bar{Q}_T. \tag{2.4.12}
$$

Then we shall show that $v_k \leq v_{k+1}$ and $w_{k+1} \leq w_k$ on $\bar{Q}_T$. Set $p = v_k - v_{k+1}$. Then $Bp = 0$ on Γ_T, $p(0,x) = 0$ on $\bar{\Omega}$ and employing (2.4.1), (2.4.2) and noting that g_u is nonincreasing in u, it follows that

$$
\begin{aligned}
\mathcal{L}p &= F(t,x,v_k;v_{k-1},w_{k-1}) - F(t,x,v_{k+1};v_k,w_k) \\
&= f(t,x,v_{k-1}) + f_u(t,x,v_{k-1})(v_k - v_{k-1})(v_k - v_{k-1}) + g(t,x,v_k) \\
&\quad + g_u(t,x,w_k)(v_{k+1} - v_k) \\
&\quad - [f(t,x,v_k) + f_u(t,x,v_k)(v_{k+1} - v_k) + g(t,x,v_k) \\
&\quad + g_u(t,x,w_k)(v_{k+1} - v_k) \\
&\leq -f_u(t,x,v_{k-1})(v_k - v_{k-1}) \\
&\quad + f_u(t,x,v_{k-1})(v_k - v_{k-1})f_u(t,x,v_k)(v_{k+1} - v_k) \\
&\quad - g_u(t,x,v_k)(v_k - v_{k-1}) + g_u(t,x,v_k)(v_k - v_{k-1}) \\
&\quad - g_u(t,x,w_k)(v_{k+1} - v_1) \\
&\leq [f_u(t,x,v_k) + g_u(t,x,w_k)]p \quad \text{in } Q_T.
\end{aligned}
$$

Corollary 2.2.1 therefore shows that $v_k \leq v_{k+1}$ on $\bar{Q}_T$. Similarly one shows that $w_{k+1} \leq w_k$ on $\bar{Q}_T$. To prove $v_{k+1} \leq w_{k+1}$ on $\bar{Q}_T$, we set

$p = v_{k+1} - w_{k+1}$ so that $Bp = 0$ on Γ_T, $p(0, x) = 0$ on $\bar{\Omega}$ and

$$
\begin{aligned}
\mathcal{L}p &= F(t, x, v_{k+1}; v_k, w_k) - G(t, x, w_{k+1}; v_k, w_k) \\
&= f(t, x, v_k) + f_u(t, x, v_k)(v_{k+1} - v_k) \\
&\quad + g(t, x, v_k) + g_u(t, x, w_k)(v_{k+1} - v_k) \\
&\quad - [f(t, x, w_k) + f_u(t, x, w_k)(w_{k+1} - w_k) \\
&\quad + g(t, x, w_k) + g_u(t, x, w_k)(w_{k+1} - w_k)] \\
&\leq -f_u(t, x, v_k)(w_k - v_k) + f_u(t, x, v_k)(v_{k+1} - v_k) \\
&\quad - f_u(t, x, v_k)(w_{k+1} - w_k) \\
&\quad - g_u(t, x, w_k)(w_k - v_k) + g_u(t, x, w_k)(v_{k+1} - v_k) \\
&\quad - g_u(t, x, w_k)(w_{k+1} - w_k) \\
&= [f_u(t, x, v_k) + g_u(t, x, w_k)]p \text{ in } Q_T.
\end{aligned}
$$

Here we have utilized (2.4.1), (2.4.2) and the monotone character of f_u. Thus Corollary 2.2.1 implies $v_{k+1} \leq w_{k+1}$ on Q_T and therefore we arrive at

$$v_{k-1} \leq v_k \leq v_{k+1} \leq w_{k+1} \leq w_k \leq w_{k-1} \text{ on } \bar{Q}_T.$$

The induction argument now assures that

$$v_0 \leq v_1 \leq \ldots \leq v_k \leq w_k \leq \ldots \leq w_1 \leq w_0 \text{ on } Q_T \text{ for all } k = 1, 2, \ldots.$$
$$(2.4.13)$$

To show that the sequences $\{v_k\}$, $\{w_k\}$ converge in $C^{1,2}[\bar{Q}_T, R]$ to (ρ, r) which are solutions of (2.3.1), we use an argument similar to that of the corresponding reasoning in Theorem 2.3.1 except that h_k is now defined as

$$h_k(t, x) = f(t, x, v_k) + g(t, x, v_k) - f_u(t, x, v_k)v_k - g_u(t, x, w_k)v_k.$$

Consequently, (ρ, r) satisfy

$$\mathcal{L}\rho = f(t, x, \rho) + g(t, x, \rho) \text{ in } Q_T, Bp = \phi \text{ on } \Gamma_T, \rho(0, x) = \phi_0 \text{ on } \bar{\Omega},$$

$$\mathcal{L}r = f(t, x, r) + g(t, x, r) \text{ in } Q_T, Br = \phi \text{ on } \Gamma_T, \; r(0, x) = \phi_0 \text{ on } \bar{\Omega}.$$

To show that $\rho = r = u$ is the unique solution of (2.3.1), we note that $\rho \leq r$ in $\bar{Q}_T$ and thus we need only to show that $r \leq \rho$ on $\bar{Q}_T$. We observe, for this purpose, that f, g satisfy the Lipschitz condition relative to $[v_0, w_0]$ with Lipschitz constants $N_1, N_2 > 0$. Hence letting $p = r - \rho$ and using the Lipschitz conditions suitably, we arrive at

$$
\begin{aligned}
\mathcal{L}p &= f(t, x, r) + g(t, x, r) - f(t, x, \rho) - g(t, x, \rho) \\
&\leq (N_1 + N_2)p \text{ on } Q_T, \\
Bp &= 0 \text{ on } \Gamma_T, p(0, x) = 0 \text{ on } \bar{\Omega}.
\end{aligned}
$$

As a result, we see that Corollary 2.2.1 gives $r \leq \rho$ on $\bar{Q}_T$ and this implies that $\rho = r = u$ is the unique solution of (2.3.1) in $[v_0, w_0]$.

We next show the quadratic convergence of $\{v_k\}$, $\{w_k\}$ to u. Let us set $p_{k+1} = u - v_{k+1}$, $q_{k+1} = w_{k+1} - u$ so that $Bp_{k+1} = 0$ on Γ_T, $Bq_{k+1} = 0$ on Γ_T, $p_{k+1}(0, x) = 0$ on $\bar{\Omega}$ and $q_{k+1}(0, x) = 0$ on $\bar{\Omega}$. Moreover,

$$
\begin{aligned}
\mathcal{L}p_{k+1} &= f(t, x, u) + g(t, x, u) - [f(t, x, v_k) + f_u(t, x, v_k)(v_{k+1} - v_k) \\
&\quad + g(t, x, v_k) + g_u(t, x, w_k)(v_{k+1} - v_1)] \\
&\leq f_u(t, x, \xi_1)p_k - f_u(t, x, v_k)(v_{k+1} - v_k) + g_u(t, x, \xi_2)p_k \\
&\quad - g_u(t, x, w_k)(v_{k+1} - v_k)
\end{aligned}
$$

where ξ_1, ξ_2 lie between v_k and u. It then follows that using the monotone nature of f_u, g_u

$$
\begin{aligned}
\mathcal{L}p_{k+1} &\leq [f_u(t, x, u) - f_u(t, x, v_k)]p_k + [f_u(t, x, v_k) + g_u(t, x, w_k)]p_{k+1} \\
&\quad + [g_u(t, x, v_k) - g_u(t, x, w_k)]p_k \\
&\leq f_{uu}(t, x, \sigma_1)p_k^2 - g_{uu}(t, x, \sigma_2)p_k(p_k + q_k) + c_2 p_{k+1},
\end{aligned}
$$

where $\mid f_u + g_u \mid \leq c_2$ on $\bar{Q}_T$. But

$$
\begin{aligned}
-g_{uu}(t, x, \sigma_2)p_k(p_k + q_k) &\leq c_3(p_k + q_k)p_k \\
&= c_3(p_k^2 + p_k q_k) \\
&\leq \frac{3}{2}c_3 p_k^2 + \frac{c_3}{2}q_k^2
\end{aligned}
$$

where $\mid g_{uu}(t, x, u) \mid \leq c_3$ on $\bar{Q}_T$. Now letting $\mid f_{uu} \mid \leq c_1$ on $\bar{Q}_T$, we obtain

$$
\begin{aligned}
\mathcal{L}p_{k+1} &\leq c_1 p_k^2 + \frac{3}{2}c_3 p_k^2 + \frac{c_3}{2}q_k^2 + c_2 p_{k+1} \\
&= c_2 p_{k+1} + (c_1 + \frac{3}{2}c_3)p_k^2 + \frac{c_3}{2}q_k^2 \text{ in } Q_T.
\end{aligned}
$$

Let $v(t)$ be the solution of the linear differential equation

$$
v' = Mv + \sigma, v(0) = 0,
$$

which is $v(t) = v_0(0)e^{Mt} + \int_0^t e^{M(t-s)}\sigma(s)ds$. Here $M = c_0 + c_2$, $c_0 \leq c(t, x)$ on $\bar{Q}_T$ and $\sigma = c_4 [\max_{\bar{Q}_T} \mid p_k \mid^2 + \max_{\bar{Q}_T} \mid q_k \mid^2]$. Since $Bp_{k_1} \leq v(t)$ on Γ_T, $p_{k+1}(0, x) \leq v(0)$ on $\bar{\Omega}$ and $v(t) \leq K[\max_{\bar{Q}_T} \mid p_k \mid^2 + \max_{\bar{Q}_T} \mid q_k \mid^2]$, we find, by Theorem 2.2.2, the estimate

$$
\max_{\bar{Q}_T} \mid p_{k+1} \mid \leq K[\max_{\bar{Q}_T} \mid p_k \mid^2 + \max_{\bar{Q}_T} \mid q_k \mid^2]
$$

which proves the quadratic convergence of $\{v_k\}$ to u. One can get a similar estimate for q_{k+1}. This completes the proof.

The following special cases that show the importance of Theorem 2.4.1 are worth noting.

Remarks 2.4.1

(i) If $g(t,x,u) \equiv 0$ in (2.3.1), we get from Theorem 2.4.1 a result when $f(t,x,u)$ is convex in u.

(ii) If $f(t,x,u) \equiv 0$ in (2.3.1), Theorem 2.4.1 gives a dual result for $g(t,x,u)$ when it is concave.

(iii) Consider the case when $g(t,x,u) \equiv 0$ and $f(t,x,u)$ is not convex in u. Suppose that $f(t,x,u) + G(t,x,u)$ is convex with $G(t,x,u)$ convex in u. Then, it would be a special case of Theorem 2.4.1 since we can write IBVP (2.3.1) as $\mathcal{L}u = \tilde{f}(t,x,u) + \tilde{g}(t,x,u)$ where $\tilde{f} = f + G$ and $\tilde{g} = -G$, and note the conditions of Theorem 2.4.1 are fulfilled. Consequently, we get the same conclusion when $f(t,x,u)$ is not convex. For example, we can choose $G(t,x,u) = Mu^2$, $M > 0$, where $-M \le \min_{\bar{Q}_T} f_{uu}(t,x,u)$, $u \in [v_0, w_0]$.

(iv) A dual situation of (iii) arises when $f(t,x,u) \equiv 0$ and $g(t,x,u)$ is not concave but $g(t,x,u) + G(t,x,u)$ is concave with $F(t,x,u)$ concave. Then (2.3.1) is valid with $\tilde{f}(t,x,u) = -F(t,x,u)$ and $\tilde{g}(t,x,u) = g(t,x,u) + F(t,x,u)$. Then the conditions of Theorem 2.4.1 are satisfied and we get the same conclusion for the case when $g(t,x,u)$ is not concave.

(v) Suppose $g(t,x,u)$ is concave in u but $f(t,x,u)$ is not convex in u in (2.3.1). Assume that $f(t,x,u) + G(t,x,u)$ is convex with $G(t,x,u)$ convex in u. Then we consider the IBVP

$$\mathcal{L}u = \tilde{f}(t,x,u) + \tilde{g}(t,x,u) \text{ in } Q_T, \qquad (2.4.14)$$

$$Bu = \phi \text{ on } \Gamma_T, \ u(0,x) = \phi_0 \text{ on } \bar{\Omega},$$

where $\tilde{f} = f + G$ and $\tilde{g} = g - G$. An application of Theorem 2.4.1 to IBVP (2.4.14) yields the same conclusion for IBVP (2.3.1).

(vi) A dual case of (v) holds as well, that is, $f(t,x,u)$ is convex in u but $g(t,x,u)$ is not concave. Then, if $\tilde{g}(t,x,u) = g(t,x,u) + F(t,x,u)$ is concave with $F(t,x,u)$ concave, we can consider the IBVP

$$\mathcal{L}u = f(t,x,u) + \tilde{g}(t,x,u) \text{ in } Q_T, \qquad (2.4.15)$$

$$Bu = \phi \text{ on } \Gamma_T, \ u(0, x) = \phi_0 \text{ on } \bar{\Omega},$$

to get the same conclusion as of Theorem 2.4.1 for IBVP (2.3.1).

(vii) Suppose now that in (2.3.1), f is not convex and g is not concave in u but $f + G$ is convex with G convex in u and $g + F$ is concave with F concave in u. Then we consider the IBVP

$$\mathcal{L}u = \tilde{f}(t, x, u) + \tilde{g}(t, x, u) \text{ in } Q_T, \tag{2.4.16}$$

$$Bu = \phi \text{ on } \Gamma_T, u(0, x) = \phi_0 \text{ on } \bar{\Omega},$$

where $\tilde{f} = f + G - F$ and $\tilde{g} = g + F - G$ and note that the conclusion of Theorem 2.4.1 holds for the IBVP (2.4.16).

2.5   Monotone Flows and Mixed Monotone Systems

We shall discuss, in this section, the parabolic system of IBVPs which possess the mixed quasimonotone property. Employing the coupled lower and upper solutions of the corresponding elliptic BVP and the theory of monotone flows, we shall investigate the convergence of the solutions of the corresponding elliptic system.

We consider the following weakly coupled parabolic IBVP in the vectorial form

$$\left[\begin{array}{rll} u_t + Lu & = g(x, u), & t > 0, \ x \in \Omega, \\ Bu & = h(x), & t > 0, \ x \in \partial\Omega, \\ u(0, x) & = u_0(x), & x \in \Omega, \end{array} \right. \tag{2.5.1}$$

and its corresponding elliptic BVP

$$\left[\begin{array}{rll} Lu & = g(x, u), & x \in \Omega, \\ Bu & = h(x), & x \in \partial\Omega, \end{array} \right. \tag{2.5.2}$$

where Ω is a bounded domain in R^n with boundary $\partial\Omega$. We suppose that Lu is elliptic and the boundary operator Bu is given by

$$L_k u^k = -\sum_{i,j=1}^{n} a_{ij}(x) u^k_{x_i x_j} + \sum_{i=1}^{n} b_i(x) u^k_{x_i},$$

$$B_k u^k = \alpha_k(x) \frac{du^k}{d\gamma} + B_k(x) u^k,$$

for $k = 1, 2, \ldots, N$ where $\frac{d}{d\gamma}$ denotes, as usual, the outward normal derivative on $\partial\Omega$. We also assume that Ω is of class $C^{2+\alpha}$ and for each k, the
coefficients of L_k are Hölder continuous in $\bar{\Omega}$, $\beta_k \in C^{1+\alpha}[\partial\Omega, R]$, and either
$\alpha_k(x) = 0$ and $\beta_k(x) = 1$ or $\alpha_k(x) = 1$ and $\beta_k(x) = 0$. The functions
$g(x, u)$, $h(x)$, $u_0(x)$ are Hölder continuous in x in their respective domains
and $u(x)$ satisfies the compatibility condition. These assumptions guarantee
the existence of a classical solution to (2.5.1).

The inequalities between vectors are understood to be componentwise. A
function $f \in C[R^N, R^N]$ is said to have the mixed monotone property if for
each k, $f_k(u_k, [u]_{p_k}, [u]_{q_k})$ is monotone nondecreasing in $[u]_{p_i}$ and monotone
nonincreasing in $[u]_{q_i}$, where for k, $1 \le k \le n$, the nonnegative integers
p_k, q_k satisfy $p_k + q_k = N - 1$ and the vector u admits a splitting into
$u = (u_i, [u]_{p_i}, [u]_{q_i})$.

We shall first prove a result which offers convergence results for the
quasimonotone parabolic IBVP (2.5.1).

Theorem 2.5.1 *Assume that*

(A_1) $\alpha_0, \beta_0 \in C^{2+\alpha}[\bar{\Omega}, R^N]$, $\alpha_0(x) \le \beta_0(x)$,

$$L\alpha_0 \le g(x, \alpha_0) \text{ in } \Omega,$$

$$B\alpha_0 \le h(x) \text{ on } \partial\Omega,$$

and

$$L\beta_0 \ge g(x, \beta_0) \text{ in } \Omega,$$

$$B\beta_0 \ge h(x) \text{ on } \partial\Omega;$$

(A_2) $g \in [\bar{\Omega} \times R^N, R^N]$, $g(x, u)$ *is Hölder continuous in* x, *is quasimonotone
nondecreasing in* u *and for* $x \in \bar{\Omega}$, $u_1, u_2 \in [\alpha_0, \beta_0]$,

$$| g(x, u_1) - g(x, u_2) | \le M \parallel u_1 - u_2 \parallel, \; M > 0.$$

Then

(i) $\alpha_0 \le u(t, x, \alpha_0) \le u(t, x, \beta_0) \le \beta_0$, *where* $u(t, x, \alpha_0)$ *and* $u(t, x, \beta_0)$ *are
solutions of* (2.5.1) *with* $u_0 = \alpha_0$ *and* $u_0 = \beta_0$ *respectively;*

(ii) $u(t, x, \alpha_0)$ *is nondecreasing in* t *and* $u(t, x, \beta_0)$ *is nonincreasing in* t,

$$\lim_{t \to \infty} u(t, x, \alpha_0) = \underline{u}(x), \; \lim_{t \to \infty} u(t, x, \beta_0) = \bar{u}(x)$$

exist and $\underline{u}(x)$, $\bar{u}(x)$ *are solutions of* (2.5.2);

(iii) *for any u_0 with $\alpha_0 \leq u_0 \leq \beta_0$, we have $\underline{u}(x) \leq u^*(x) \leq \bar{u}(x)$ on $\bar{\Omega}$, if* $\lim_{t\to\infty} u(t, x, u_0) = u^*(x)$ *on $\bar{\Omega}$.*

Proof It is easy to see that $\alpha_0(x)$, $\beta_0(x)$ are also lower and upper solutions of (2.5.1) provided $\alpha_0(x) \leq u_0(x) \leq \beta_0(x)$. Since $g(x, u)$ is Lipschitzian in u, it follows that there exists a unique solution $u(t, x)$ of (2.5.1) such that $\alpha_0(x) \leq u(t, x, u_0) \leq \beta_0(x)$ on $R \times \bar{\Omega}$. Using the comparison theorem 2.2.1, we see that $\alpha_0(x) \leq u(t, x, \alpha_0)$, where $u(t, x, \alpha_0)$ is the unique solution of (2.5.1) with $u(0, x) = \alpha_0(x)$. Similarly we get $\beta_0(x) \geq u(t, x, \beta_0)$, where $u(t, x, \beta_0)$ is the unique solution of (2.5.1) with $u(0, x) = \beta_0(x)$. This proves (i).

To prove (ii), we let, for fixed $s > 0$, $v(t, x) = u(t+s, x, \beta_0)$ and $u(t, x) = u(t, x, \beta_0)$. Then using (i), we have

$$\left[\begin{array}{lll} v_t + Lv & = g(x, v), & (t > 0, x \in \Omega), \\ Bv & = h & (t > 0, x \in \partial\Omega), \\ v(0, x) & = u(s, x, \beta_0) \leq \beta_0(x), & (x \in \Omega) \end{array}\right. \tag{2.5.3}$$

and

$$\left[\begin{array}{lll} u_t + Lu & = g(x, u), & (t > 0, x \in \Omega), \\ Bu & = h, & (t > 0, x \in \Omega), \\ u(0, x) & = \beta_0(x), & (x \in \Omega). \end{array}\right. \tag{2.5.4}$$

Hence by applying the Comparison Theorem 2.2.1, we obtain that $v(t, x) \leq u(t, x)$, which implies $u(t+s, x, \beta_0) \leq u(t, x, \beta_0)$ on $R_+ \times \bar{\Omega}$. This proves $u(t, x, \beta_0)$ is nonincreasing in t. Similarly, one can prove $u(t, x, \alpha_0)$ is nondecreasing in t. We therefore have the estimate

$$\alpha_0(x) \leq u(t, x, \alpha_0) \leq u(t, x, \beta_0) \leq \beta_0(x) \text{ on } R_+ \times \bar{\Omega}.$$

Hence the limits $\lim_{t\to\infty} u(t, x, \alpha_0) = \underline{u}(x)$ and $\lim_{t\to\infty} u(t, x, \beta_0) = \bar{u}(x)$ exist on $\bar{\Omega}$. In order to prove that the limits $\bar{u}(x)$ and $\underline{u}(x)$ are solutions of (2.5.2), we proceed with the following arguments.

Let $w(x)$ be the generalized solutions of

$$\left[\begin{array}{ll} Lw + cw & = G(x, \bar{u}) \text{ in } \Omega, \\ Bw & = h(x), \text{ on } \partial\Omega, \end{array}\right. \tag{2.5.5}$$

where c is a positive constant and

$$G(x, u) = cu + g(x, u).$$

Now setting $W(t,x) = u(t,x,\beta_0) - w(x)$, it is easy to see that $W(t,x)$ satisfies

$$(W)_t + LW + cW = q(t,x)$$

$$BW = h(t,x) \equiv 0,$$

$$W(0,x) = W_0(x),$$

where $q(t,x) = cw(t,x) + g(x,u(t,x,\beta_0)) - g(x,\bar{u}(x))$, and

$$W(0,x) = \bar{u}(x) - w(x).$$

Since $\lim_{t\to\infty} u(t,x,\beta_0) = \bar{u}(x)$, we have $q(t,x) \to 0$ as $t \to \infty$. By Theorem A.3.1, we get $W(t,x) \to 0$ in $L_2(\Omega)$ as $t \to \infty$. This proves $\bar{u}(x)$ is the generalized solution of (2.5.2) which immediately shows that $\bar{u}(x)$ is the generalized solution of (2.5.2). Now using Lemma A.3.1, $\bar{u} \in C^{1+\alpha}(\bar{\Omega})$ for some $\alpha \in (0,1)$. This implies $g(x,\bar{u}(x))$ is $C^\alpha(\bar{\Omega})$. Moreover, using Theorem A.3.1, we see that $\bar{u} \in C^{2+\alpha}(\bar{\Omega})$ and therefore $\bar{u}(x)$ is the solution of (2.5.2). Similarly one can show that $\underline{u}(x) = \lim_{t\to\infty} u(t,x,\alpha_0)$ is the solution of (2.5.2). Thus we have proved (ii).

To prove (iii), let $\alpha_0(x) \leq u_0(x) \leq \beta_0(x)$ and consider the solution $u(t,x,u_0)$ of (2.5.1). By repeated application of Comparison Theorem 2.2.1, it follows that

$$u(t,x,\alpha_0) \leq u(t,x,u_0) \leq u(t,x,\beta_0) \text{ on } R_+ \times \bar{\Omega}.$$

Hence if $\lim_{t\to\infty} u(t,x,u_0) = u^*(x)$ exists, then because of (ii), we obtain

$$\underline{u}(x) \leq u^*(x) \leq \bar{u}(x) \text{ on } \bar{\Omega},$$

proving (iii). The proof is complete.

Next, we consider the mixed monotone parabolic system of IBVP given by

$$\left[\begin{array}{ll} u_t + Lu = H(x,u), & (t > 0, x \in \Omega), \\ \quad\quad Bu = h(x), & (t > 0, x \in \partial\Omega), \\ \quad u(0,x) = u_0(x), & (x \in \Omega), \end{array}\right. \tag{2.5.6}$$

where H satisfies the same assumptions as g in (2.5.1) except that $H(x,u)$ is now assumed to be mixed monotone in u. The corresponding steady state system is

$$\left[\begin{array}{ll} Lu & = H(x,u), \text{ on } \bar{\Omega}, \\ Bu & = h(x) \text{ on } \partial\Omega. \end{array}\right. \tag{2.5.7}$$

We can now prove the following results relative to (2.5.6) and (2.5.7).

Theorem 2.5.2 *Assume that*

(B_0) *there exists a function $F \in C[\Omega \times R^N \times R^N, R^N]$ such that $F(x, u, u)$*
$= H(x, u)$, $F(x, u, v)$ is Hölder continuous in x, $F(x, u, v)$ is quasi-
monotone nondecreasing in u and monotone nonincreasing in v, and
$F(x, u, v)$ is Lipschitzian in u and v;

(B_1) *$\alpha_0, \beta_0 \in C^{2+\alpha}[\bar{\Omega}, R^n]$, $\alpha_0(x) \le \beta_0(x)$ on $\bar{\Omega}$,*

$$L\alpha_0 \le F(x, \alpha_0, \beta_0), \ x \in \bar{\Omega},$$

$$B\alpha_0 \le h(x), \ x \in \partial\Omega,$$

and

$$L\beta_0 \ge F(x, \beta_0, \alpha_0), \ x \in \bar{\Omega},$$

$$B\beta_0 \ge h(x), \ x \in \partial\Omega.$$

  Then

(i) *$\alpha_0(x) \le u(t, x, \alpha_0) \le u(t, x, \beta_0) \le \beta_0(x)$, $(t, x) \in R_+ \times \bar{\Omega}$, where*
$u(t, x, \alpha_0)$, $u(t, x, \beta_0)$ are the solutions of (2.5.6);

(ii) *$u(t, x, \alpha_0)$ is nondecreasing in t, $u(t, x, \beta_0)$ is nonincreasing in t, $\lim_{t\to\infty}$*
$u(t, x, \alpha_0) = \underline{u}$ and $\lim_{t\to\infty} u(t, x, \beta_0) = \bar{u}$ exist and $\underline{u}$, $\bar{u}$ are solutions
of (2.5.7);

(iii) *if $u(t, x, u_0)$ is the solution of (2.5.6) with $\alpha_0 \le u_0 \le \beta_0$ then*

$$u(t, x, \alpha_0) \le u(t, x, u_0) \le u(t, x, \beta_0)$$

and if $\lim_{t\to\infty} u(t, x, u_0) = u^*(x)$ exists, then $\underline{u}(x) \le u^*(x) \le \bar{u}(x)$ on
$\bar{\Omega}$.

Proof   We define $g \in C[\bar{\Omega} \times R^{2N}, R^{2N}]$ by

$$g(x, z) = g(x, u, v) = (F(x, u, v), F(x, v, u)).$$

Introduce the ordering in $K = R_+^N \times \{-R_+^N\}$ by $Z_1 \le Z_2$, where $Z_1 = (u_1, v_1)$, $Z_2 = (u_2, v_2)$ iff $u_1 \le u_2$, $v_1 \ge v_2$.

  Define the reflection operator $P : R^{2N} \to R^{2N}$ by

$$P(u, v) = (v, u),$$

and let $\xi_0 = (\alpha_0, \beta_0)$, $\eta_0 = P\xi_0$, $Z = (u, v)$ and $Z(t, x, \xi_0) = (u(t, x, \alpha_0), u(t, x, \beta_0))$. Then $g(x, z)$ satisfies the assumptions of Theorem 2.5.1. As a result, conclusions (i), (ii) follow from Theorem 2.5.1. Since $\eta_0 = P\xi_0$, $Z(t, x, \eta_0) = PZ(t, x, \xi_0)$, we get $P\underline{Z} = \bar{Z}$, where $\underline{Z} = (\bar{u}, \underline{u})$. To prove (iii), let $Z_0 = (u_0, u_0)$ so that $\xi_0 \leq Z_0 \leq \eta_0$, which implies from Theorem 2.5.1 that $u(t, x, \alpha_0) \leq u(t, x, u_0) \leq u(t, x, \beta_0)$, which provides (iii). Hence the proof of the theorem is complete.

Remark 2.5.1 Since $H(x, u)$ is mixed quasimonotone and satisfies a Lipschitz condition in u, in Theorem 2.5.2, it follows that $\tilde{H}_k(x, u) \equiv H_k(x, u) + Mu_k$, where $M > 0$, is mixed monotone. As a result, one can define $F(x, u, v)$ whose existence is assumed in Theorem 2.5.2 in the following ways:

(i) Suppose that $\tilde{H}(x, u)$ admits a splitting

$$\tilde{H}(x, u) = H_1(x, u) + H_2(x, u)$$

where $H_1(x, u)$ is nondecreasing and $H_2(x, u)$ is nonincreasing in u. Then F can be taken as

$$F(x, u, v) = H_1(x, u) + H_2(x, v).$$

(ii) Suppose that $H(x, u)$ is mixed quasimonotone in U as defined earlier that is, $H_k(t, x, u_k, [u]_{p_k}, [u]_{q_k})$ where $p_k + q_k = N - 1$, H is nondecreasing in $[u]_{p_k}$ and nonincreasing in $[u]_{q_k}$. Then, as in (i)

$$\tilde{H}(x, u) = H(x, [u]_{p_k+1}, [u]_{q_k})$$

and in this case, (H_1), (H_2) in (1) take the form

$$H_1(x, u) = H[x, [u]_{p_k+1}, [0]_{q_k}],$$

$$H_2(x, u) = H[x, [0]_{p_k+1}, [u]_{q_k}],$$

so that $F(x, u, v)$ can be defined as before

$$F(x, u, v) = H_1(x, u) + H_2(x, v).$$

In each of the foregoing situations, we need to consider the IBVPs

$$\left[\begin{array}{rl} u_t^k + L_k u^k + Mu^k & = \tilde{H}_k(x, u), \quad t > 0, x \in \Omega, \\ Bu = h(x), & T > 0, x \in \partial\Omega, \\ u(0, x) & = u_0(x), \quad\quad x \in \Omega, \end{array} \right. \tag{2.5.8}$$

with a corresponding change in (2.5.7) designated as (2.5.8). An application of Theorem 2.5.2 to (2.5.7) yields the same conclusion as in Theorem 2.5.2 for the IBVP (2.5.6).

2.6   General Comparison Results (GCRs) for Weakly Coupled Systems

We shall extend the comparison result (2.2.1) to weakly coupled parabolic systems in a very general framework. We consider the general IBVP

$$
\left[
\begin{array}{lll}
u_t^k & = f_k(t, x, u, u_x^k, u_{xx}^k) & \text{in } Q_T, \\
Bu & = \phi(t, x) & \text{on } \Gamma_T, \\
u(0, x) & = u_0(x) & \text{on } \bar{\Omega},
\end{array}
\right.
\tag{2.6.1}
$$

where for each $k = 1, 2, \ldots, N$, $f_k \in C[\bar{Q}_T \times R^N \times R^n \times R^{n^2}, R]$ is elliptic in Q_T and $B_k u^k$ is the boundary operator defined as before, namely,

$$
B_k u^k = p_k(t, x) u^k + q_k(t, x) \frac{du^k}{d\gamma}.
$$

Here p_k, $q_k \in C[\Gamma_T, R]$ are such that $p_k(t, x) \geq 0$, $q_k(t, x) \geq 0$ and $p_k(t, x) + q_k(t, x) > 0$. Without further mention, here and in the next section all vectorial inequalities mean that the same inequalities hold between their corresponding components. Also, $\phi \in C[\Gamma_T, R^N]$ and $u_0 \in C[\bar{\Omega}, R^N]$.

For each $k = 1, 2, \ldots, N$, the function f_k is said to be elliptic at a point (t_1, x_1) if, for any $u, P^k, R_{ij}^k, S_{ij}^k, i, j = 1, 2, \ldots, n$, the quadratic form

$$
\sum_{i,j=1}^{n} (R_{ij}^k - S_{ij}^k) \lambda_i \lambda_j \leq 0, \ \lambda \in R^n,
$$

implies that

$$
f_k(t_1, x_1, u, P, R) \leq f_k(t_1, x_1, u, P, S).
$$

If this property holds at every point $(t, x) \in Q_T$, then we say that $f(t, x, u, P, R)$ is elliptic in Q_T.

The basic comparison result relative to IBVP (2.6.1) may now be proved.

Theorem 2.6.1 *Assume that*

(i) *$v_k, w_k \in C^{1,2}$, $f_k \in C[\bar{Q}_T \times R^N \times R^n \times R^{n2}]$, it is elliptic and $(v_k)_t \leq f_k(t, x, v, (v_k)_x, (v_k)_{xx})$, $(w_k)_t \geq f_k(t, x, w, (w_k)_x, (w_k)_{xx})$ on Q_T, for $k \in I$;*

(ii) (a) *$v(0, x) < w(0, x)$ on $\bar{\Omega}$;*

  (b) *$Bv(t, x) < Bw(t, x)$ on Γ_T;*

(iii) $f(t, x, u, P, Q)$ is quasimonotone nondecreasing in u for fixed (t, x, P, Q), where $t, x \in \bar{Q}_T$, $P \in R^n$, $Q \in R^{n2}$.

Then $v(t, x) < w(t, x)$ on $\bar{Q}_T$, if one of the inequalities of (i) is strict.

Proof   We assume that one of the inequalities in (i) is strict and consider the function $m(t, x) = v(t, x) - w(t, x)$. If the conclusion is not true, there exists an index k and $t_1 > 0$, $x_1 \in \bar{\Omega}$, such that

$$m_j(t, x) < 0 \text{ on } [0, t_1) \times \bar{\Omega}, \ j \neq k$$

$$m_k(t, x) < 0 \text{ on } [0, t_1) \times \bar{\Omega} \text{ and } m_k(t_1, x_1) = 0.$$

It is easy to observe that $m_k(t, x)$ has its maximum at x_1 which is equal to zero. Clearly, $t_1 > 0$ because of (ii)(a). Also $(t_1, x_1) \notin \Gamma_T$, and therefore we have

$$\lim_{h \to 0} \frac{m_k(t_1, x_1) - m_k(t_1, x_1 - h\gamma)}{h} \geq 0.$$

This would contradict (ii)(b). Hence $(t_1, x_1) \in Q_T$. Consequently, we have $(m_k)_{x_i}(t_1, x_1) = 0$, $i = 1, 2, \ldots, n$, and the quadratic form

$$\sum_{i,j=1}^{m} (m_k)_{x_i x_j}(t_1, x_1)\lambda_k, \lambda_k \leq 0$$

for an arbitrary vector $\lambda \in R^m$. This implies that

$$v_j(t_1, x_1) \leq w_j(t_1, x_1), \quad j = k$$
$$v_k(t_1, x_1) = w_k(t_1, x_1)$$
$$(v_k)_{x_i}(t_1, x_1) = (w_k)_{x_i}(t_1, x_1), \quad i = 1, 2, \ldots, m$$

and

$$\sum_{i,j=1}^{m} \frac{\partial^2}{\partial x_i \partial x_j}[v_k(t_1, x_1) - w_k(t_1, x_1)]\lambda_i \lambda_k \leq 0.$$

Also $m_k(t_1 - h, x_1) < 0$ for $h > 0$ sufficiently small. It therefore follows that $(m_k)_t \geq 0$. However, by (i) and using (iii), the quasimonotonicity of f and ellipticity of f_k, we have

$$\begin{aligned}
0 \leq \ & (m_k)_t \\
< \ & f_k(t_1, x_1, v(t_1, x_1), (v_k)_x(t_1, x_1), (v_k)_{xx}(t_1, x_1)) \\
& - f_k(t_1, x_1, w(t_1, x_1), (w_k)_x(t_1, x_1), (w_k)_{xx}(t_1, x_1)) \\
\leq \ & 0,
\end{aligned}$$

which is a contradiction. This completes the proof of the theorem.

The conclusion of Theorem 2.6.1 is valid even if the strict inequality in (i) is not satisfied provided f satisfies the following condition:

(C_0) $z_k \in C^{1,2}$, $z_k > 0$ on $\bar{Q}_T$, $\frac{dz_k(t,x)}{d\gamma} \geq \gamma_k > 0$ on Γ_k and for sufficiently small $\epsilon > 0$, either

(a) $\epsilon z_{kt} > f_k(t, x, w + \epsilon z, (w_k)_x + \epsilon(z_k)_x, (w_k)_{xx} + \epsilon(z_k)_{xx})$
$- f_k(t, x, w, (w_k)_x, (w_k)_{xx})$ or

(b) $\epsilon z_{kt} > f_k(t, x, v, (v_k)_x, (v_k)_{xx}) - f_k(t, x, v - \epsilon z, v_{kx} - \epsilon(z_k)_x, (v_k)_{xx} - \epsilon(z_k)_{xx})$ on Q_T, where $v_k, w_k \in C^{1,2}$, for $k \in I$.

We can dispense with the strict inequality required in Theorem 2.6.1 which is the content of the following result.

Theorem 2.6.2 *Let the assumptions* (i)–(iii) *of Theorem 2.6.1 hold. Suppose further that the condition* (C_0) *is satisfied. Then the relations*

$$v(0, x) \leq w(0, x) \ for \ x \in \bar{\Omega},$$

$$Bv(t, x,) \leq Bw(t, x) \ on \ \Gamma_T,$$

imply

$$v(t, x) \leq w(t, x) \ on \ \bar{Q}_T.$$

Proof Assume that the condition (a) of (C_0) holds. Consider $(\tilde{w}_k)_t = (w_k)_t + \epsilon(z_k)_t$ for $k \in I$. We then have

$$\begin{aligned}
(\tilde{w}_k)_t &= (w_k)_t + \epsilon(z_k)_t \\
&\geq f_k(t, x, w, (w_k)_x, (w_k)_{xx}) + \epsilon(z_k)_t \\
&> f_k(t, x, w + \epsilon z, (w_k)_x + \epsilon(z_k)_x, (w_k)_{xx} + \epsilon(z_k)_{xx}) \\
&= f(t, x, \tilde{w}, (\tilde{w}_k)_x, (\tilde{w}_k)_{xx}) \ \text{on} \ Q_T.
\end{aligned}$$

Also $v_k(0, x) < \tilde{w}_k(0, x)$ for $x \in \bar{\Omega}$ and

$$B_k \tilde{w}_k = B_k w_k + \epsilon B_k z_k \geq B_k v_k + \epsilon B_k z_k > B_k v_k \ \text{on} \ \Gamma_T \ \text{by} \ (C_0).$$

Thus the functions $v, \tilde{w}$ satisfy the assumptions of Theorem 2.6.1 and hence $v_k(t, x) < \tilde{w}_k(t, x)$ on $\bar{Q}_T$ for $k \in I$. Taking the limit as $\epsilon \to 0$ yields the required result and this completes the proof.

Remark 2.6.1 If $q_k(t,x) = 0$ for $k \in I$, then the assumption (C_0) in Theorem 2.6.2 can be replaced by a weaker hypothesis, namely a one-sided Lipschitz condition of the form

$$f_k(t,x,u,P,R) - f_k(t,x,v,P,R) \leq L \left(\sum_{j=1}^{n} (u_j - v_j) \right)$$

whenever $u \geq v$. In this case, it is enough to set $\tilde{w}_k = w_k + \epsilon e^{(N+1)Lt}$ so that $v_k(0,x) < \tilde{w}_k(0,x)$, $x \in \bar{\Omega}$, and $B_k v_k(t,x) < B_k \tilde{w}_k(t,x)$ on Γ_T. Also we have

$$
\begin{aligned}
(\tilde{w}_k)_t &\geq f_k(t,x,w,(w_k)_x,(w_k)_{xx}) + (N+1)\epsilon e^{(N+1)Lt} \\
&\geq f_k(t,x,\tilde{w},(\tilde{w}_k)_x,(\tilde{w}_k)_{xx}) - n\epsilon e^{(N+1)Lt} + (N+1)\epsilon e^{(N+1)Lt} \\
&> f_k(t,x,\tilde{w},(\tilde{w}_k)_x,(\tilde{w}_k)_{xx}) \text{ on } Q_T \text{ for } k \in I.
\end{aligned}
$$

As in the scalar case if $p_k(t,x) = 0$ for $k \in I$, then condition (C_0) becomes essential.

As an example of Theorem 2.6.2, consider the interesting special case

$$f_k(t,x,u,u_x^k,u_{xx}^k) = a^k u_{xx}^k + b^k u_x^k + F_k(t,x,u) \qquad (2.6.2)$$

where

$$a^k u_{xx}^k = \sum_{i,j=1}^{n} a_{ij}^k u_{x_i x_j}^k, \quad b^k u_x^k = \sum_{j=1}^{n} b_j^k u_{x_j}^k, \quad k = 1, 2, \dots, N$$

and F is Lipschitzian and quasimonotone nondecreasing in u. That is, F satisfies

$$| F_k(t,x,u) - F_k(t,x,v) | \leq L \sum_{\mu=1}^{n} | u_\mu - v_\mu |, \quad (t,x) \in H.$$

Assume also that the boundary H is smooth enough, that is, there exists a $h \in C^2[\Omega, R_+]$ such that $\frac{\partial h}{\partial \gamma} \geq 1$ on ∂H_1 and h_x, h_{xx} are bounded. Let $M > 1$ and define $H(x) = \exp((mLh(x))$, $Z(x) = (\exp(N_0 t)) H(x)$ and $\tilde{Z}(x) = \tilde{e} Z(x)$, where $\tilde{e} = (1,1,\dots,1)$, $N_0 = MLN + A$, L is the Lipschitz constant for F and $| a^k H_{xx} + b^k H_x | \leq A_k \leq A$, $k = 1, 2, \dots, N$. Then

$$\frac{dZ}{d\gamma} = ML \left(\frac{d(x)}{d\gamma} \right) \tilde{e} \geq ML\tilde{e} > 0 \text{ on } \partial H_1$$

and

$$\epsilon(\tilde{Z}_t^k - a^k \tilde{Z}_{xx}^k - b^k \tilde{Z}_x^k) \geq \epsilon(N_1 - A)\tilde{Z}^k = \epsilon M L N \tilde{Z}^k > \epsilon L N Z.$$

Using the Lipschitz condition for F, we have

$$\mid F^k(t,x,w+\epsilon\tilde{Z}) - F^k(t,x,w) \mid \leq L \sum_{\mu=1}^{n} \epsilon\tilde{Z}_\mu = \epsilon L N Z$$

and consequently we get for $\epsilon > 0$,

$$\epsilon\tilde{Z}_t > \epsilon[a\tilde{Z}_{xx} + b\tilde{Z}_{xx}] + F(t,x,w+\epsilon\tilde{Z}) - F(t,x,w)$$

which is exactly condition (a) of (C_0). Similarly (b) of (C_0) is also satisfied.

2.7   Stability and Vector Lyapunov Functions

We shall investigate, in this section, the stability properties of the trivial solution of weakly coupled system of IBVPs by means of the method of vector Lyapunov functions and show that this effective approach is a natural setting for the discussion of such systems.

Let us consider the weakly coupled system

$$\left[\begin{array}{ll} u_t = Lu + f(t,x,u) & \text{in } J \times \Omega, \quad J = [t_0, \infty), t_0 \geq 0, \\ \frac{du}{d\gamma}(t,x) = 0 & \text{on } J \times \partial\Omega, \\ u(0,x) = \phi_0(x) & \text{in } \bar{\Omega}, \end{array} \right. \tag{2.7.1}$$

where the elliptic operator is such that

$$L_k u^k = \sum_{i,j=1}^{n} a_{ij}^k(t,x) u_{x_i x_j}^k + \sum_{i=1}^{n} b_i(t,x)u^k$$

$$\sum_{i,j=1}^{n} a_{ij}^k(t,x)\lambda_i\lambda_j \geq \beta \mid \lambda \mid^2, \ \lambda \in R^n,$$

and $f \in C[R_+ \times \bar{\Omega} \times R^N, R^N]$. Here $\Omega \subset R^n$ is assumed to be a bounded domain equipped with a smooth boundary. We also assume the existence and uniqueness of solutions $u(t,x)$ of (2.7.1) for $J \times \bar{\Omega}$.

We can now extend the method of vector Lyapunov functions to study the stability properties of the solution of (2.7.1).

Let us first define the stability notions, assuming that (2.7.1) has the trivial solution. We denote $\parallel u(t,\cdot) \parallel_0 = \max_{\bar{\Omega}} \parallel u(t,x) \parallel$, where $u(t,x)$ is the solution of (2.7.1).

Definition 2.7.1 *The trivial solution of* (2.7.1) *is said to be*

(i) *Stable, if given $\epsilon > 0$ and $t_0 \in R_+$, there exists a $\delta = \delta(t_0, \epsilon) > 0$ such that $\| u_0 \|_0 < \delta$ implies $\| u(t, \cdot) \|_0 < \epsilon$, $t \geq t_0$;*

(ii) *uniformly stable, if δ in (i) is independent of t_0;*

(iii) *quasi-asymptotically stable, if, given $\epsilon > 0$ and $t_0 \in R_+$, there exist $\delta_0 = \delta_0(t_0) > 0$, $T = T(t_0, \epsilon) > 0$ such that*

$$\| u_0 \|_0 < \delta_0 \ implies \ \| u(t, \cdot) \|_0 < \epsilon, t \geq t_0 + T;$$

(iv) *uniformly quasi-asymptotically stable, if in (iii) δ_0, T are independent of t_0;*

(v) *asymptotically stable, if (i) and (iii) hold simultaneously;*

(vi) *uniformly asymptotically stable if (ii) and (iv) are satisfied at the same time.*

We can now provide sufficient conditions, in terms of vector Lyapunov-like functions, for the stability properties of the trivial solution of (2.7.1).

Theorem 2.7.1 *Assume that*

(i) $V \in C^1[R_+ \times R^N, R_+^N]$, $V_u(t, u) \, Lu \leq LV(t, u)$, and $V_t(t, u) + V_u(t, u)f(t, x, u) \leq g(t, V(t, u))$ on $R_+ \times \Omega \times R^N$, where $g \in C[R_+ \times R_+^N, R_+^N]$, $g(t, u)$ is quasimonotone nondecreasing and locally Lipschitzian in u;

(ii) $f(t, x, 0) \equiv 0$ and $g(t, 0) \equiv 0$;

(iii) on $R_+ \times R^N$, $b(\| u \|) \leq \sum_{i=1}^N V_i(t, u) \leq a(\| u \|)$, where $a, b \in K$, $K = \{\phi \in C[R_+, R_+] : \phi(0) = 0 \ and \ \phi(s) \ is \ strictly \ increasing \ in \ s\}$.

Then the stability properties of the trivial solution of either

(a)
$$y' = g(t, y), y(t_0) = y_0 \geq 0, \tag{2.7.2}$$

   or

(b)
$$\left[\begin{array}{l} v_t = Lv + g(t, v) \quad in \ J \times \Omega \\ v(t_0, x) = \psi_0(x) \geq 0 \quad in \ \bar{\Omega}, \frac{dv(t,x)}{dv} = 0 \ on \ J \times \partial\Omega, \end{array} \right. \tag{2.7.3}$$

imply the corresponding stability properties of the trivial solution of (2.7.1).

Proof   Let $u(t,x)$ be any solution of (2.7.1). Setting $m(t,x) = V(t, u(t,x))$ and using assumption (i), we get

$$
\left[
\begin{array}{l}
m_t \leq Lm + g(t,m) \quad \text{in } J \times \Omega \\
m(t_0, x) = V(t_0, \phi_0(x)) \quad \text{in } \bar{\Omega}, \ \frac{dm(t,x)}{d\gamma} = 0 \text{ in } J \times \partial\Omega.
\end{array}
\right.
\tag{2.7.4}
$$

Let $y(t)$ and $r(t,x)$ be the solutions of (2.7.2) and (2.7.3) respectively existing for $t \geq t_0$ and $x \in \bar{\Omega}$. Then we have

$$
y' = g(t,y), y(t_0) = y_0 \geq m(t_0, x) \text{ in } \bar{\Omega}
\tag{2.7.5}
$$

and

$$
\left[
\begin{array}{l}
r_t = Lr + g(t,r) \quad \text{in } J \times \Omega \\
r(t_0, x) \geq m(t_0, x) \quad \text{in } \bar{\Omega}, \ \frac{dm(t,x)}{d\gamma} = 0 \text{ in } J \times \partial\Omega.
\end{array}
\right.
\tag{2.7.6}
$$

Consequently, applying Theorem 2.6.1 yields with $v = m$ and $w = y$ or $w = r$ the estimates $V(t, u(t,x)) \leq y(t)$, or $V(t, u(t,x)) \leq r(t,x)$ in $J \times \bar{\Omega}$.

Let $0 < \epsilon$ and $t_0 \in R_+$ be given.

Suppose that the trivial solution of (2.7.3) is stable, which implies that, given $b(\epsilon) > 0$ and $t_0 \in R_+$, there exists a $\delta_1 = \delta_1(t_0, \epsilon) > 0$ such that

$$
\| \psi_0 \|_0 = \max_{\bar{\Omega}} \sum_{i=1}^{N} \| \psi_{0i}(x) \| < \delta_1 \text{ implies } \| r(t, \cdot) \|_0 = \max_{\bar{\Omega}} \sum_{i=1}^{N} r_i(t,x) < b(\epsilon)
\tag{2.7.7}
$$

for $t \geq t_0$.

From the estimate $V(t, u(t,x)) \leq r(t,x)$, $t \geq t_0$, $x \in \bar{\Omega}$, we find, in view of (ii)

$$
b(\| u(t,x) \|) \leq \sum_{i=1}^{N} V_i(t, u(t,x)) \leq \sum_{i=1}^{N} r_i(t,x), t > t_0, x \in \bar{\Omega}.
\tag{2.7.8}
$$

Choose $\delta = \delta(t_0, \epsilon) > 0$ such that

$$
a(\delta) < \delta_1;
\tag{2.7.9}
$$

and let $\| \phi_0 \|_0 < \delta$. Then, it follows that $\sum_{i=1}^{N} r_i(t_0, x) = \sum_{i=1}^{N} V_i(t_0, \phi_0(x)) \leq a(\| \phi_0(x) \|) \leq a(\| \phi_0 \|_0) < a(\delta) < \delta_1$. Consequently, this implies from (2.7.7), (2.7.8) that

$$
b(\| u(t,x) \|) \leq \sum_{i=1}^{N} r_i(t,x) < b(\epsilon), \ t \geq t_0, x \in \bar{\Omega},
$$

which yields the relation

$$\| u(t, \cdot) \|_0 < \epsilon, \ t \geq t_0,$$

proving that the trivial solution of (2.7.1) is stable.

Other stability notions can be proved similarly using the standard arguments in Lyapunov stability theory related to vector Lyapunov functions. This proves Theorem 2.7.1.

Remark 2.7.1 If we have the same operator L, that is, the same diffusion law for all the components of u in (2.7.1), then one can use a single Lyapunov function. On the other hand, if the system (2.7.1) does not enjoy this luxury, then it is not possible to employ a single Lyapunov function. We note also that if each component of V is convex in u then $V_u(t, u)Lu \leq LV$ clearly holds. Thus it is obvious that for general weakly coupled systems, utilizing vector Lyapunov functions is natural and advantageous.

Let us next demonstrate the result with some typical examples. Before that, let us note the following.

Let $\psi(x)$ be the solution of the steady state problem

$$\left[\begin{array}{ll} L\psi + f(x, \psi) = 0 & \text{in } \Omega. \\ \frac{d\psi(x)}{d\gamma} = 0 & \text{on } \partial\Omega. \end{array} \right. \tag{2.7.10}$$

Then, setting $w = u - \psi$, we see that w satisfies

$$\left[\begin{array}{lll} w_t & = Lw + F(x, w) & \text{in } J \times \Omega, \\ w(t_0, x) & = \phi_0(x) - \psi(x) = \bar{\psi}_0(x) & \text{in } \bar{\Omega}, \\ \frac{dw(t,x)}{d\gamma} & = 0 & \text{in } J \times \partial\Omega, \end{array} \right. \tag{2.7.11}$$

where $F(x, w) = f(x, w + \psi) - f(x, \psi)$. Noting that $F(x, 0) = 0$, we observe that the stability properties of the trivial solution of (2.7.11) imply the corresponding stability properties of solutions of (2.7.1) relative to the steady state solution $\psi(x)$.

Now, for the purpose of illustration, let us consider a typical comparison system

$$\left[\begin{array}{lll} v_t & = Av_{xx} - bv_x + g(t, v), & 0 < x < 1, t \geq 0 \\ v(0, x) & = \psi_0(x) \geq 0, & 0 \leq x \leq 1, \\ v_x(t, 0) & = v_x(t, 1) = 0, & t \geq 0, \end{array} \right. \tag{2.7.12}$$

where $A > 0$ is a diagonal matrix, $b > 0$ and $g(t, v)$ satisfies

$$g(t, u_1) - g(t, u_2) \leq L(u_1 - u_2), \text{ whenever } 0 \leq u_2 \leq u_1 \leq Q, \tag{2.7.13}$$

for some $Q > 0$, where L is an $N \times N$ matrix. We shall consider two cases.

(i) L is a positive matrix with

$$\max_i \sum_{k=1}^{n} L_{ik} = \tilde{L}, \ 1 \leq i \leq N;$$

(ii) L is a Metzeler matrix satisfying the dominant diagonal condition.
That is,

$$L_{ij} \geq 0 \text{ for } i \neq j \text{ and } -L_{ii} > \sum_{\substack{j=1 \\ i \neq j}}^{n} L_{ij} L_{ij}.$$

Define $R(t,x) = (Ke^{-\alpha t + \beta(1-x)})\tilde{e}$, $\alpha, K > 0$ and $\beta \in R$ to be chosen and $\tilde{e} = (1, 1, \ldots, 1)$. Substituting R in (2.7.12), we get

$$R_t - AR_{xx} + bR_x - g(t, R) \geq R(-\alpha - a\beta^2 - b\beta) - g(t, R), \qquad (2.7.14)$$

where $a = \max A_{ii}$. Let $Q = \max_i(\max_{0 \leq x \leq 1} \psi_{0i}(x))$ and $\alpha = L_0 - \tilde{L} > 0$. In case (i), we now have

$$R_t - AR_{xx} + bR_x - g(t, R) \geq -R[L_0 - \tilde{L} + a\beta^2 + b\beta + \tilde{L}] = 0,$$

if β is a root of $a\beta^2 + b\beta + L_0 = 0$. But since $\beta - (-b \pm \sqrt{b^2 - 4aL_0})/2a$, if we suppose that $0 < a < (b^2)/(4L_0)$, then β is negative. It is easy to check that R satisfies initial and boundary conditions by choosing $K = Qe^{-\beta}$. As a result, it follows from Theorem 2.6.1 with $v = r$ and $w = r$ that

$$0 \leq r(t, x) \leq R(t, x), \ t \geq 0, \ 0 \leq x \leq 1,$$

which implies exponential asymptotic stability of the trivial solution of (2.7.12).

In case (ii), set

$$-\gamma_i = L_{ii} + \sum_{\substack{j=1 \\ i \neq j}}^{n} L_{ij} \text{ and } \tilde{L} = \min_t \gamma_i.$$

Then, we get from (2.7.14),

$$R_t - AR_{xx} + bR_x - g(t, R) \geq -R[a\beta^2 + b\beta + L_0] = 0,$$

with $L_0 = \alpha - \tilde{L}$. We again have two cases: $\alpha > L$ and $0 < \alpha \leq \tilde{L}$. If $\alpha > L$, then as in the previous situation, β is negative if we assume $0 < a < b^2/4L_0$ and we obtain the same conclusion as before. If, on the other hand, $0 < \alpha \leq \tilde{L}$ so that $L_0 \leq 0$, then $b^2 - 4aL_0$ is always nonnegative and hence we have one negative root β and consequently the conclusion remains the same in the previous case.

It is clear that in case (i) diffusion and convection terms are contributing to stability and in case (ii) reaction terms are also playing a role. From these two cases, one can obtain several possibilities for the coefficients.

2.8 Notes and Comments

The standard results of Section 2.2 are taken from Ladde et al. [47]. See also Lakshmikantham and Leela [55] for differential inequalities. The results of Sections 2.3 and 2.4 are based on the work of McRae [69], [68]. The contents of Section 2.5 are due to Lakshmikantham and Vatsala [61]. The comparison theorems for weakly coupled systems given in Section 2.6 are adapted from Ladde et al. [47]. See also Lakshmikantham [51] for a comparison result in arbitrary cones. The results of Section 2.7, dealing with the study of stability theory via the method of vector Lyapunov functions, are due to Lakshmikantham and Leela [56]. For parabolic systems in unbounded domains relative to existence, the monotone method and asymptotic behavior, see Pao [77]. For mixed quasimonotone parabolic systems relative to coupled quasi-extremal solutions by the monotone method, see Liu Zhenhai [84]. See Byszewski [11] for the monotone method for a system of nonlocal parabolic IBVPs and Chabrowski [19] for nonlocal parabolic IBVPs with respect to basic existence and comparison results. For results analogous to those in Section 2.5 utilizing the monotone method for mixed monotone parabolic systems, see Pao [76].

For related results see Chandra et al. [20], Carl [13], Carl and Grossman [14], Pao [74], Lakshmikantham and Vatsala [59], J. Wang [80], Vatsala and Wang [79].

Chapter 3

Impulsive Parabolic Equations

3.1 Introduction

In this chapter, we introduce second-order parabolic IBVPs with impulses. The theory of impulsive ordinary differential equations has recently become very popular and the extension of impulsive effects to parabolic IBVPs is a natural development.

Section 3.2 incorporates the necessary framework for studying impulsive parabolic IBVPs, and proves suitable comparison results in a similar general set-up as in Section 2.6. This section also deals with flow invariance by utilizing the method of lower and upper solutions, and obtains bounds for solutions of impulsive parabolic IBVPs in terms of the solutions of impulsive ordinary differential equations. In Section 3.3, various comparison theorems for different types of coupled lower and upper solutions are derived, which are essential to consider the corresponding situations in the method of generalized quasilinearization, which is discussed in Section 3.4. A mathematical model for the growth of populations with impulses is presented in detail in Section 3.5.

3.2 Comparison Results for Impulsive Parabolic Systems (IPSs)

We shall introduce, in this section, systems of impulsive parabolic IBVPs, develop necessary basic notation, and prove comparison results in the present

framework. The contents of this section are therefore natural extensions
of the results in Section 2.6. In fact, we shall discuss much more, such as
flow invariance by employing the method of lower and upper solutions and
then using these results to obtain bounds to the solutions of impulsive parabolic IBVPs in terms of the solutions of ordinary differential equations with
impulse effects.

Let Ω be a smooth bounded domain in R^N, $Q_T = (0,T] \times \Omega$ for some
$T > 0$, $\Gamma_T = (0,T) \times \partial\Omega$ and γ be a unit outward normal vector field on
Γ_T. For a given partition $0 < \tau_1 < \tau_2 < \ldots < \tau_p < T$ of the interval $[0,T]$
we introduce the following notation:

$$M_i = \{(\tau_i, x) : \tau_i \in (0,T), \ x \in \Omega\} \text{ and } M = \bigcup_{i=1}^{p} M_i,$$

$$N_i = \{(\tau_i, x) : \tau_i \in (0,T), \ x \in \partial\Omega\} \text{ and } N = \bigcup_{i=1}^{p} N_i.$$

We denote by $\mathcal{C}$ the set of functions $u(t,x)$, $u : [0,T] \times \bar{\Omega} \to R^N$ with
the following properties:

(i) $u(t,x)$ is of class $C_{t,x}^{1,1}$ for $(t,x) \in \bar{Q}_T/(M \cup N)$;

(ii) there exists $\partial^2 u(t,x)/\partial x^2$ which is continuous for $(t,x) \in Q_T/M$;

(iii) there exist the following limits:

$$\lim_{t \to \tau_k^+} v(t,x) = v(\tau_k, x)$$

and

$$\lim_{t \to \tau_k^-} v(t,x) = v(\tau_k^-, x) < \infty \text{ for } x \in \bar{\Omega},$$

where $v(t,x)$ stands for $(u(t,x), u_t(t,x), u_x(t,x), u_{xx}(t,x))$, $u_t = \frac{\partial u}{\partial t}$,
$u_x(\frac{\partial u}{\partial x_1}, \ldots \frac{\partial u}{\partial x_n})$, $u_{xx} = (\frac{\partial^2 u}{\partial x_1}, \frac{\partial^2 u}{\partial x_1 \partial x_2}, \ldots, \frac{\partial^2 u}{\partial x_n^2})$.

A function $f \in C[\bar{Q}_T \times R^N \times R^n \times R^{n^2}, R^N]$ is said to be *quasimonotone
nondecreasing* in u if for some i such that $1 \le i \le n$, $u \le v$ and $u_i = v_i$,
the inequality $f_i(t,x,u,P,Q) \le f_i(t,x,v,P,Q)$ holds, where $(t,x) \in \bar{Q}_T$,
$P \in R^n$, $Q \in R^{n^2}$.

A function $f \in C[\bar{Q}_T \times R^N \times R^n \times R^{n^2}, R^N]$ is said to be *elliptic at the point* (t_1, x_1) if for any u, P, R_{ik}, S_{ik} $(i, k = 1, 2, \ldots, m)$ the quadratic form

$$\sum_{i,k=1}^{n} (R_{ik} - S_{ik}) \mu_i \mu_k \leq 0$$

for arbitrary vector $\mu \in R^n$ implies $f(t_1, x_1, u, P, R) \leq f(t_1, x_1, u, P, S)$. If this property holds for any $(t, x) \in Q_T$, then $f(t, x, u, P, R)$ is said to be *elliptic in* Q_T in which case the differential operator $T(u) = u_t - f(t, x, u, u_x, u_{xx})$ is called *parabolic*.

We are concerned with a system of initial boundary value problems (IBVPs) of parabolic type with impulse perturbations at fixed moments of time. Namely, we consider the system of equations

$$u_{kt} = f_k(t, x, u, (u_k)_x, (u_k)_{xx}) \text{ for } (t, x) \in Q_T/M \tag{3.2.1}$$

subject to the initial condition

$$u(0, x) = u_0(x) \text{ for } x \in \bar{\Omega}, \tag{3.2.2}$$

and boundary condition

$$Bu = \phi(t, x) \text{ for } (t, x) \in \Gamma_T/N, \tag{3.2.3}$$

and impulse perturbations at fixed moments of time τ_i, $i = 1, 2, \ldots, p$

$$\Delta u_k(\tau_i, x) \equiv u_k(t_i, x) - u_k(\tau_i^-, x) = g_{ki}(x, u_k(\tau_k^-, x)) \text{ for } x \in \bar{\Omega}. \tag{3.2.4}$$

In what follows, we assume that

(i) for each $k \in I = \{1, 2, \ldots, N\}$ the function $f_k(t, x, u, (u_k)_x, (u_k)_{xx}) \in C[\bar{Q}_T \times R^N \times R^n \times R^{n^2}, R^N]$ is elliptic in Q_T;

(ii) for each $k \in I$, $B_k : C \to C[\Gamma_T/N, R^n]$ is a boundary operator defined by

$$B_k u_k(t, x) = p_k(t, x) u_k + q_k(t, x) \frac{du_k}{d\gamma}$$

with $p_k(t, x)$, $q_k(t, x) \in C[\Gamma_T, R^n]$ such that $p_k(t, x) \geq 0$, $q_k(t, x) \geq 0$ and $p_k(t, x) + q_k(t, x) > 0$ on Γ_T;

(iii) for each $k \in I$ the function $g_{ki}(x, u) \in C[\Omega \times R^n, R^n]$;

(iv) the function $\phi(t, x) \in C[\Gamma_T, R^N]$.

The function $u(t, x) \in C$ satisfying (3.2.1)–(3.2.4) is called a *solution of IBVP* (3.2.1)–(3.2.4).

The function $u(t, x) \in C$ is called an *upper (lower) solution of IBVP* (3.2.1)–(3.2.4) if it satisfies the following inequalities

$$(u_k)_t \geq f_k(t, x, u, (u_k)_x, (u_k)_{xx}) \quad ((u_k)_t \leq f_k(t, x, u, (u_k)_x, (u_k)_{xx}))$$

for $(t, x) \in Q_T/M$,

$$u(0, x) \geq u_0(x) \quad (u(0, x) \leq u_0(x)) \text{ for } x \in \bar{\Omega},$$

$$Bu \geq \phi(t, x) \quad (Bu \leq \phi(t, x)) \text{ for } (t, x) \in \Gamma_T/N,$$

$$u_k(\tau_i, x) - u_k(\tau_i^-, x) \geq g_{ki}(x, u_k(\tau_i^-, x))$$

$$(u_k(\tau_i, x) - u_k(\tau_i^-, x) \leq g_{ki}(x, u_k(\tau_i^-, x))) \text{ for } x \in \bar{\Omega}.$$

We shall now give some basic theorems on upper and lower solutions of IBVP (3.2.1)–(3.2.4) along with the flow invariance results.

Theorem 3.2.1 *Assume that the following conditions hold:*

(i) $v_k, w_k \in C$, $f_k \in C[\bar{Q}_T \times R^N \times R^n \times R^{n2}, R^N]$ *is elliptic and*

$$(v_k)_t \leq f_k(t, x, v, (v_k)_x, (v_k)_{xx}) \text{ for } (t, x) \in Q_T/M,$$

$$v_k(\tau_i, x) - v_k(\tau_i^-, x) \leq g_{ki}(x, v_k(\tau_i^-, x)) \text{ for } x \in \bar{\Omega},$$

$$(w_k)_t \geq f_k(t, x, w, (w_k)_x, (w_k)_x x) \text{ for } (t, x) \in Q_T/M,$$

$$w_k(\tau_i, x) - w_k(\tau_i^-, x) \geq g_{ki}(x, w_k(\tau_i^-, x)) \text{ for } x \in \bar{\Omega}$$

for any $k \in I$, $i = 1, 2, \ldots, p$;

(ii)   (a) $v(0, x) < w(0, x)$ *for $x \in \bar{\Omega}$;*

(b) $Bv(t, x) < Bw(t, x)$ *for $(t, x) \in \Gamma_T/N$;*

(iii) $f(t, x, u, P, Q)$ *is quasimonotone nondecreasing in u for fixed (t, x, P, Q), where $(t, x) \in \bar{Q}_T$, $P \in R^n$, $Q \in R^{n2}$;*

(iv) $g_{ki}(x, u)$ *is quasimonotone nondecreasing in u for fixed $x \in \bar{\Omega}$.*

Then if one of the inequalities in (i) is strict, $v(t, x) < w(t, x)$ for $(t, x) \in Q_T$.

Proof   We assume that one of the inequalities in (i) is strict and define the function m by the formula $m(t, x) = v(t, x) - w(t, x)$. If the conclusion of the theorem is not true, there exist an index k and $t_1 \geq 0$, $x_1 \in \bar{\Omega}$ such that one of the following two cases holds:

Case 1 For $t_1 \neq \tau_i$, $i = 1, 2, \ldots, p$

$$m_j(t, x) < 0 \text{ for } (t, x) \in [0, t_1) \times \bar{\Omega}, \ j \neq k,$$

$$m_k(t, x) < 0 \text{ for } (t, x) \in [0, t_1) \times \bar{\Omega} \text{ and } m_k(t_1, x_1) = 0.$$

Case 2 For some $i, i = 1, 2, \ldots, p$, $\ t_1 = \tau_i$ and

$$m_j(t, x) < 0 \text{ for } (t, x) \in [0, t_1) \times \bar{\Omega}, \ j \neq k,$$

$$m_k(t, x) < 0 \text{ for } (t, x) \in [0, t_1) \times \bar{\Omega} \text{ and } m_k(t_1, x_1) \geq 0.$$

For the first case, the proof can be carried out as in Theorem 2.6.1. For the second case, we have

$$
\begin{aligned}
0 \ &\leq \ m_k(\tau_i, x_1) \\
&= \ v_k(\tau_i, x_1) - w_k(\tau_i, x_1) \\
&= \ v_k(\tau_i^-, x_1) - w_k(\tau_i^-, x_1) + g_{ki}(x_1, v_k(\tau_i^-, x_1)) - g_{ki}(x_1, w_k(\tau_i^-, x_1)) \\
&= \ m_k(\tau_i^-, x_1) + [g_{ki}(x_1, v_k(\tau_i^-, x_1)) - g_{ki}(x_1, w_k(\tau_i^-, x_1))].
\end{aligned}
$$

According to our assumptions $m_k(\tau_i^-, x_1) < 0$ and the summand in the brackets is nonpositive due to condition (iv) of the theorem. That leads to the contradiction $0 \leq m_k(\tau_i, x_1) < 0$ which completes the proof of the theorem.

It is possible to dispense with the strict inequality in (i) in Theorem 3.2.1 and the conclusion of the theorem remains valid provided the functions f and g_i satisfy the following conditions:

(A) there exists a function $z_k \in \mathcal{B}$ such that

$$z_k(t, x) > 0 \text{ on } Q_T/M, \quad \frac{\partial z_k(t, x)}{\partial \gamma} \geq \gamma_k > 0 \text{ on } \Gamma_T/N$$

and for sufficiently small $\epsilon > 0$ and for two arbitrary given functions v_k, $w_k \in \mathcal{B}$ either

(a) $\epsilon z_{kt} > f_k(t, x, w + \epsilon z, (w_k)_x + \epsilon (z_k)_x, (w_k)_{xx} + \epsilon (z_k)_{xx})$

$$- f_k(t, x, w, (w_k)_x, (w_k)_{xx})$$

on Q_T/M and

(b) $\epsilon[z_k(\tau_i, x) - z_k(\tau_i^-, x)]$

$$\geq g_{ki}(x, w_k(\tau_i^-, x) + \epsilon z_k(\tau_i^-, x)) - g_{ki}(x, w_k(\tau_i^-, x))$$

for $x \in \bar{\Omega}$, $k \in I$ or

(c) $\epsilon z_{kt} > f_k(t, x, v, v_{kx}, v_{kxx}) - f_k(t, x, v - \epsilon z, v_{kx} - \epsilon z_{kx}, v_{kxx} - \epsilon z_{kxx})$
on Q_T/M and

(d) $\epsilon[z_k(\tau_i, x) - z_k(\tau_i^-, x)]$

$$\geq g_{ki}(x, v_k(\tau_i^-, x)) - g_{ki}(x, v_k(\tau_i^-, x) - \epsilon z_k(\tau_i^-, x))$$

for $x \in \bar{\Omega}$, $k \in I$.

Theorem 3.2.2 *Let the assumptions* (i)–(iv) *of Theorem 3.2.1 hold and let the condition* (A) *be satisfied. Then the relations*

$$v(0, x) \leq w(0, x) \text{ for } x \in \bar{\Omega}, \ Bv(t, x) \leq Bw(t, x) \text{ on } \Gamma_T/N$$

imply $v(t, x) \leq w(t, x)$ *on* $\bar{Q}_T$.

Proof   Assume that the conditions (a) and (c) of (A) hold and let us define the function $\tilde{w}_k = w_k + \epsilon z_k$ for $k \in I$. Then one can show that the functions v and $\tilde{w}$ satisfy the assumptions of Theorem 3.2.1 and hence $v_k(t, x) < \tilde{w}_k(t, x) = w_k(t, x) + \epsilon z_k(t, x)$ on $\bar{Q}_T$ for $k \in I$.

Taking the limit as $\epsilon \to 0^+$ we get the desired inequality $v_k(t, x) \leq w_k(t, x)$ on $\bar{Q}_T$.

For the case when the conditions (b) and (d) of (A) hold, it is necessary to define the function $\tilde{v}_k = v_k - \epsilon z_k$ for $k \in I$ and to show that the functions $\tilde{v}$ and w satisfy the assumptions of Theorem 3.2.1 which completes the proof of the theorem.

Let us now assume that solutions of the IBVP (3.2.1)–(3.2.4) exist on $\bar{Q}_T$.

A closed set $F \subset R^N$ is said to be *flow invariant relative to IBVP* (3.2.1)–(3.2.4) if, for every solution $u(t, x)$ of (3.2.1)–(3.2.4), $u_0(x) \in F$ on $\bar{\Omega}$ implies $u(t, x) \in F$ on $\bar{Q}_T$.

The following results on flow invariance can be useful in obtaining bounds for the solutions of the IBVP (3.2.1)–(3.2.4).

Theorem 3.2.3 *Assume that the following conditions hold:*

(i)   (a) $f_k(t, x, 0, 0, 0) \geq 0$;

(b) $g_{ki}(x,0) \geq 0$ *and conditions (a) and (c) of (A) are satisfied with*
$v = u$, *where* $u = u(t,x)$ *is any solution of IBVP* (3.2.1)–(3.2.4)
for $\phi(t,x) \geq 0$;

(ii) *conditions (iii) and (iv) of Theorem 3.2.1 are satisfied.*

Then the closed set $\bar{U}$, *where* $U = \{u \in R^N, u > 0\}$, *is flow invariant relative to the system* (3.2.1)–(3.2.4).

Proof   We define the function $m(t,x) = u(t,x) + \epsilon z(t,x)$, where $u(t,x)$ is the arbitrary solution of the IBVP (3.2.1)–(3.2.4) such that $u_0(x) \in \bar{U}$ and $\epsilon > 0$ is sufficiently small. In view of conditions (a) and (c) of (A), we have $m(t,x) > 0$ on Q_T and $B_m(t,x) > 0$ on Γ_T/N.

Actually, if this conclusion is not true, there exist an index k and $t_1 \geq 0$, $x_1 \in \bar{\Omega}$ such that one of the following two cases holds:

Case 1 For $t_1 \neq \tau_i$, $i = 1, 2, \ldots, p$

$$m_j(t,x) > 0 \text{ for } (t,x) \in [0,t_1) \times \bar{\Omega}, \ j \neq k,$$

$$m_k(t,x) > 0 \text{ for } (t,x) \in [0,t_1) \times \bar{\Omega} \text{ and } m_k(t_1,x_1) = 0.$$

Case 2 For some i, $i = 1, 2, \ldots, p$, $t_1 = \tau_i$ and

$$m_j(t,x) > 0 \text{ for } (t,x) \in [0,t_1) \times \bar{\Omega}, \ j \neq k,$$

$$m_k(t,x) > 0 \text{ for } (t,x) \in [0,t_1) \times \bar{\Omega} \text{ and } m_k(t_1,x_1) \leq 0.$$

For the first case, one can show as in Theorem 2.6.1 that $(t_1,x_1) \in Q_T$ and $m(t,x) = u(t,x) + \epsilon z(t,x) > 0$ on Q_T/M.

For Case 2, using the condition (c) of (A) and the assumption (i) (b) of the theorem we get

$$
\begin{aligned}
0 \geq\ & m_k(\tau_i, x_1) \\
=\ & u_k(\tau_i, x_1) + \epsilon z_k(\tau_i, x_1) \\
=\ & u_k(\tau_i^-, x_1) + g_{ki}(x_1, u_k(\tau_i^-, x_1)) + \epsilon z_k(\tau_i, x_1) \\
\geq\ & u_k(\tau_i^-, x_1) + g_{ki}(x_1, u_k(\tau_i^-, x_1)) + \epsilon z_k(\tau_i^-, x_1) \\
& + g_{ki}(x_1, u_k(\tau_i^-, x_1)) + g_{ki}(x_1, u_k(\tau_i^-, x_1) + \epsilon z_k(\tau_i^-, x_1)) \\
=\ & m_k(\tau_i^-, x_1) + g_{ki}(x_1, m_k(\tau_i^-, x_1)).
\end{aligned}
$$

The assumption (ii) of the theorem along with $m_k(\tau_i^-, x_1) > 0$ implies $g_{ki}(x_1, m_k(\tau_i^-, x_1)) \geq g_{ki}(x_1, 0)$ and thus we arrive at the following contradiction:

$$0 \geq m_k(\tau_i, x_1) \geq m_k(\tau_i^-, x_1) + g_{ki}(x_1, 0) > 0.$$

Thus we can conclude that $m(t, x) = u(t, x) + \epsilon z(t, x) > 0$ on $\bar{Q}_T$ and hence taking the limit as $\epsilon \to 0^+$ we get the flow invariance of the set $\bar{U}$ relative to IBVP (3.2.1)–(3.2.4) which completes the proof of the theorem.

Corollary 3.2.1 *Suppose that the following conditions hold:*

(i) (a) $f_k(t, x, 0, 0, 0) \leq 0;$

   (b) $g_{ki}(x, 0) \leq 0$ *and conditions* (b) *and* (d) *of* (A) *are satisfied with* $w = u$, *where* $u = u(t, x)$ *is any solution of IBVP* (3.2.1)–(3.2.4);

(ii) *condition* (ii) *of Theorem 3.2.3 is satisfied.*

Then the closed set $\bar{U}$, where $U = \{u \in R^n, u < 0\}$, is flow invariant relative to the IBVP (3.2.1)–(3.2.4).

The proof of the corollary is similar to that of Theorem 3.2.3 with the evident choice $m(t, x) = u(t, x) - \epsilon z(t, x)$, and natural changes of the signs in the inequalities.

Combining Theorem 3.2.3 with Corollary 3.2.1, we get the following result.

Corollary 3.2.2 *Suppose that the condition* (A) *is satisfied with $v = w = u$, where $u = u(t, x)$ is any solution of IBVP* (3.2.1)–(3.2.4). *Suppose further that*

$$f_k(t, x, a, 0, 0) \leq 0, \quad g_{ki}(a, 0) \leq 0;$$
$$f_k(t, x, b, 0, 0) \geq 0, \quad g_{ki}(b, 0) \geq 0;$$

and the condition (ii) *of Theorem 3.2.3 holds.*

Then the closed set $\bar{W}$, where $W = \{u \in R^n, a < u < b\}$, is flow invariant relative to the IBVP (3.2.1)–(3.2.4).

Next we consider the result concerning upper and lower bounds for the solutions of IBVP (3.2.1)–(3.2.4) in terms of solutions of the corresponding impulsive ordinary differential equations.

Theorem 3.2.4 *Suppose that the following conditions are fulfilled:*

(i) *condition* (A) *holds with $v = w = u$, where $u = u(t, x)$ is an arbitrary solution of IBVP* (3.2.1)–(3.2.4);

(ii) *there are functions $f_1, f_2 \in C[[0, T] \times R^N, R^N]$ such that*

$$f_{2k}(t, u) \leq f_k(t, x, u, 0, 0) \leq f_{1k}(t, u) \text{ for } k \in I$$

and f_1, f_2 are quasimonotone nondecreasing in u for fixed $t \in [0, T];$

(iii) *there are functions $g_{1i}, g_{2i} \in C[R^N, R^N]$ such that*

$$g_{2ki}(u) \le g_{ki}(x, u) \le g_{1ki}(u) \ for \ k \in I$$

and g_{1i}, g_{2i} are quasimonotone nondecreasing in u;

(iv) *$r(t), \rho(t)$ are solutions of the following impulsive ordinary differential equations*

$$\frac{dr}{dt} = f_1(t, x), \ t \ne \tau_i, \ \Delta r(\tau_i) = g_{1i}(t), \ r(0) = r_0 \qquad (3.2.5)$$

and

$$\frac{d\rho}{dt} = f_2(t, \rho), \ t \ne \tau_i, \ \Delta \rho(\tau_i) = g_{2i}(\rho), \ \rho(0) = \rho_0 \qquad (3.2.6)$$

which exist on $[0, T]$ and such that $\rho_0 \le u_0 \le r_0$ on $\bar{\Omega}$ and $p(t, x)\rho(t)$ $\le Bu(t, x) \le p(t, x)r(t)$ on Γ_T/N.

Then $\rho(t) \le u(t, x) \le r(t)$ on $\bar{Q}_T$.

Proof Let us define the function $m_k(t, x) = u_k(t, x) - r_k(t)$. Then $m_k(t, x)$ satisfies the following system:

$$\left[\begin{array}{ll} (m_k)_t = F_k(t, x, m, (m_k)_x, (m_k)_{xx}) & \text{on } Q_T/M, \\ m(0, x) = u_0(x) - r(0) = u_0(x) - r_0 & \text{on } \bar{\Omega}, \\ B_k m_k(t, x) = B_k u_k(t, x) - p_k(t, x)r_k(t) \le 0 & \text{on } \Gamma_T/N, \\ \Delta m_k(\tau_i) = G_{ki}(x, m_k(\tau_i^-)), \end{array} \right. \qquad (3.2.7)$$

where

$$F_k(t, x, m, (m_k)_x, (m_k)_{xx}) = f_k(t, x, m + r, (u_k)_x, (u_k)_{xx}) - f_{1k}(t, r),$$

$$G_{ki}(x, m) = g_{ki}(x, m + r) - g_{1ki}(x, r).$$

One can show that (3.2.7) satisfies the assumptions of Theorem 3.2.3 and hence $u(t, x) \le r(t)$ on $\bar{Q}_T$.

Making use of Corollary 3.2.1 one can show that the inequality $\rho(t) \le u(t, x)$ holds on $\bar{Q}_T$. Now the proof of the theorem is complete.

If we know a priori that the solution u of the IBVP (3.2.1)–(3.2.4) is such that $a \le u \le b$, then the following result may be useful.

Corollary 3.2.3 *If the closed set $\bar{W}$, where $W = \{u \in R^n : a < u < b, a, b \in R^N\}$, is flow invariant relative to the IBVP (3.2.1)–(3.2.4), then there exist functions f_1, f_2, g_{1i}, g_{2i} satisfying the assumptions of Theorem 3.2.4 provided f_k is elliptic for each $k \in I$.*

Proof   The functions f_1, f_2, g_{1i}, g_{2i} are constructed in the following way:

$$
\begin{aligned}
f_{1k}(t,x) &= \sup\{f_k(t,x,v,0,0),\ x \in \bar{\Omega},\ a \le v \le u,\ v_i = u_i\},\\
f_{2k}(t,u) &= \inf\{f_k(t,x,v,0,0),\ x \in \bar{\Omega},\ a \le v \le u,\ v_i = u_i\},\\
g_{1ki}(u) &= \sup\{g_{ki}(x,v),\ x \in \bar{\Omega},\ a \le v \le u,\ v_i = u_i\},\\
g_{2ki}(u) &= \inf\{g_{ki}(x,v),\ x \in \bar{\Omega},\ a \le v \le u,\ v_i = u_i\}.
\end{aligned}
$$

Finally, we present a comparison result related to IBVP (3.2.1)–(3.2.4).

Theorem 3.2.5 *Suppose that the following conditions are satisfied:*

(i) *The functions $v_k, w_k \in C$ are such that*

$$v_{kt} \le f_k(t,x,\sigma,(v_k)_x,(v_k)_{xx}) \text{ and } v_k(\tau_i,x) - v_i(\tau_i^-,x) \le g_{ki}(x,\sigma)$$

for all σ such that $v(t,x) \le \sigma \le w(t,x)$ and $\sigma_i = v_i(t,x)$;

$$w_{kt} \ge f_k(t,x,\sigma,(w_k)_x,(w_k)_{xx}) \text{ and } w_k(\tau_i,x) - w_k(\tau_i^-,x) \ge g_{ki}(x,\sigma)$$

for all σ such that $v(t,x) \le \sigma \le w(t,x)$ and $\sigma_i = w_i(t,x)$;

(ii) (a) $f_k(t,x,\bar{\sigma},(\bar{\sigma}_k)_x,(\bar{\sigma}_k)_{xx}) - f_k(t,x,\sigma,(\sigma_k)_x,(\sigma_k)_{xx})$

$$\le \Phi_k(t,x,|\bar{\sigma}_1 - \sigma_1|,\ldots,\bar{\sigma}_k - \sigma_k,\ldots,|\bar{\sigma}_n - \sigma_n|,$$

$$(\bar{\sigma}_k)_x - (\sigma_k)_x,(\bar{\sigma}_k)_{xx} - (\sigma_k)_{xx});$$

   (b) $g_{ki}(x,\bar{\sigma}) - g_{ki}(x,\sigma)$

$$\le \Psi_{ki}(x,|\bar{\sigma}_1 - \sigma_1|,\ldots,\bar{\sigma}_k - \sigma_k,\ldots,|\bar{\sigma}_n - \sigma_n|),$$

whenever $\bar{\sigma}_k \ge \sigma_k$, where $\Phi_k \in C[\bar{Q}_T \times R^N \times R^n \times R^{n^2}, R^N]$ and $\Psi_{ki} \in C[\bar{\Omega} \times R^N, R^N]$ are quasimonotone nondecreasing in u and there exists a function $z_k \in B$, $z_k > 0$ on Q_T/M, $\frac{\partial z_k(t,x)}{\partial \gamma} \ge \delta_k > 0$ on Γ_T/N for $k \in I$ such that for sufficiently small $\epsilon > 0$,

   (c) $\epsilon z_{kt} > \Phi_k(t,x,\epsilon z,\epsilon(z_k)_x,\epsilon(z_k)_{xx})$ *on Q_T/M and*

   (d) $\epsilon[z_k(\tau_i,x) - z_k(\tau_i^-,x)] \ge \Psi_{ki}(x,\epsilon z)$ *for $x \in \bar{\Omega}$, $k \in I$;*

(iii) *$u(t,x)$ is any solution of the IBVP (3.2.1)–(3.2.4) such that $v(0,x) \le u_0 \le w(0,x)$ for $x \in \bar{\Omega}$ and $B_k v_k \le B_k u_k \le B_k w_k$ on Γ_T/N.*

Then $v(t,x) \le u(t,x) \le w(t,x)$ on $\bar{Q}_T$.

Proof To begin with, we prove the conclusion of the theorem under the assumption that v, w satisfy strict inequalities in (i), (iii). Let us define the functions m and n by $m = u - w$ and $n = u - v$ on $\bar{Q}_T$. If the conclusion is not true, there exist an index $k \in I$ and the point $(t_1, x_1) \in \dot{Q}_T$ such that one of the following two cases holds:

Case 1 For $t_1 \neq \tau_i$, $i = 1, 2, \ldots, p$

$$m_j(t, x) < 0 < n_j(t, x) \text{ for } (t, x) \in [0, t_1) \times \bar{\Omega}, j \neq k$$

and

$$m_k(t, x) < 0 < n_k(t, x) \text{ for } (t, x) \in [0, t_1) \times \bar{\Omega}$$

and either $m_k(t_1, x_1) = 0$ or $n_k(t_1, x_1) = 0$.

Case 2 For some i, $i = 1, 2, \ldots, p$, $t_1 = \tau_i$ and

$$m_j(t, x) < 0 < n_j(t, x) \text{ for } (t, x) \in [0, t_1) \times \bar{\Omega}, j \neq k$$

and

$$m_k(t, x) < 0 < n_k(t, x) \text{ for } (t, x) \in [0, t_1) \times \bar{\Omega}$$

and either $m_k(t_1, x_1) \geq 0$ or $n_k(t_1, x_1) \leq 0$. The first case is handled in the same way as in Theorem 2.6.1. For the second case, we have either $m_k(\tau_i, x_1) \geq 0$ or $n_k(\tau_i, x_1) \leq 0$. Let us suppose that the last inequality holds. Then we easily arrive at the following contradiction:

$$
\begin{aligned}
0 &\geq n_k(\tau_i, x_1) \\
&= u_k(\tau_i, x_1) - v_k(\tau_i, x_1) \\
&\geq u_k(\tau_i^-, x_1) \\
&= v_k(\tau_i^-, x_1) + g_{ki}(x_1, u_k(\tau_i^-, x_1)) - g_{ki}(x_1, \sigma) \\
&= n_k(\tau_i^-, x_1) + g_{ki}(x_1, u_k) - g_{ki}(x_1, \sigma) \\
&> 0 \quad .
\end{aligned}
$$

since $n_k(\tau_i^-, x_1) > 0$ and $v(t_1, x_1) \leq u(t_1, x_1) \leq w(t_1, x_1)$. Thus the conclusion of the theorem is true for the strict inequalities in (i) and (iii).

To finish the proof for the general case, let us define the functions $\tilde{w}_k$ and $\tilde{v}_k$ setting $\tilde{w}_k = w_k + \epsilon z_k$, $\tilde{v}_k = v_k - \epsilon z_k$ on Q_T/M.

Evidently we have $\tilde{v}(0, x) < u_0(x) < \tilde{w}(0, x)$ on $\bar{\Omega}$ and $B_k \tilde{v}_k < B_k u_k$ $< B_k \tilde{w}_k$ on Γ_T/N.

Define the function p_k by

$$p_k(t, x, \bar{\sigma}) = \max\{v_k(t, x), \min\{\bar{\sigma}_k, w_k(t, x)\}\}.$$

Then for $\bar{\sigma}$, $\tilde{v} \leq \bar{\sigma} \leq \tilde{w}$ and $\bar{\sigma}_k = \tilde{w}$, it follows that the function $\sigma = p(t, x, \bar{\sigma})$ satisfies the inequality $v \leq \sigma \leq w$ and $\sigma_k = w_k$. Making use of the assumptions (i), (ii) of the theorem we get by the quasimonotonicity of Φ_k

$$
\begin{aligned}
(\tilde{w}_k)_t &= (w_k)_t + \epsilon(z_k)_t \geq f_k(t, x, \sigma, (w_k)_x, (w_k)_{xx}) + \epsilon z_{kt} \\
&\geq f_k(t, x, \bar{\sigma}, (\tilde{w}_k)_x, (\tilde{w}_k)_{xx}) + \epsilon(z_k)_t \\
&\quad - \Phi_k(t, x, \mid \bar{\sigma}_1 - \sigma_1 \mid, \ldots, \epsilon z_k, \ldots, \mid \bar{\sigma}_n - \sigma_n \mid, \\
&\quad \epsilon(z_k)_x, \epsilon(z_k)_{xx}) \\
&> f_k(t, x, \bar{\sigma}, (\tilde{w}_k)_x, (\tilde{w}_k)_{xx})
\end{aligned}
$$

since $\mid \bar{\sigma}_k - \sigma_k \mid \leq \epsilon z_k$ for all $k \in I$.

Moreover,

$$
\begin{aligned}
\tilde{w}_k(\tau_i, x) &= w_k(\tau_i, x) + \epsilon z_k(\tau_i, x) \\
&\geq w_k(\tau_i^-, x) + g_{ki}(x, \sigma) + \epsilon z_k(\tau_i, x) \\
&\geq w_k(\tau_i^-, x) + g_{ki}(x, \bar{\sigma}) \\
&\quad - \Psi_{ki}(x, \mid \bar{\sigma}_1 + \sigma_1 \mid, \ldots, \epsilon z_k, \ldots \mid \bar{\sigma}_n - \sigma_n \mid) + \epsilon z_k(\tau_i, x) \\
&\geq w_k(\tau_i - 0, x) + \epsilon z_k(\tau_i^-, x) + g_{ki}(x, \bar{\sigma}) \\
&= \tilde{w}_k(\tau_0^-, x) + g_{ki}(x, \bar{\sigma}).
\end{aligned}
$$

Thus the functions $\tilde{w}_k$ and $\tilde{v}_k$ satisfy all the conditions required for the strict inequality result proved above and we get

$$v_k(t, x) - \epsilon z_k(t, x) < u_k(t, x) < w_k(t, x) + \epsilon z_k(t, x)$$

on $\bar{Q}_T$ for arbitrary sufficiently small $\epsilon > 0$ and $k \in I$. Taking the limit as $\epsilon \to 0^+$ we get $v(t, x) \leq u(t, x) \leq w(t, x)$ on $\bar{Q}_T$. Now the proof of the theorem is complete.

3.3 Coupled Lower and Upper Solutions

In order to extend the method of generalized quasilinearization to the impulsive parabolic IBVP, we require various comparison results for different forms of coupled lower and upper solutions. Since we would like to provide, at least once, the various possibilities in the method of generalized quasilinearization, we have chosen to develop these results relative to the impulsive parabolic IBVP, which we shall describe in the next section. To prove the quadratic convergence of the obtained monotone sequences to the unique solution of the IBVP in question, we need to use a comparison result which enables us to avoid the computational complexity.

Let us consider the following IBVP of parabolic type with impulsive perturbations at fixed moments of time:

$$\left[\begin{array}{rcll} \mathcal{L}u &=& f(t,x,u) + g(t,x,u) & \text{in } Q_T/M, \\ Bu &=& \phi & \text{on } \Gamma_T/N, \\ u(0,x) &=& u_0 & \text{on } \bar{\Omega} \\ u(t_k^+, x) &=& I_k(t_k, x) & \text{in } \bar{\Omega} \end{array} \right. \tag{3.3.1}$$

for each $k = 1, 2, \ldots, p$ and $0 < t_1 < t_2 < \ldots < t_p < T$. Here $\mathcal{L}$ is the second-order differential operator defined by

$$\mathcal{L} = \frac{\partial}{\partial t} - L, \quad L = \sum_{i,j=1}^{n} a_{ij}(t,x) \frac{\partial^2}{\partial x_i \partial x_j} + \sum_{i=1}^{n} b_i(t,x) \frac{\partial}{\partial x_i},$$

and B is the boundary operator given by

$$Bu = p(t,x)u + q(t,x) \frac{du}{d\gamma}$$

where $\frac{du}{d\gamma}$ denotes the normal derivative of u and $v(t,x)$ is the unit outward normal vector field on $\partial\Omega$ for $(t,x) \in \Gamma_T$. We list the following assumptions for convenience.

(A_0) (i) For each $i, j \in I = \{1, 2, \ldots, n\}$, $a_{ij}, b_j \in C^{\alpha/2, \alpha}[\bar{Q}_T, R]$.

  (ii) The operator $T(u) = u_t - Lu - [f(t,x,u) + g(t,x,u)]$ is strictly uniformly parabolic for $(t,x) \in \bar{Q}_T$.

  (iii) $p, q \in C^{1+\frac{\alpha}{2}, 1+\alpha}[\bar{\Gamma}_T, R]$, $p(t,x) \geq 0$, $q(t,x) \geq 0$, and $p(t,x) + q(t,x) > 0$ on Γ_T.

  (iv) $I_k \in C^{2+\alpha}[R, R]$.

  (v) $\phi \in C^{1+\frac{\alpha}{2}, 1+\alpha}[\bar{\Gamma}_T, R]$, $u_0(x) \in C^{2+\alpha}[\bar{\Omega}, R]$.

  (vi) $f, g \in C^{\frac{\alpha}{2}, \alpha}[\bar{Q}_T \times R, R]$ that is $f(t,x,u)$ and $g(t,x,u)$ are Hölder continuous in t and (x,u) with exponent $\frac{\alpha}{2}$ and α, respectively.

  (vii) $\partial\Omega$ belongs to $C^{2+\alpha}$.

  (viii) The IBVP (3.3.1) satisfies the compatibility condition of order $\frac{\alpha+1}{2}$.

Definition 3.3.1 A function $v(t, x) \in C$ is called a *lower solution* of (3.2.1) if

$$\mathcal{L}v \leq f(t, x, v) + g(t, x, v) \text{ in } Q_T/M,$$

$$Bv \leq \phi \text{ on } \Gamma_T/N,$$

$$v(0, x) \leq u_0(x) \text{ in } \bar{\Omega},$$

$$v(t_k^+, x) \leq I_k(v(t_k, x)) \text{ in } \bar{\Omega}.$$

If the inequalities are reversed then $v(t, x)$ is called an *upper solution*.

This is the natural definition of upper and lower solutions. However, in this situation, it is possible to define coupled upper and lower solutions in several ways which we shall see in the comparison theorem proved below. For this purpose, it is convenient to set

$$F(t, x, v, w) = f(t, x, v) + g(t, x, w).$$

Theorem 3.3.1 *Let $v, w \in C$, $F \in C[\bar{Q}_T \times R^2, R]$. Further, assume that any one of the following conditions hold:*

(H_1) (a) $\mathcal{L}v \leq F(t, x, v, v), \quad \mathcal{L}w \geq F(t, x, w, w);$

   (b) $F(t, x, y_1, y_2) - F(t, x, \tilde{y}_1, \tilde{y}_2) \leq L[(y_1 - \tilde{y}_1) + (y_2 - \tilde{y}_2)]$ *whenever* $y_1 \geq \tilde{y}_1$ *and* $y_2 \geq \tilde{y}_2;$

   (c) $v(0, x) \leq w(0, x)$, $Bv(t, x) \leq Bw(t, x)$,$v(t_k^+, x) \leq I_k(v(t_k, x))$ $w(t_k^+, x) \geq I_k(w(t_k, x))$ *and* $I_k(u_1) - I_k(u_2) \leq M(u_1 - u_2)$ *for* $M > 0$ $(M < 1)$ *and* $u_1 \geq u_2;$

   Also $I_k(u)$ is nondecreasing in u;

(H_2) (a) $\mathcal{L}v \leq F(t, x, v, w), \mathcal{L}w \geq F(t, x, w, v);$

   (b) $F(t, x, y_1, y_2) - f(t, x, \tilde{y}_1, \tilde{y}_2) \leq L(y_1 - \tilde{y}_1), \ L > 0, \ y_1 \geq \tilde{y}_1;$ $F(t, x, y_1, y_2) - F(t, x, y_1, \tilde{y}_2) \geq -L(y_2 - \tilde{y}_2),$ *whenever* $y_2 \geq \tilde{y}_2;$ *and (c) of (H_1) hold;*

(H_3) (a) $\mathcal{L}v \leq F(t, x, w, v), \mathcal{L}w \geq F(t, x, v, w);$

   (b) $F(t, x, y_1, y_2) - F(t, x, y_1, \tilde{y}_2) \leq L(y_2 - \tilde{y}_2), \ L > 0, \ y_2 \geq \tilde{y}_2;$ $F(t, x, y_1, y_2) - F(t, x, \tilde{y}_1, y_2) \geq -L(y_1 - \tilde{y}_1), \ L > 0, \ y_1 \geq \tilde{y}_1;$ *and (c) of (H_1) holds;*

(H_4) (a) $\mathcal{L}v \leq F(t, x, w, w), \mathcal{L}w \geq F(t, x, v, v);$

   (b) $F(t, x, y_1, y_2) - f(t, x, \tilde{y}_1, \tilde{y}_2) \leq L[(y_1 - \tilde{y}_1) + (y_2 - \tilde{y}_2)], \ L > 0,$ *and (c) of (H_1) holds.*

Then $v(t,x) \leq w(t,x)$ on $\bar{Q}_T$.

Proof The conclusion relative to (H_1) is a scalar version of Theorem 3.3.2. Here we provide the proof of (H_1) for completeness. We also provide the proof for (H_2). The conclusions relative to (H_3) and (H_4) can be proven along the same lines. Consider (H_1) (a)–(c), when all the inequalities are strict. Define the function $m(t,x) = v(t,x) = w(t,x)$. If the conclusion of the theorem is not true, there exists an index k and a $t_1^* > 0$, $x_1 \in \bar{\Omega}$ such that one of the following two cases hold:

Case 1 For $t_1^* \neq t_i$, $i = 1, \ldots, p$, $m(t,x) < 0$ for $(t,x) \in [0, t_1^*) \times \bar{\Omega}$, and $m(t_1^*, x) = 0$.

Case 2 For some i, $i = 1, \ldots, p$, $t_1^* = t_k$ and $m(t,x) < 0$ for $(t,x) \in [0, t_k) \times \bar{\Omega}$ and $m(t_k^+, x) \geq 0$.

In the first case, the proof can be carried out in analogy with Theorem 2.6.1 and by induction. For the second case, we have

$$0 \leq m(t_k^+, x) = v(t_k^+, x) - w(t_k^+, x) < I_k(v(t_k, x)) - I_k(w(t_k, x)) \leq 0,$$

which is a contradiction using the nondecreasing nature of I_k and the continuity of $m(t,x)$ to the left of t_k. This completes the situation when the inequalities in (H_1) are all strict. In order to handle the nonstrict inequality, let $\tilde{v} = v - \epsilon e^{3Lt}$, $\tilde{w} = w + \epsilon e^{3Lt}$, where $\epsilon > 0$ is arbitrarily small and L is the Lipschitz constant of (H_1) (b). Certainly, $\tilde{v} < v$ and $\tilde{w} > w$. Then

$$
\begin{aligned}
\mathcal{L}\tilde{v} &= \mathcal{L}v - 3L\epsilon e^{3Lt} \\
&\leq F(t,x,v,v) - 3L\epsilon e^{3Lt} \\
&\leq F(t,x,\tilde{v},\tilde{v}) + 2L\epsilon e^{3Lt} - 3L\epsilon e^{3Lt} \\
&< F(t,x,\tilde{v},\tilde{v}),
\end{aligned}
$$

using (H_1) (b). Similarly, one can prove that

$$\mathcal{L}\tilde{w} > F(t,x,\tilde{w},\tilde{w}).$$

Also

$$\tilde{v}(0,x) = v(0,x) - \epsilon < v(0,x) \leq w(0,x) < w(0,x) + \epsilon = \tilde{w}(0,x),$$

that is, $\tilde{v}(0,x) < \tilde{w}(0,x)$.

Further,

$$
\begin{aligned}
B\tilde{v}(t,x) &= Bv(t,x) \\
&= p(t,x)\epsilon e^{3Lt} \\
&< Bv(t,x) \\
&\leq Bw(t,x) \\
&< Bw(t,x) + p(t,x)\epsilon e^{3Lt} \\
&= B\tilde{w}(t,x),
\end{aligned}
$$

since $p(t,x) > 0$ on Q_T. Also

$$
\begin{aligned}
\tilde{v}(t_k^+,x) &= v(t_k^+,x) - \epsilon e^{3Lt} \\
&\leq I_k(v(t_k,x)) - \epsilon e^{3Lt_k} \\
&\leq I_k(\tilde{v}(t_k,x)) + (M-1)\epsilon e^{3Lt_k} \\
&< I_k(\tilde{v}(t_k,x)),
\end{aligned}
$$

since $M < 1$. Similarly, it follows that

$$
\begin{aligned}
\tilde{w}(t_k^+,x) &= w(t_k^+,x) + \epsilon e^{3Lt_k} \\
&\geq I_k(w(t_k,x)) + \epsilon w^{3Lt_k} \\
&\geq I_k(\tilde{w}(t_k,x)) + (1-M)\epsilon e^{3Lt_k} \\
&> I_k(\tilde{w}(t_k,x)).
\end{aligned}
$$

This proves $\tilde{v}$, $\tilde{w}$ satisfies (H_1) (a), (c) with strict inequality. Now using the strict inequality result, we have $\tilde{v}(t,x) < \tilde{w}(t,x)$ on $\bar{Q}_T$. Taking the limit as $\epsilon \to 0$, we have $v(t,x) \leq w(t,x)$.

Next, we consider the situation when (H_2) holds. The proof for strict inequality follows on the same lines as in (H_1) for strict inequalities. In order to prove the nonstrict inequality case, let $\tilde{v}(t,x)$, $\tilde{w}(t,x)$ be as before except that L, the Lipschitz constant, is the same as in (H_2) (b). Obviously (H_2) (c) is satisfied by $\tilde{v}$, and $\tilde{w}$ with strict inequality. It is enough to show that $\tilde{v}$, $\tilde{w}$ satisfy (H_2) (a) with strict inequality. In order to prove that, consider

$$
\begin{aligned}
\mathcal{L}\tilde{v} &= \mathcal{L}v - 3L\epsilon e^{3Lt} \\
&\leq F(t,x,v,w) - 3L\epsilon e^{3Lt} \\
&\leq F(t,x,\tilde{v},w) + L\epsilon e^{3Lt} - 3L\epsilon e^{3Lt} \\
&\leq F(t,x,\tilde{v},\tilde{w}) + 2L\epsilon e^{3Lt} - 3L\epsilon e^{3Lt} \\
&< f(t,x,\tilde{v},\tilde{w});
\end{aligned}
$$

using (H_2) (b), since $\tilde{v} < v$ and $\tilde{w} > w$.

Similarly one can prove that $\mathcal{L}\tilde{w} > F(t, x, \tilde{w}, \tilde{v})$ using (H_2) (b). Now, using the strict inequality result, it follows that $\tilde{v}(t, x) < \tilde{w}(t, x)$. Making $\epsilon \to 0$, we get $v(t, x) \le w(t, x)$. The conclusion for (H_3) and (H_4) follows on the same lines.

In view of the definition of $F(t, x, v, w)$ in terms of $f(t, x, v)$, $g(t, x, w)$, we also note that (H_2), (H_3) and (H_4) gives the definition of all possible coupled upper and lower solutions of (3.3.1).

The next two results are the existence of the solution for IBVP (3.3.1) and a comparison theorem. The impulsive case can be proved by induction. Also the comparison theorem we state below is a special case of Theorem 3.2.4. Hence we merely state the theorems here without proof. Both of the theorems are needed to prove our main results.

For that purpose we consider the weakly coupled impulsive parabolic IBVP

$$
\left[
\begin{array}{ll}
\mathcal{L}u_i = H_i(t, x, u_1, \ldots, u_n) & \text{in } Q_T/M, \\
Bu = \phi & \text{on } \Gamma_T/N, \\
u_i(0, x) = u_{0i}(x) & \text{on } \bar{\Omega}, \\
u_i(t_k^+, x) = I_k(u_i(t_k, x)) & \text{in } \bar{\Omega},
\end{array}
\right.
\tag{3.3.2}
$$

for each $i = 1, \ldots, N$ and for each $k = 1, \ldots, p$ where $\mathcal{L}$ and B are as before.

We first state the existence theorem. For that purpose, we define Müller-type upper and lower solutions for (3.3.2).

(A_1) Let $v = (v_1, v_2, \ldots, v_n)$, $w = (w_1, w_2, \ldots, w_n)$ be such that $v_i, w_i \in C$ for each $i = 1, \ldots, n$ and $v(t, x) \le w(t, x)$ on $\bar{Q}_T$ and

$$
\left[
\begin{array}{ll}
\mathcal{L}v_i \le H_I(t, x, \sigma) & \text{in } Q_T/M, \\
Bv \le \phi & \text{on } \Gamma_T/N, \\
v(0, x) \le u_0(x) & \text{in } \bar{\Omega}, \\
v(t_k^+, x) \le I_k(v(t_k, x)) & \text{in } \bar{\Omega},
\end{array}
\right.
\tag{3.3.3}
$$

for all σ such that $\sigma_i = v_i(t, x)$ and $v(t, x) \le \sigma \le w(t, x)$ in Q_T/M;

$$
\left[
\begin{array}{ll}
\mathcal{L}w_i \ge H_i(t, x, \sigma) & \text{in } Q_T/M, \\
Bw \ge \phi & \text{on } \Gamma_T/N, \\
w(0, x) \ge u_0(x) & \text{in } \bar{\Omega}, \\
w(t_k^+, x) \ge I_k(w(t_k, x)) & \text{in } \bar{\Omega},
\end{array}
\right.
\tag{3.3.4}
$$

for all σ such that $\sigma = w(t, x)$ and $v(t, x) \le \sigma \le w(t, x)$ in $\bar{Q}_T/M$. Then v, w are called Müller-type lower and upper solutions.

Note that if H_i is mixed quasimonotone, then one can define coupled upper and lower solutions of (3.3.2). Further note that if H_i is mixed quasimonotone and v, w are coupled upper and lower solutions, then v, w are Müller-type lower and upper solutions also. That is the reason we state the existence theorem in terms of Müller-type upper and lower solutions. The next result is precisely this.

Theorem 3.3.2 *Suppose that the assumptions of (A_0) and (A_1) hold except (A_0) (ii). Instead of $(A_0)(ii)$, assume the operator $Lu^i - H(t, x, u_1, \ldots, u_n)$ be strictly uniformly parabolic in $\bar{Q}_T$. Further, let $I_k(u_i) \in C^{2+\alpha}$ be such that $I_k(u_i)$ is nondecreasing in u_i for each $i = 1, \ldots, n$, $k = 1, \ldots, p$. Then the weakly coupled reaction diffusion system (3.3.2) possesses a solution $u \in C^{1+\alpha/2, 2+\alpha}$ on $\bar{Q}_T / M$. Further the solution is unique if $H_i(t, x, u)$ is Lipschitzian in u.*

Proof   The existence on the interval $[0, t_1]$ follows from Theorem A.4.1. Now using (A_0) (iv), and the nature of $I_k(u_i)$, it follows that $u(t_1^+) \in C^{1+\alpha/2, 2+\alpha}$ on $\bar{\Omega}$. Again, the existence on $[t_1, t_2]$ can be proven using Theorem A.4.1. Using the method of mathematical induction, the existence of the solution of (3.3.1) can be proven on $[0, T]$.

Next we state a comparison theorem which is useful in the quadratic convergence of a sequence of solutions which converges to the solution of (3.3.1).

Theorem 3.3.3 *Suppose that*

(i) $m \in C^{1,2}[\bar{Q}_T, R]$ be such that $\mathcal{L}m \leq G(t, x, m)$, $t \neq t_k$ where $G(t, x, u) \in C[Q_T \times R, R]$ where the operator $\mathcal{L}$ is parabolic, $I_k(u)$ is nondecreasing in u, for $k = 1, \ldots, p$.

(ii) $h \in C[[0, T] \times R_+, R_+]$ and let $r(t, 0, y_0) \geq 0$ be the maximal solution of the system of ordinary impulsive differential equation,

$$y' = h(t, y), \ t \neq t_k$$

$$y(0) = y_0 \geq 0$$

$$y(t_k^+) = \bar{I}_k(y(t_k))$$

existing for $t \geq 0$ and $\bar{I}_k(y(t_k))$ is nondecreasing in y. Further

$$H(t, x, z) \leq h(t, z), \ for \ z \geq 0$$

$$I_k(z(t_k, x)) \leq \bar{I}_k(z(t_k)).$$

(iii) $m(0,x) \leq r(0,0,y_0)$ for $x \in \bar{\Omega}$, $Bm \leq p(t,x)r(t)$ on Γ_T/N.

Then $m(t,x) \leq r(t,0,y_0)$ on $\bar{Q}_T$.

As mentioned earlier, this theorem is a special case of Theorem 3.2.4.

3.4   Generalized Quasilinearization

We shall now extend the method of generalized quasilinearization to the impulsive parabolic IBVP (3.3.1) when $f(t,x,u)$ is convex in u and $g(t,x,u)$ is concave in u. So far we have investigated this technique only under the assumption of natural lower and upper solutions. We shall discuss, in this section, other possible situations when the lower and upper solutions are of coupled type. Of course, when the impulses are absent, these results reduce to those of parabolic IBVPs and therefore we get new results even in this special case. Let us start with the case of natural lower and upper solutions.

Theorem 3.4.1 *Assume that all assumptions of* (A_0) *of Section 3.3 hold except* (vi). *Suppose further that*

(A_1) $v_0, w_0 \in C$ *be natural lower and upper solutions of (3.3.1) such that* $v_0(t,x) \leq w_0(t,x)$ *on* $\bar{Q}_T$;

(A_2) $f, f_u, g, g_u \in C^{\frac{\alpha}{2},\alpha}[\bar{Q}_T \times R, R]$ f_{uu}, g_{uu} *exist and are continuous, and* $f_{uu} \geq 0$, $g_{uu} \leq 0$ *for* $v_0(t,x) \leq u \leq w_0(t,x)$ *and* $(t,x) \in Q_T$;

(A_3) $I_{ku}(u) \in C^{2+\alpha}[R,R]$ *for* $k = 1,\ldots,p$ *be such that* $I_k(u)$ *is nondecreasing in* u, *and* $I_{ku} \leq M(< 1)$ *and* $I_{kuu} \geq 0$ *on* $[v_0,w_0]$.

Then there exist monotone sequences $\{v_n(t,x)\}$, $\{w_n(t,x)\}$ *which converge uniformly to the unique solution* $u(t,x)$ *of (3.3.1) on* $\bar{Q}_T/M$, *and the convergence is quadratic.*

Proof   Since $f_{uu}(t,x) \geq 0$ and $g_{uu}(t,x) \leq 0$, we get the inequalities

$$f(t,x,u) \geq f(t,x,v) + f_u(t,x,v)(u-v) \qquad (3.4.1)$$

and

$$g(t,x,u) \geq g(t,x,v) + g_u(t,x,u)(u-v) \qquad (3.4.2)$$

for $u \geq v$. Similarly from (A_3), we get

$$I_k(u) \geq I_k(v) + I_{ku}(v)(u-v) \qquad (3.4.3)$$

whenever $u \geq v$, for $k = 1, \ldots, p$. Now consider the following set of two linear impulsive reaction diffusion equations,

$$
\left[
\begin{aligned}
\mathcal{L}u &= F(t, x, v_0, w_0; u) \\
&= f(t, x, v_0) + f_u(t, x, v_0)(u - v_0) \\
&\quad + g(t, x, v_0) + g_u(t, x, w_0)(u - v_0) \text{ on } Q_T/M, \\
Bu(t, x) &= \Phi(t, x) \text{ on } \Gamma_T/N, \\
u(0, x) &= u_0(x) \text{ in } \bar{\Omega}, \\
u(t_k^+, x) &= I_k(v_0(t_k, x)) + I_{ku}(v_0(t_k, x))(u(t_k, x) - v_0(t, x)) \\
&= \tilde{I}_{k1}(v_0(t_k, x); u(t_k, x)), \text{ on } \bar{\Omega}
\end{aligned}
\right.
\tag{3.4.4}
$$

for each $k = 1, 2, \ldots, p$ and

$$
\left[
\begin{aligned}
\mathcal{L}v &= G(t, x, v_0, w_0; v) \\
&= f(t, x, w_0) + f_u(t, x, w_0)(v - w_0) \\
&\quad + g(t, x, w_0) + g_u(t, x, w_0)(v - w_0) \text{ on } Q_T/M; \\
Bv(t, x) &= \Phi(t, x) \text{ on } \Gamma_T/N, \\
v(0, x) &= u_0(x) \text{ on } \bar{\Omega}, \\
v(t_k^+, x) &= I_k(w_0(t_k, x)) + I_{ku}(v_0(t_k, x))(v(t_k, x) - w_0(t_k, x)) \\
&= \tilde{I}_{k2}(v_0(t_k, x), w_0(t_k, x); v(t_k, x)) \text{ on } \bar{\Omega}.
\end{aligned}
\right.
\tag{3.4.5}
$$

It is easy to see that the solutions $u(t, x)$, $v(t, x)$ of (3.4.5) exists on $[0, t_1)$. Using the continuity of these solutions to the left of t_1 and the fact that $I_k(u) \in C^{2+\alpha}$ it is easy to show that $u(t_1^+, x)\ v(t_1^+, x) \in C^{2+\alpha}$. From this, it follows that the solutions of (3.4.4) and (3.4.5) exist and are unique on $[t_1, t_2)$. Using mathematical induction, one can complete the solutions of (3.4.4) and (3.4.5) respectively on $\bar{Q}_T$. Further the solution is unique, since they are solutions of linear equations.

Since v_0, w_0 are natural lower and upper solutions of (3.3.1), we have

$$\mathcal{L}v_0 \leq f(t, x, v_0) + g(t, x, v_0) = F(t, x, v_0, w_0; v_0) \text{ on } Q_T/M,$$

$$Bv_0 \leq \Phi(t, x) \text{ on } \Gamma_T/N,$$

$$v(0, x) \leq u_0(x) \text{ on } \bar{\Omega},$$

$$v_0(t_k^+, x) \leq I_k(v_0(t_k, x)) \equiv \tilde{I}_{k1}(v_0(t_k, x); v_0(t_k, x)) \text{ on } \bar{\Omega},$$

for each $k = 1, \ldots, p$. This proves that v_0 is a lower solution of (3.4.4). Similarly,

$$
\left[
\begin{aligned}
\mathcal{L}w_0 &\geq f(t, x, w_0) + g(t, x, w_0) = G(t, x, v_0, w_0; w_0) \text{ on } Q_T/M, \\
Bw_0 &\geq \Phi(t, x) \text{ on } \Gamma_T/N, \\
w(0, x) &\geq u_0(x) \text{ on } \bar{\Omega}, \\
w_0(t_k^+, x) &\geq I_k(w_0(t_k, x)) \geq \tilde{I}_{k1}(v_0(t_k, x); w_0(t_k, x)) \text{ on } \bar{\Omega},
\end{aligned}
\right.
$$

$$\tag{3.4.6}$$

using (3.4.3), since $v_0(t,x) \leq w_0(t,x)$. This proves that v_0, w_0 are upper and lower solutions of (3.4.4). Using (H_1) of Theorem 3.3.1, it is easy to conclude $v_0(t,x) \leq v_1(t,x) \leq w_0(t,x)$, where $v_1(t,x)$ is the unique solution of the linear impulsive initial boundary value problem (3.4.4). Similarly, one can prove that $v_0(t,x)$, $w_0(t,x)$ are lower and upper solutions of (3.4.5). As before, using (H_1) of Theorem 3.3.1, we get $v_0(t,x) \leq w_1(t,x) \leq w_0(t,x)$ on $\bar{Q}_T$. Now using (3.4.1), (3.4.2) and (3.4.3), we get

$$\left[\begin{array}{ll} \mathcal{L}v_1 & \leq f(t,x,v_1) + g(t,x,v_1) \text{ on } Q_T/M, \\ Bv_1(t,x) & \leq \Phi(t,x) \text{ on } \Gamma_T/N, \\ v_1(0,x) & = u_0(x) \text{ on } \bar{\Omega}, \\ v_1(t_k^+, x) & \geq I_k(v_1(t_k,x)) \text{ on } \bar{\Omega}, \end{array} \right. \tag{3.4.7}$$

since $v_0(t,x) \leq v_1(t,x) \leq w_0(t,x)$. Similarly, using $v_0(t,x) \leq w_1(t,x) \leq w_0(t,x)$ and (3.4.1)–(3.4.3), we obtain

$$\mathcal{L}w_1 \geq f(t,x,w_1) + g(t,x,w_1) \text{ on } Q_T/M,$$

$$Bw_1(t,x) = \Phi(t,x) \text{ on } \Gamma_T/N,$$

$$w_1(0,x) = u_0(x) \text{ on } \bar{\Omega},$$

$$w_1(t_k^+, x) \geq I_k(w_1(t_k,x)) \text{ on } \bar{\Omega}.$$

Using (H_1) of Theorem 3.3.1, (3.4.5) and (3.4.6), we can conclude that $v_1(t,x) \leq w_1(t,x)$ on $\bar{Q}_T$, since f and g satisfy the Lipschitz condition. This proves

$$v_0(t,x) \leq v_1(t,x) \leq w_1(t,x) \leq w_0(t,x) \text{ on } \bar{Q}_T. \tag{3.4.8}$$

Now assume $v_l(t,x)$ and $w_l(t,x)$ as solutions of the linear impulsive initial boundary value problems (3.4.4) and (3.4.5), respectively where v_0, w_0 are replaced by v_{l-1}, w_{l-1}, respectively. It is clear that v_l and w_l exist and are unique on $\bar{Q}_T$. Now assume for some $l > 1$,

$$v_0(t,x) \leq v_1(t,x) \leq \ldots \leq v_l(t,x) \leq w_l(t,x) \leq \ldots \\ \leq w_1(t,x) \leq w_0(t,x) \text{ on } \bar{Q}_T. \tag{3.4.9}$$

From (3.4.8), it follows that (3.4.9) is certainly true for $l = 1$. Using the method of mathematical induction and (3.4.1)–(3.4.3), we can prove that (3.4.9) is true for all l. Employing standard arguments as in the proof of Theorems 2.3.1 and 2.4.1 to each subinterval $(t_k, t_{k+1}) \subseteq [0,T]$ we can conclude that $\{v_n(t,x)\}$, $\{w_n(t,x)\}$ converge uniformly and monotonically to the unique solution of (3.3.1) on $\bar{Q}_T$.

We now claim that the convergence of the sequences $\{v_n(t,x)\}$ and $\{w_n(t,x)\}$ to the unique solution of (3.3.1) is quadratic. In order to prove quadratic convergence set $p_{n+1}(t,x) = u(t,x) - v_{n+1}(t,x)$ and $q_n(t,x) = w_{n+1}(t,x) - u(t,x)$, where $u(t,x)$ is the unique solution of (3.3.1). Here we prove the quadratic convergence of $p_n(t,x)$. On the same lines, one can prove the quadratic convergence of $q_n(t,x)$. Since $f_{uu} \geq 0$ and $g_{uu} \leq 0$, it follows that

$$
\begin{aligned}
\mathcal{L}p_{n+1} &= \mathcal{L}u - \mathcal{L}v_{n+1} \\
&= f(t,x,u) + g(t,x,u) \\
&\quad -[f(t,x,v_n) + f_u(t,x,v_n)(v_{n+1} - v_n) \\
&\quad + g(t,x,v_n) + g_u(t,x,w_n)(v_{n+1} - v_n)] \\
&\leq f(t,x,v_n) + f_u(t,x,v_n)(u - v_n) + f_{uu}(t,x,\xi)\frac{(u - v_n)^2}{2!} \\
&\quad - f(t,x,v_n) - f_u(t,x,v_n)(v_{n+1} - v_n) \\
&\quad + g_u(t,x,\eta)(u - v_n) - g_u(t,x,w_n)(u - v_n) + g_u(t,x,w_n)p_{n+1} \\
&\leq f_u(t,x,v_n)p_{n+1} + g_u(t,x,w_n)p_{n+1} + f_{uu}(t,x,\xi)\frac{p_n^2}{2!} \\
&\quad - g_{uu}(t,x,\delta)(u - v_n)(w_n - v_n) \\
&\leq [f_u(t,x,v_n) + g_u(t,x,w_n)]p_{n+1} \\
&\quad + [f_{uu}(t,x,\xi) - \frac{3}{2}g_{uu}(t,x,\delta)]p_n^2 - \frac{1}{2}g_{uu}(t,x,\delta)q_n^2,
\end{aligned}
$$
$$ t \neq t_k $$

for some $v_n \leq \xi \leq u$, $v_n \leq \eta \leq u$ and $\eta \leq \delta \leq w_n$. Thus we have

$$
\mathcal{L}p_{n+1} \leq (\frac{R}{2} + \frac{3}{2}Q)p_n^2 + \frac{Q}{2}q_n^2 + (M + N)p_{n+1},
$$

where $|f_u(t,x,u)| \leq M$, $|g_u(t,x,u)| \leq N$, $|f_{uu}(t,x,u)| \leq R$ and $|g_{uu}(t,x,u)| \leq Q$ on $\bar{Q}_T \times R^N$. Also we have $Bp_{n+1} = Bu-$

$Bv_{n+1} = 0$, $p_{n+1}(0, x) = 0$. At $t = t_k$, we have

$$
\begin{aligned}
p_{n+1}(t_k^+, x) &= u(t_k^+, x) - v_{n+1}(t_k^+, x) \\
&= I_k(u(t_k, x)) - [I_k(v_n(t_k, x)) + I_{ku}(v_n(t_k, x))(v_{n+1}(t_k, x) \\
&\quad - v_n(t_k, x))] \\
&= I_k(v_n(t_k, x)) + I_{ku}(v_n(t_k, x))(u(t_k, x) - v_n(t_k, x)) \\
&\quad + \frac{1}{2} I_{kuu}(\cdot) p_n^2(t_k, x) - I_k(v_n(t_k, x)) \\
&\quad + I_{ku}(v_n(t_k, x))[p_{n_1}(t_k, x) - p_n(t_k, x)] \\
&= I_{ku}(v_n(t_k, x)) p_{n+1}(t_k, x) + \frac{1}{2!} I_{kuu}(\cdot) p_n^2(t_k, x) \\
&\leq M_{1k} p_{n+1}(t_k, x) + \frac{N_{1k}}{2} p_n^2(t_k, x),
\end{aligned}
$$

where $|\ I_{ku}(u)\ | \leq M_{1k}$ and $|\ I_{kuu}(\cdot)\ | \leq N_{1k}$ on $\bar{\Omega}$. Thus, we have the following inequalities,

$$
\begin{bmatrix}
\mathcal{L} p_{n+1} & \leq (M+N) p_{n+1} + \frac{1}{2}(R+3Q) p_n^2 + \frac{Q}{2} q_n^2 \text{ on } Q_T/M \\
B p_{n+1}(t, x) & \leq 0 \text{ on } \Gamma_T/N \\
p_{n+1}(0, x) & \leq 0 \text{ on } \bar{\Omega}, \\
p_{n+1}(t_k^+, x) & \leq M_{1k} p_{n+1}(t_k^+, x) + \frac{N_{1k}}{2} p_n^2(t_k, x) \text{ on } \bar{\Omega}.
\end{bmatrix}
$$

$$(3.4.10)$$

Compare these inequalities with the ordinary impulsive differential equation

$$
\begin{bmatrix}
y' & = h(t, y) \\
y(0) & = y_0, \ y_0 \geq 0 \\
y(t_k^+) & = \tilde{I}_k(y(t_k))
\end{bmatrix}
$$

$$(3.4.11)$$

with $h(t, y) = (M+N)y + R_1$ where

$$
R_1 = \frac{1}{2}(R+3Q) \sup_{\bar{Q}_T}(p_n^2) + \frac{Q}{2} \sup_{\bar{Q}_T}(q_n^2).
$$

Also $\tilde{I}_k(y(t_k)) = M_{1k} y(t_k) + \tilde{R}_k$, where $\tilde{R}_k = \frac{N_{1k}}{2} \sup_{\bar{Q}_T} p_n^2(t_k, x)$. It is easy to observe that $y(t) \geq 0$, since $I_k(y)$ is nondecreasing in y. It easily follows that $B p_{n+1} \leq 0 \leq p(t, x) r(t)$, where $r(t, 0, y_0)$ is the maximal solution of (3.4.11). Now solving the linear problem (3.4.11) and applying Theorem

3.3.1, we get

$$
\begin{aligned}
p_{n+1}(t,x) \;\le\; & \frac{1}{(MN)}\left[\sum_{t_0<t_k<t}\prod_{t_0<t_j<t} M_{1k}e^{(M+N)(t-t_k)}\tilde{R}_k \right.\\
& \left. + \int_0^t \prod_{s<t_k<t} M_{1k}e^{(M+N)(t-s)}R_1 ds\right]\\
\le\; & \frac{e^{(M_N)T}}{M+N}\left[\sum_{t_0<t_k<t}\prod_{t_0<t_j<t} M_{1k}e^{-(M+N)t_k}\tilde{R}_k \right.\\
& \left. + \int_0^T \prod_{s<t_k<t} M_{1k}e^{-(M+N)s}R_1 ds\right]\\
\le\; & A_1\sup_{\bar{Q}_T}(p_n^2) + A_2\sup_{\bar{Q}_T}(q_n^2),
\end{aligned}
$$

where $A_1 = \frac{e^{(M+N)T}}{2(M+N)}[\sum 3\Omega_k N_{1k}e^{-(M+N)t_k}(R+3Q)]$, $\Omega_k = \prod_{t_k<t_j<t} M_{1k}$, $\Lambda_0 = \prod_k N_{1k}$, and $A_2 = \frac{e^{(M+N)T}}{2(M+N)}[\sum \Omega_k N_{1k}\,e^{-(M+N)t_k} + \Lambda_0 Q]$. This proves the sequences $\{v_n(t,x)\}$ converge quadratically to the unique solution $u(t,x)$ of the impulsive reaction–diffusion equation (3.3.1). Similarly, one can show that the sequence $\{w_n(t,x)\}$ converges quadratically to the unique solution $u(t,x)$ of (3.3.1).

In the next result, we assume the existence of coupled upper and lower solutions in the form of (H_2) of Theorem 3.3.1. As a result, we need some extra assumptions.

Theorem 3.4.2 *Let all the assumptions of Theorem 3.4.1 hold except (A_1). In place of (A_1), assume*

(A_1^*) *$(v_0, w_0) \in C$ be coupled lower and upper solutions of (3.3.1) as in (H_2) of Theorem 3.3.1 such that $v_0(t,x) \le w_0(t,x)$ on $\bar{Q}_T$.*

Further in (A_2) assume also that $g_u \le 0$ on $\bar{Q}_T \times R^N$. Then the conclusion of Theorem 3.4.1 holds.

Proof   The assumptions $f_{uu}(t,x,u) \ge 0$ and $g_{uu}(t,x,u) \le 0$ yield the following inequalities that will be useful in the present situation.

$$f(t,x,u) \ge f(t,x,v) + f_u(t,x,v)(u-v) \tag{3.4.12}$$

$$g(t,x,u) \le g(t,x,v) + g_u(t,x,v) \text{ for } u \ge v. \tag{3.4.13}$$

It is also clear that f and g satisfy the Lipschitz condition for some positive constant for u, v such that $v_0 \leq u, v \leq w_0$ on $\bar{Q}_T$. In addition, since $I_{kuu} \geq 0$, we get the inequality

$$I_k(u(t_k, x)) \geq I_k(v(t_k, x)) + I_{ku}(v(t_k, x))(u(t_k, x) - v(t_k, x)). \quad (3.4.14)$$

We develop two sequences which converge quadratically to the unique solution of (3.3.1). Consider the coupled set of linear impulsive reaction diffusion equations

$$
\begin{bmatrix}
\begin{aligned}
\mathcal{L}u &= F(t, x, v_0, w_0; u, v) \\
&= f(t, x, v_0) + f_u(t, x, v_0)(u - v_0) \\
&\quad + g(t, x, w_0) + g_u(t, x, v_0)(v - w_0) \text{ on } Q_T/M \\
\mathcal{L}v &= G(t, x, v_0, w_0; u, v) \\
&= f(t, x, w_0) + f_u(t, x, v_0)(v - w_0) \\
&\quad + g(t, x, v_0) + g_u(t, x, v_0)(u - v_0) \text{ on } Q_T/M \\
Bu(t, x) &= \Phi(t, x), \\
Bv(t, x) &= \Phi(t, x) \text{ on } \Gamma_T/N, \\
u(0, x) &= u_0(x) \\
v(0, x) &= u_0(x) \text{ on } \bar{\Omega}, \\
u(t_k^+, x) &= I_k(v_0(t_k, x)) + [I_{ku}(v_0(t_k, x))](u(t_k, x) - v_0(t_k, x)) \\
&= \bar{I}_{k1}(v_0(t_k, x); u(t_k, x)) \\
v(t_k^+, x) &= I_k(w_0(t_k, x)) + [I_{ku}(v_0(t_k, x))](v(t_k, x) - w_0(t_k, x)) \\
&= \bar{I}_{k2}(v_0(t_k, x), w_0(t_k, x); v(t_k, x)),
\end{aligned}
\end{bmatrix}
$$
$$(3.4.15)$$

for each $k = 1, \ldots, p$.

Using the inequalities (3.4.12), (3.4.13) and the coupled lower and upper solutions of (3.3.1) of the hypothesis, it follows that

$$\mathcal{L}v_0 \leq f(t, x, v_0) + g(t, x, w_0) = F(t, x, v_0, w_0; v_0, w_0) \text{ on } Q_T/M,$$

and

$$
\begin{aligned}
\mathcal{L}w_0 &\geq f(t, x, w_0) + g(t, x, v_0) \\
&\geq f(t, x, v_0) + f_u(t, x, v_0)(w_0 - v_0) \\
&\quad + g(t, x, w_0) + g_u(t, x, v_0)(v_0 - w_0) \\
&= f(t, x, v_0, w_0; w_0, v_0), \text{ on } Q_T/M
\end{aligned}
$$

also

$$\begin{aligned}
\mathcal{L}v_0 &\leq f(t,x,v_0) + g(t,x,w_0) \\
&\leq f(t,x,w_0) + f_u(t,x,v_0)(v_0 - w_0) \\
&\quad + g(t,x,v_0) + g_u(t,x,v_0)(w_0 - v_0) \\
&\equiv G(t,x,v_0,w_0;v_0,w_0) \text{ on } Q_T/M,
\end{aligned}$$

and

$$\begin{aligned}
\mathcal{L}w_0 &\geq f(t,x,w_0) + g(t,x,v_0) \\
&\equiv G(t,x,v_0,w_0;w_0,v_0) \text{ on } Q_T/M.
\end{aligned}$$

Further Bv_0, Bw_0, $v_0(0,x)$, $w_0(0,x)$, $v_0(t_k^+,x)$, $w_0(t_k^+,x)$ are as in the definition of natural lower and upper solutions.

Since $g_u \leq 0$, it is easy to see that $F(t,x,v_0,w_0;u,v)$ is nonincreasing in v and $G(t,x,v_0,w_0;u,v)$ is nonincreasing in u respectively. This proves that (v_0,w_0) are also coupled lower and upper solutions of (3.4.15). Since $F(t,x,v_0,w_0;u,v)$ and $G(t,x,v_0,w_0;u,v)$ are Lipschitzian in u and v, it follows from Theorem 3.3.2 that the linear system (3.4.15) has (v_1,w_1) as the unique solution such that $v_0 \leq v_1$, $w_1 \leq w_0$ on $\bar{Q}_T$. Next we prove $v_1 \leq w_1$ on $\bar{Q}_T$. Using (3.4.12), (3.4.13) and (3.4.14) and the fact that $v_0 \leq v_1$, $w_1 \leq w_0$, it follows that

$$\mathcal{L}v_1 \leq f(t,x,v_1) + g(t,x,w_1), \text{ on } Q_T/M$$

$$\mathcal{L}w_1 \geq f(t,x,w_1) + g(t,x,w_1), \text{ on } Q_T/M$$

$$v_1(0,x) = u_0(x), \text{ on } \bar{\Omega}$$

$$w_1(0,x) = u_0(x), \text{ on } \bar{\Omega}$$

$$Bv_1(t,x) = \Phi(t,x) \text{ on } \Gamma_T/N$$

$$Bw_1(t,x) = \Phi(t,x) \text{ on } \Gamma_T/N$$

$$v_1(t_k^+,x) \leq I_k(v_1(t^k,x)), \text{ on } \bar{\Omega}$$

$$w_1(t_k^+,x) \geq I_k(w_1(t^k,x)), \text{ on } \bar{\Omega}.$$

This proves (v_1,w_1) are coupled lower and upper solutions of (3.3.1) of the type (H_2) of Theorem 3.3.1. Hence using Theorem 3.3.1, we get $v_1(t,x) \leq w_1(t,x)$ on $\bar{Q}_T$, since $I_{ku} \leq M < 1$, from the hypothesis. This proves $v_0(t,x) \leq v_1(t,x) \leq w_1(t,x) \leq w_0(t,x)$ on $\bar{Q}_T$. Now assume that $(v_l(t,x),w_l(t,x))$ is the coupled solution of the linear system of impulsive initial boundary value problems (3.4.15), where (v_0, w_0) are replaced by (v_{l-1}, w_{l-1}) for some $l \geq 1$. For $l = 1$, we already have that (v_l, w_l) exist and are unique such that

$$v_0(t, x) \leq v_1(t, x) \leq w_1(t, x) \leq w_0(t, x) \text{ on } \bar{Q}_T. \tag{3.4.16}$$

Now assume that for some $l > 1$, (v_l, w_l) exist such that

$$v_0(t, x) \leq v_1(t, x) \leq \ldots \leq v_l(t, x) \leq w_l(t, x) \leq \ldots \tag{3.4.17}$$

$$\leq w_1(t, x) \leq w_0(t, x) \text{ on } \bar{Q}_T.$$

From (3.4.16) it follows that (3.4.17) is certainly true for $l = 1$. Using the method of mathematical induction and using the inequalities (3.4.12)–(3.4.14), we can prove that (3.4.17) is true for all l. Employing standard arguments as in Theorem 2.3.1 to each subinterval $(t_k, t_{k+1}) \subseteq [0, T]$, we can conclude that $\{v_n(t, x)\}$, $\{w_n(t, x)\}$ converge uniformly and monotonically to $v(t, x)$, $w(t, x)$ such that $v(t, x) = u(t, x) = w(t, x)$, where $u(t, x)$ is the unique solution of (3.3.1).

We now claim that the convergence of the coupled sequences $\{v_n(t, x)\}$, $\{w_n(t, x)\}$ to the unique solution of (3.3.1) is quadratic. In order to prove quadratic convergence, set $p_{n+1}(t, x) = u(t, x) - v_{n+1}(t, x)$, $q_{n+1}(t, x) = w_{n+1}(t, x) - u(t, x)$. It is easy to observe that $p_{n+1}(t, x) \geq 0$ and $q_{n+1}(t, x) \geq 0$ on $\bar{Q}_T$. Since $f_{uu} \geq 0$ and $g_{uu} \leq 0$, it follows that

$$
\begin{aligned}
\mathcal{L}p_{n+1} &= \mathcal{L}u + \mathcal{L}v_{n+1} \\
&= f(t, x, u) + g(t, x, u) - [f(t, x, v_n) + f_u(t, x, v_n)(v_{n+1} - v_n) \\
&\quad + g(t, x, w_n) + g_u(t, x, v_n)(w_{n+1} - w_n)] \\
&\leq f(t, x, v_n) + f_u(t, x, v_n)(u - v_n) + f_{uu}(t, x, \xi)\frac{(u - u_n)^2}{2!} \\
&\quad - f(t, x, v_n) - f_u(t, x, v_n)(v_{n+1} - v_n) \\
&\quad + g_u(t, x, \eta)(u - w_n) - g_u(t, x, v_n)(w_{n+1} - w_n)
\end{aligned}
$$

where $v_n < \xi < u$ and $u < \eta < w_n$. Since $g_u(t, x, u)$ is decreasing in u and $u \leq w_n$, we have

$$g_u(t, x, \eta)(u - w_n) \leq g_u(t, x, w_n)(u - w_n).$$

Using this, we have

$$
\begin{aligned}
\mathcal{L}p_{n+1} \;\le\; & f_u(t,x,v_n)p_{n+1} + f_{uu}(t,x,\xi)p_n^2 \\
& -g_u(t,x,w_n)q_n - g_u(t,x,v_n)(q_{n+1}-q_n) \\
\le\; & f_u(t,x,v_n)p_{n+1} - g_u(t,x,v_n)q_{n+1} + f_{uu}(t,x,\xi)p_n^2 \\
& +g_{uu}(t,x,\delta)(w_n - u + u - v_n)q_n \\
\le\; & f_u(t,x,v_n)p_{n+1} - g_u(t,x,v_n)q_{n+1} \\
& +g_{uu}(t,x,\delta)(p_n q_n + q_n^2) + f_{uu}t,x,\xi)p_n^2,
\end{aligned}
$$

where $v_n < \delta < w_n$. Now using the estimates of $\mid f_u \mid$, $\mid g_u \mid$ and $\mid g_{uu} \mid$ of Theorem 3.4.1, it follows that

$$
\mathcal{L}p_{n+1} \le PM_{n+1} + Iq_{n+1} + PR_n^2 + Q(p_n q_n + q_n^2),
$$

where M, N, R, Q are as in Theorem 3.4.1. At $t = t_k^+$, the estimate for $p_{n+1}(t_k^+, x)$ is as in Theorem 3.4.1.

Similarly for q_{n+1}, using the mean value theorem and the nature of f_{uu}, g_{uu}, we have

$$
\begin{aligned}
\mathcal{L}q_{n+1} \;=\; & \mathcal{L}w_{n+1} - \mathcal{L}u \\
=\; & f(t,x,w_n) + f_u(t,x,v_n)(w_{n+1}-w_n) \\
& +g(t,x,v_n) + g_u(t,x,v_n)(v_{n+1}-v_n) \\
& -[f(t,x,u) + g(t,x,u)] \\
\le\; & f_u(t,x,w_n)q_n + f_u(t,x,v_n)(q_{n+1}-q_n) \\
& +g_u(t,x,v_n)(-p_{n+1}+p_n) - g_u(t,x,u)p_n \\
\le\; & f_u(t,x,w_n)q_{n+1} - g_u(t,x,v_n)p_{n+1} \\
& +f_{uu}(t,x,\delta)q_n(q_n+p_n) - g_{uu}(t,x,\eta)p_n^2,
\end{aligned}
$$

where $v_n < \delta < w_n$ and $v_n < \eta < u$ on $\bar{Q}_T/M$. Now using the same estimate on $\mid f_{uu} \mid$, $\mid g_{uu} \mid$, $\mid f_u \mid$, and $\mid g_u \mid$ as in Theorem 3.4.1, we get

$$
\mathcal{L}q_{n+1} \le Np_{n+1} + IQ_{n+1} + Rp_n q_n + Rq_n^2 + Qp_n^2.
$$

At $t = t_k$, for q_{n+1}, using the hypothesis $I_{kuu} \geq 0$, we have

$$
\begin{aligned}
q_{n+1}(t_k^+, x) &= w_{n+1}(t_k^+, x) \\
&= u(t_k^+, x)I_k(w_n(t_k, x)) \\
&\quad + [I_{k,u}(v_n(t_k, x))][w_{n+1}(t_k, x) - w_n(t_k, x)] - I_k(u(t_k, x)) \\
&\leq I_{k,u}(w_n(t_k, x))q_n(t_k, x) - I_{k,u}(v_n(t_k, x))q_n(t_k, x) \\
&\quad + I_{k,u}(v_n(t_k, x))q_{n+1}(t_k, x) \\
&\leq I_{k,u}(\gamma)(w_n(t_k, x) - v_n(t_k, x))q_n(t_k, x) \\
&\quad + I_{k,u}(v_n(t_k, x))q_{n+1}(t_k, x) \\
&\leq I_{k,u}(\gamma)(q_n + p_n)q_n + I_{k,u}(v_n(t_k, x))q_{n+1},
\end{aligned}
$$

where $v_n(t_k, x) < \gamma < w_n(t_k, x)$. Now using the estimate on $\mid I_{k.u}(\cdot) \mid$, and $\mid I_{k,uu}(\cdot) \mid$ as in Theorem 3.4.1, we have

$$
q_{n+1}(t_k^+, x) \leq M_{1k}q_{n+1} + N_{1k}(p_n q_n + q_n^2) \text{ on } \bar{\Omega}.
$$

Setting $H_{n+1}(t, x) = p_{n+1}(t, x) + q_{n+1}(t, x)$, and adding the inequalities related to p_{n+1} and q_{n+1}, we get the following inequalities:

$$
\left[
\begin{array}{ll}
\mathcal{L}H_{n+1}(t, x) & \leq (M + N)H_{n+1}(t, x) + (R + Q)H_n^2(t, x) \text{ on } \quad Q_T/M, \\
BH_{n+1}(t, x) & = Bp_{n+1}(t, x) + Bq_{n+1}(t, x) = 0 \quad \text{on } \Gamma_T/N, \\
H_{n+1}(0, x) & = p_{n+1}(0, x) + q_{n+1}(0, x) = 0 \text{ on } \quad \bar{\Omega}, \\
H_{n+1}(t_k^+, x) & \leq M_{1k}H_{n+1}(t_k, x) + N_{1k}H_n^2(t_k, x).
\end{array}
\right.
$$

$$(3.4.18)$$

We note that the estimate of $p_{n+1}(t_k^+, x)$ of Theorem 3.4.1 is used to obtain the estimate of $H_{n+1}(t_k^+, x)$. One can now easily show the quadratic convergence of H_{n+1} as in Theorem 3.4.1 by comparing (3.4.18) with the ordinary impulsive differential equation with initial condition as in (3.4.11), where now

$$
h(t, y) = (M_N)y + R_1
$$

where $R_1 = (R + Q)\sup_{\bar{Q}_T}[p_n(t, x) + q_n(t, x)]^2$ and $I_k(y(t_k)) = M_{1k}y(t_k) + \tilde{R}_k$, where $\tilde{R}_k = N_{1k}\sup_{\bar{Q}_T}[p_n(t_k, x) + q_n(t_k, x)]^2$. Hence the proof is complete.

Remark 3.4.1 Note that one can also prove quadratic convergence of p_{n+1}, q_{n+1} using the corresponding estimate of the system of ordinary impulsive initial value problems. For simplicity, here we have used the scalar version of the comparison Theorem 3.3.3.

Next we merely state the results related to other two types of coupled lower and upper solutions of (3.3.1). The coupled lower and upper solutions of (3.3.1), considered in the next two theorems, are of type (H_3) and (H_4) from Theorem 3.3.1 respectively.

Theorem 3.4.3 *Let all assumptions of Theorem 3.4.1 hold except (A_1). In place of (A_1), assume*

(A_1^{**}) *$(v_0, w_0) \in C$ be coupled lower and upper solutions of (3.3.1) as in (H_3) of Theorem 3.3.1 such that $v_0(t,x) \le w_0(t,x)$ on $\bar{Q}_T$.*

Further in (A_2), assume also that $f_u(t,x,u) \le 0$ on $[v_0, w_0]$. Then the conclusion of Theorem 3.4.1 holds.

Theorem 3.4.4 *Let all the assumptions of Theorem 3.4.1 hold except (A_1). In place of (A_1), assume*

(A_1^{***}) *$(v_0, w_0) \in C$ be coupled lower and upper solutions of (3.3.1) as in (H_4) of Theorem 3.3.1, such that $v_0(t,x) \le w_0(t,x)$ on $\bar{Q}_T$.*

Further in (A_2), assume also that $f_u(t,x,u) \le 0$ and $g_u(t,x,u) \le 0$ on $[v_0, w_0]$. Then the conclusion of Theorem 3.4.1 holds.

All theorems of this section include several special cases of interest. For example, Theorem 3.4.1 covers all the corresponding particular results mentioned following Theorem 2.4.1 in Remarks 2.4.1. We do not repeat to avoid monotony.

3.5 Population Dynamics with Impulses

In this section, we shall consider a mathematical model for the growth of populations of two competing species under the presence of abrupt harvesting which is defined by the system of parabolic equations

$$\left[\begin{array}{l} u_t - D_1 \nabla^2 u = f_1(u,v) \equiv u(a_1 - b_1 u - c_1 v), \\ v_t - D_2 \nabla^2 v = f_2(u,v) \equiv v(a_2 - b_2 u - c_2 v) \end{array} \right. \tag{3.5.1}$$

on Q_T/M along with the boundary conditions

$$\frac{\partial u}{\partial \gamma} = 0, \quad \frac{\partial v}{\partial \gamma} = 0, \tag{3.5.2}$$

on Γ_T/N, and the initial conditions

$$u(0, x) = u_0(x), \quad v(0, x) = v_0(x) \tag{3.5.3}$$

in $\bar{\Omega}$ and the equations of impulsive perturbations

$$\left[\begin{array}{ll} \Delta u(\tau_i, x) = g_{1i}(u, v) & \equiv u(\tau_i^-, x)(-d_1 + e_1 v(\tau_i^-, x)) \\ \Delta v(\tau_i, x) = g_{2i}(u, v) & \equiv v(\tau_i^-, x)(-d_2 + e_2 u(\tau_i^-, x)) \end{array} \right. \tag{3.5.4}$$

in $\bar{\Omega}$, where u and v are the population densities of two competing species which are continuously distributed throughout a bounded habitat Ω in R^2, a_i, b_i, c_i, d_i, e_i, D_i $(i = 1, 2,)$ are positive constants and $u_0(x)$, $v_0(x) \geq 0$ on $\bar{\Omega}$. We denote by $\frac{\partial}{\partial \gamma}$ the outward normal derivative on Γ_T/N and condition (3.5.2) implies that there is no migration across the boundary of Ω. Further, $J = \{t_i\}$ is a nonempty set of real numbers such that

$$card([a, a + 1] \cap J) < \infty \text{ for all } a \in R, \tag{3.5.5}$$

where $card\ M$ denotes the number of elements of the set M. We also suppose that $t_i < t_j$ for $i < j$ and $\lim_{i \to \infty} t_i = +\infty$. Equations (3.5.1) mean that in the absence of competition $(c_1 = c_2 = 0)$ each population grows according to a Malthusian law, while under competition the growth rate of each population is reduced at a rate proportional to its size and the size of its competitor.

Equations (3.5.4) mean that each population is harvested at a rate proportional to its size and to the size of its competitor and the duration of harvesting is small in comparison with the duration of the process of population growth. It is natural that two competing species can coexist, both extinguish, or one of them wipes out the other, and we will be especially interested in the existence and stability of nonnegative steady-state solutions to the problem (3.5.1)–(3.5.4).

We observe that $f_1(0, v) \geq 0$, $f_2(u, 0) \geq 0$, $g_{1i}(0, v) \geq 0$, $g_{2i}(u, 0) \geq 0$, and the functions f_1, f_2, g_{1i}, g_{2i} satisfy a local Lipschitz condition with the constants

$$M_1 = |-a_1 - a_2 \frac{c_1}{c_2}|, \quad M_2 = |-a_2 - a_1 \frac{b_2}{b_1}|,$$

$$L_1 = |-d_1 + e_1|, \quad L_2 = |-d_2 + e_2|,$$

respectively. Hence we can conclude that $u(t, x) \geq 0$ on $\bar{Q}_T$. Moreover, due to Theorem 3.2.4, the following estimate for the solutions of the system (3.5.1)–(3.5.4) is valid:

$$0 \leq u(t, x) \leq r(t) \text{ on } \bar{Q}_T,$$

where $r(t)$ is the solution of the following system of impulsive ordinary differential equations

$$\left[\begin{array}{ll} \frac{du}{dt} = u(a_1 - b_1 u - c_1 v), & t \neq \tau_i, \\ \frac{dv}{dt} = v(a_2 - b_2 u - c_2 v), & t \neq \tau_i \\ \Delta u(\tau_i) = u(-d_1 + e_1 v), & \\ \Delta v(\tau_i) = v(-d_2 + e_2 u), & \\ u(0) = u_0, & v(0) = v_0. \end{array} \right. \tag{3.5.6}$$

Now we are interested in the behavior of the solutions of the system of impulsive differential equations (3.5.6), and steady state solutions. It follows from (3.5.6) that there are two stationary solutions to the system of impulsive differential equations (3.5.6), the trivial solution $u = 0$, $v = 0$ and the nontrivial one

$$u_* = \frac{a_1 c_2 - a_2 c_1}{b_1 c_2 - b_2 c_1}, \quad v_* = \frac{a_2 b_1 - a_1 b_2}{b_1 c_2 - b_2 c_1},$$

provided

$$d_1 = e_1 \frac{a_2 b_1 - a_1 b_2}{b_1 c_2 - b_2 c_1}, d_2 = e_2 \frac{a_1 c_2 - a_2 c_1}{b_1 c_2 - b_2 c_1}, (b_1 c_2 - b_2 c_1 \neq 0). \tag{3.5.7}$$

Naturally, we are interested in stability properties of the nontrivial solution, so in what follows, we suppose that condition (3.5.7) is fulfilled. Taking into account that (3.5.4) implies that the coefficients e_1, e_2 from the biological point of view must be rather small, e_1, $e_2 \ll 1$, we conclude from (3.5.7) that the same concerns also the coefficients d_1, d_2, so the harvesting rate cannot be very high.

Linearizing the system of impulsive differential equations (3.5.6) in the neighborhood of the stationary point u_*, v_*, we get

$$\frac{dx}{dt} = Ax, \ x \neq \tau_i$$

$$\Delta x(\tau_i) = Bx,$$

where

$$A = \left(\begin{array}{cc} -b_1 u_* & -c_1 u_* \\ -b_2 v_* & -c_2 v_* \end{array} \right), \ B = \left(\begin{array}{cc} 0 & e_1 u_* \\ e_2 v_* & 0 \end{array} \right).$$

It is known that the stationary solution u_*, v_* of the system (3.5.6) is asymptotically stable if the following condition holds:

$$\alpha + p \ln \beta < 0, \tag{3.5.8}$$

where

$$\alpha = \mathrm{Re}\lambda_{\max}(A), \quad \beta^2 = \lambda_{\max}((I+B)^T(I+B)),$$

and

$$p = \lim_{t\to\infty} \frac{card J_{(t,t+T)}}{T}, \quad J_{(t,t+T)} = J \cap [t, t+T].$$

It turns out that the eigenvalues of the matrix A are real and they are given by the formula

$$\lambda_{1,2} = \frac{1}{2}\left[-b_1 u_* - c_2 v_* \pm \sqrt{(b_1 u_* - c_2 v_*)^2 + 4b_2 c_1 u_* v_*}\right].$$

Straightforward calculations give us also two real eigenvalues for the matrix $(I+B)^T(I+B)$:

$$\lambda_{1,2} = 1 + \frac{1}{2}\left[(e_1 u_*)^2 + (e_2 v_*)^2 \pm \sqrt{((e_1 u_*)^2 - (e_2 v_*)^2)^2 + 4(e_1 u_* + e_2 v_*)^2}\right],$$

so we get $\alpha^2 = 1 + R$, where

$$R = \frac{1}{2}\left[(e_1 u_*)^2 + (e_2 v_*)^2 + \sqrt{((e_1 u_*)^2 - (e_2 v_*)^2)^2 + 4(e_1 u_* + e_2 v_*)^2}\right].$$

Thus condition (3.5.8) which guarantees the asymptotic stability of steady-state solution u_*, v_* for our case can be written in the following form:

$$b_1 u_* + c_2 v_* - Q - p\ln(1+R) > 0, \tag{3.5.9}$$

where R is as above and

$$Q = \sqrt{(b_1 u_* - c_2 v_*)^2 + 4b_2 c_1 u_* v_*}.$$

Summing up, we can now formulate the following result.

Theorem 3.5.1 *Assume that condition (3.5.9) holds. Then the steady-state solution u_*, v_* of the reaction–diffusion system with abrupt harvesting (3.5.1)–(3.5.4) is globally asymptotically stable. This means that for any solution of the system*

$$\lim_{t\to\infty} u(t, x) = u_*, \quad \lim_{t\to\infty} v(t, x) = v_*$$

on $\bar{\Omega}$.

Taking into account that condition (3.5.8) means both roots of the characteristic polynomial for the matrix A must be less than $-p\ln(1+R)$, we can also obtain the following sufficient conditions for the asymptotic stability of the steady-state solution u_*, v_*:

$$\left[\begin{array}{l} \frac{p^2}{4}(\ln(1+R))^2 \quad -\frac{p}{2}(b_1u_* + c_2v_*)\ln(1+R) + (b_1c_2 - b_2c_1)u_*v_* > 0 \\ \quad b_1u_* + c_2v_* \quad > p\ln(1+R). \end{array}\right.$$

$$(3.5.10)$$

To conclude, we give some comments concerning conditions (3.5.10). As we mentioned above, the harvesting rate is rather small due to the nature of the system and condition (3.5.7). So, we conclude $R \ll 1$ and consequently $\ln(1+R) = o(R)$. Furthermore, condition (3.5.5) implies that the upper limit

$$\limsup_{s\to\infty} \frac{card J_{(t,t+s)}}{s} = q < \infty$$

exists uniformly with respect to $t \in R$. Thus, the steady-state solution u_*, v_* of the impulsive system (3.5.1)–(3.5.4) is asymptotically stable if the following conditions hold:

$$\left[\begin{array}{l} (b_1c_2 - b_2c_1)u_*v_* \geq \delta > 0, \\ \quad b_1u_* + c_2v_* \geq \eta > 0 \end{array}\right.$$

$$(3.5.11)$$

with some δ, η which can be rather small. Evidently the first condition of (3.5.11) implies the fulfillment of condition (3.5.9) while the second condition is almost always satisfied. It can be said that harvesting at a reasonable rate preserves the asymptotic stability of steady-state solution of the competition model without harvesting.

3.6   Notes and Comments

The development of impulsive parabolic IBVPs introduced in Section 3.2 is due to Kirane and Rogovchenko [40]. The contents of Sections 3.3 and 3.4, related to the method of generalized quasilinearization for impulsive parabolic IBVPs discussed in several possible ways depending on the definition of lower and upper solutions, are taken from Drici et al. [28]. See also Drici et al. [29] for special cases. The discussion of the population model under the presence of abrupt harvesting is due to Kirane and Rogovchenko [40]. See also, Erbe et al. [31], Kirane and Rogovchenko [40] and Lakshmikantham and Drici [53] for particular special cases. For details of impulsive differential equations, refer to Lakshmikantham, Bainov and Simeonov [52], where impulsive comparison theorems are also found.

Chapter 4

Hyperbolic Equations

4.1 Introduction

This chapter deals with hyperbolic partial differential equations relative to
monotone iterates and fast convergence. Since the approaches utilized for
elliptic and parabolic equations do not work for hyperbolic equations, we
shall first proceed to consider, in Section 4.2, linear second-order hyperbolic
problems and develop the variation of parameters formula using Laplace invariants. We then derive several comparison results of interest and consider
also the important case of constant coefficients. In Section 4.3, we shall
introduce suitable compact notation to reduce the complexity that is generated in the investigation of nonlinear hyperbolic IBVPs. We then address
the extension of the monotone iterative technique in the general framework
so that one can derive several special cases of interest. Section 4.4 is dedicated to the method of generalized quasilinearization which complicates the
process even further and necessitates the use of the compact notation of
Section 4.3.

4.2 Variation of Parameters (VP) and Compari-
son Results

The approach employed in deriving comparison results in elliptic and parabolic differential equations does not extend to hyperbolic equations. As we
have seen, comparison results are crucial in order to investigate monotone
iterative techniques and the application of the method of generalized quasilinearization to nonlinear hyperbolic partial differential equations.

In this section, we shall obtain variation of parameters formulas for a general linear hyperbolic partial differential equation of the type

$$u_{xy} = a(x, y)u + b(x, y)u_x + c(x, y)u_y + h(x, y)$$

with variable coefficients a, b, c and then deduce some comparison theorems of interest. The formulas obtained are explicit, in the sense that they yield the solution in closed form.

Let $a, b \in R$, $a, b > 0$ and let I_a, I_b and I_{ab} denote respectively the interval $[0, a]$, $[0, b]$ and the rectangle $I_a \times I_b$. The linear second-order hyperbolic partial differential operator of interest is given by

$$L[u] = u_{xy} - a(x, y)u - b(x, y)u_x - c(x, y)u_y = h(x, y), \qquad (4.2.1)$$

where the coefficients $a, b, c \in C^1[R_{ab}, \mathbb{R}]$, and the forcing term $h \in C[R_{ab}, \mathbb{R}]$ The adjoint L^* of L is given by

$$L^*[v] = v_{xy} + (b_x + c_y - a)v + bv_x + cv_y.$$

Even though $L^{**} = L$ is a property enjoyed by any second-order linear operator, L is self-adjoint (that is, $L^* = L$) only when $b \equiv c \equiv 0$, in which case, (4.2.1) reduces to the telegraph equation

$$u_{xy} - au = h. \qquad (4.2.2)$$

For every (x_0, y_0) regarded as fixed, the Riemann function $R_L = R_L(x_0, y_0; x, y)$ for L has, among others, the following properties:

$$\frac{\partial}{\partial x} R_L(x_0, y_0; x, y_0) - c(x, y_0) R_L(x_0, y_0; x, y_0) = 0,$$

and

$$\frac{\partial}{\partial y} R_L(x_0, y_0; x_0, y) - b(x_0, y) R_L(x_0, y_0; x_0, y) = 0,$$

which respectively imply

$$R_L(x_0, y_0; x, y_0) = \exp\left(\int_{x_0}^x c(s, y_0)ds\right), \qquad (4.2.3)$$

and

$$R_L(x_0, y_0; x_0, y) = \exp\left(\int_{y_0}^y b(x_0, t)dt\right). \qquad (4.2.4)$$

We have the following representation theorem for (4.2.1).

Theorem 4.2.1 *Any solution of* (4.2.1) *on* R_{ab} *is given by*

$$
\begin{aligned}
u(x,y) \;=\;& u(x,0)R_L(0,0;x,y) \\
&+ \int_0^x [u_x(s,0) - c(s,0)u(s,0)]R_L(s,0;x,y)ds \\
&+ \int_0^y [u_y(0,t) - b(0,t)u(0,t)]R_L(0,t;x,y)dt \\
&+ \int_0^x ds \int_0^y h(s,t)R_L(s,t;x,y)dt .
\end{aligned}
$$

In general, it is difficult to express the Riemann function R_L for (4.2.1) in terms of the coefficients a, b and c. This restricts the applicability of Theorem 4.2.1 while deriving the comparison theorems (max-min principles) in their explicit form. In special cases, however, we can obtain an explicit expression for R_L. In the self-adjoint case (4.2.2), when a is of separable variables type, i.e., $a(x,y) = a_1(x)a_2(y)$, we have

$$
R_L(x_0, y_0; x, y) = \sum_{i=0}^{\infty} \frac{1}{i!} \left(\int_{x_0}^x a_1(t)dt \right)^i \cdot \left(\int_{y_0}^y a_2(t)dt \right)^i .
$$

In particular, when a is a constant,

$$
R_L(x_0, y_0; x, y) = J_0 \left(2[-a(x - x_0)(y - y_0)]^{1/2} \right),
$$

where $J_0(z)$, the Bessel function of order 0, is the holomorphic function given by

$$
J_0(z) = \sum_{i=0}^{\infty} \frac{(-1)^i}{(i!)^2} \left(\frac{z}{2} \right)^{2i} . \tag{4.2.5}
$$

We shall obtain several variation of parameters formulas of interest, for the linear, nonself-adjoint, scalar hyperbolic partial differential equation

$$
L[u] = h \tag{4.2.6}
$$

which express the (unique) solution of (4.2.6) explicitly in terms of the coefficients a, b, c and h. For any $z \in C^2[R_{ab}, \mathbb{R}]$, $z(x,y) \neq 0$, $(x,y) \in R_{ab}$, the change of dependent variable

$$
u(x,y) = v(x,y)z(x,y) \tag{4.2.7}
$$

transforms (4.2.6) into

$$
v_{xy} = A(x,y)v + B(x,y)v_x + C(x,y)v_y + H(x,y), \tag{4.2.8}
$$

where $A = (az + bz_x + cz_y - z_{xy})/z$, $B = (bz - z_y)/z$, $C = (cz - z_x)/z$ and $H = h/z$. Notice that $A + BC - B_x = a + bc - b_x$ and $A + BC - C_y = a_bc - c_y$, a property of the two Laplace invariants of (4.2.6), relative to the substitution (4.2.7). The choice

$$z(x,y) = \lambda(y) \exp\left(\int_0^x c(s,y)ds\right),$$
(4.2.9)

with any never vanishing function λ of y alone, reduces (4.2.8) to

$$v_{xy} = L_1 v + m v_x + h_1,$$
(4.2.10)

where

$$L_1(x,y) = a(x,y) + b(x,y)c(x,y) - c_y(x,y),$$
(4.2.11)

is one of the Laplace invariants,

$$m(x,y) = b(x,y) - \int_0^x c_y(s,y)ds - \lambda'(y)/\lambda(y),$$
(4.2.12)

and

$$h_1(x,y) = h(x,y)\exp\left(-\int_0^x c(s,y)ds\right)/\lambda(y).$$
(4.2.13)

Similarly, for any never vanishing function μ of x alone, the choice

$$z(x,y) = \mu(x)\exp\left(\int_0^y b(x,t)dt\right),$$
(4.2.14)

reduces (4.2.8) to

$$v_{xy} = L_2 v + n v_y + h_2,$$
(4.2.15)

where

$$L_2(x,y) = a(x,y) + b(x,y)c(x,y) - b_x(x,y),$$
(4.2.16)

$$n(x,y) = c(x,y) - \int_0^y b(x,t)dt - \mu'(x)/\mu(x),$$
(4.2.17)

and

$$h_2(x,y) = h(x,y)\exp\left(-\int_0^y b(x,t)dt\right)/\mu(x).$$
(4.2.18)

Equation (4.2.10) simplifies to

$$v_x(x,y) = v_x(x,0)\exp\left(\int_0^y m(x,t)dt\right)$$
$$+ \int_0^y [L_1(x,t)v(x,t)$$
$$+ h_1(x,t)]\exp\left(\int_t^y m(x,\tau)d\tau\right)dt,$$
(4.2.19)

and finally to the Volterra integral equation

$$v(x,y) = w_1(x,y) + \int_0^x \int_0^y k_1(x,y;s,t)v(s,t)dtds, \qquad (4.2.20)$$

where

$$k_1(x,y;s,t) = L_1(s,t)\exp\left(\int_t^y m(s,\tau)d\tau\right), \qquad (4.2.21)$$

and

$$\begin{aligned}
w_1(x,y) &= v(0,y) + \int_0^x v_x(s,0)\exp\left(\int_0^y m(s,t)dt\right)ds \quad (4.2.22)\\
&+ \int_0^x \int_0^y h_1(s,t)\exp\left(\int_t^y m(s,\tau)d\tau\right)dtds,
\end{aligned}$$

that is,

$$\begin{aligned}
w_1(x,y) &= v(0,y) + v(x,0)\exp\left(\int_0^y m(x,t)dt\right) \quad (4.2.23)\\
&- v(0,0)\exp\left(\int_0^y m(0,t)dt\right)\\
&- \int_0^x v(s,0)\exp\left(\int_0^y m(s,t)dt\right)[\int_0^y m_x(s,t)dt]ds\\
&+ \int_0^x \int_0^y h_1(s,t)\exp\left(\int_t^y m(s,\tau)d\tau\right)dtds.
\end{aligned}$$

By the standard resolvent kernel method, the (unique) solution of (4.2.20) is given by

$$v(x,y) = w_1(x,y) + \int_0^x \int_0^y K(x,y;s,t)w_1(s,t)dtds, \qquad (4.2.24)$$

where

$$K(x,y;s,t) = \sum_{i=1}^\infty K_i(x,y;s,t), \qquad (4.2.25)$$

with

$$K_1(x,y;s,t) = k_1(x,y;s,t), \text{ and for } i \geq 2,$$

$$K_i(x,y;s,t) = \int_s^x \int_t^y K_1(x,y;\sigma,\tau)K_{i-1}(\sigma,\tau;s,t)d\tau d\sigma. \qquad (4.2.26)$$

Similarly, solving (4.2.10), we have another set of expressions:

$$v_y(x,y) = v_y(0,y)\exp\left(\int_0^x n(s,y)ds\right) \tag{4.2.27}$$

$$+ \int_0^x [L_2(s,y)v(s,y) + h_2(s,y)]\exp\left(\int_s^x n(\sigma,y)d\sigma\right)ds,$$

$$v(x,y) = w_2(x,y) + \int_0^x\int_0^y k_2(x,y;s,t)v(s,t)dtds, \tag{4.2.28}$$

where

$$k_2(x,y;s,t) = L_2(s,t)\exp\left(\int_s^x n(\sigma,t)d\sigma\right), \tag{4.2.29}$$

and

$$w_2(x,y) = v(x,0) + \int_0^y v_y(0,t)\exp\left(\int_0^x n(s,t)ds\right)dt \tag{4.2.30}$$

$$+ \int_0^x\int_0^y h_2(s,t)\exp\left(\int_s^x n(\sigma,t)d\sigma\right)dtds.$$

Thus

$$w_2(x,y) = v(x,0) + v(0,y)exp\left(\int_0^x n(s,y)ds\right) \tag{4.2.31}$$

$$- v(0,0)\exp\left(\int_0^x n(s,0)ds\right)$$

$$- \int_0^y v(0,t)\exp\left(\int_0^x n(s,t)ds\right)[\int_0^x n_y(s,t)ds]dt$$

$$+ \int_0^x\int_0^y h_2(s,t)\exp\left(\int_s^x n(\sigma,t)d\sigma\right)dtds.$$

The unique solution of (4.2.28) is given by

$$v(x,y) = w_2(x,y) + \int_0^x\int_0^y M(x,y;s,t)w_2(s,t)dtds, \tag{4.2.32}$$

where

$$M(x,y;s,t) = \sum_{i=1}^\infty M_i(x,y;s,t), \tag{4.2.33}$$

with

$$M_1(x,y;s,t) = k_2(x,y;s,t),$$

and for $i \geq 2$,

$$M_i(x, y; s, t) = \int_s^x \int_t^y M_1(x, y; \sigma, \tau) M_{i-1}(\sigma, \tau; s, t) d\tau d\sigma. \qquad (4.2.34)$$

The above considerations lead to:

Theorem 4.2.2 (Variation of Parameters Formula) *Let* $a, b, c \in C^1[R_{ab}, \mathbb{R}]$ *and* $h \in C[R_{ab}, \mathbb{R}]$. *Then the unique solution* u *of* (4.2.6) *on* R_{ab} *is given by the formula*

$$u(x, y) = \lambda(y) \exp \left(\int_0^x c(s, y) ds \right) \{ w_1(x, y) \qquad (4.2.35)$$

$$+ \int_0^x \int_0^y K(x, y; s, t) w_1(s, t) dt ds \},$$

where $\lambda \in C^1[I_b, \mathbb{R}]$ *is any never vanishing function and* w_1, K *are respectively as in* (4.2.23) *and* (4.2.25); *or by the formula*

$$u(x, y) = \mu(x) \exp \left(\int_0^y b(x, t) dt \right) \{ w_2(x, y) \qquad (4.2.36)$$

$$+ \int_0^x \int_0^y M(x, y; s, t) w_2(s, t) dt ds \},$$

where $\mu \in C^1[I_a, \mathbb{R}]$ *is any never vanishing function and* w_2, M *are respectively as in* (4.2.31) *and* (4.2.33).

Some special cases of Theorem 4.2.2 which are of interest, are in order. Firstly, motivated by the properties (4.2.3) and (4.2.4) of the Riemann function R_L, we make the following choices for the functions $\lambda(y)$ and $\mu(x)$:

$$\lambda(y) = \exp \left(\int_0^y b(0, t) dt \right), \mu(x) = \exp \left(\int_0^x c(s, 0) ds \right). \qquad (4.2.37)$$

The functions m in (4.2.12) and n in (4.2.17) then, respectively, take the forms

$$m_1(x, y) = \int_0^x [b_x(s, y) - c_y(s, y)] ds = \int_0^x [L_1(x, y) - L_2(s, y)] ds, \quad (4.2.38)$$

$$n_1(x, y) = \int_0^y [c_y(x, t) - b_x(x, t)] dt = \int_0^y [L_2(x, t) - L_1(x, t)] dt. \quad (4.2.39)$$

Let

$$R_1(x,y) = \int_0^y b(0,t)dt + \int_0^x c(s,y)ds, \qquad (4.2.40)$$

$$R_2(x,y) = \int_0^x c(s,0)ds + \int_0^y b(x,t)dt. \qquad (4.2.41)$$

Then w_1 in (4.2.22), (4.2.23) and w_2 in (4.2.30), (4.2.31) respectively become

$$
\begin{aligned}
z_1(x,y) \;=\;& u(0,y)\exp(-R_1(0,y)) \qquad\qquad\qquad (4.2.42)\\
&+ \int_0^x \exp\left(-R_1(s,0) + \int_0^y m_1(s,t)dt\right.\\
&\times [u_x(s,0) - c(s,0)u(s,0)]\big)\, ds\\
&+ \int_0^x \int_0^y h(s,t) \times \exp\left(-R_1(s,t) + \int_t^y m_1(s,\tau)d\tau\right) dtds.
\end{aligned}
$$

Hence

$$
\begin{aligned}
z_1(x,y) \;=\;& u(0,y)\exp(-R_1(0,y)) \qquad\qquad\qquad (4.2.43)\\
&+u(x,0)\exp\left(-R_1(x,0) + \int_0^y m_1(x,t)dt\right)\\
&-u(0,0) - \int_0^x u(s,0)\exp\\
&\times\left(-R_1(s,0) + \int_0^y m_1(s,t)dt\right)[\int_0^y m_{1x}(s,t)dt]ds\\
&+ \int_0^x \int_0^y h(s,t)\exp\left(-R_1(s,t) + \int_t^y m_1(s,\tau)d\tau\right) dtds,
\end{aligned}
$$

and

$$
\begin{aligned}
z_2(x,y) \;=\;& u(x,0)\exp(-R_2(x,0)) \qquad\qquad\qquad (4.2.44)\\
&+ \int_0^y \exp\left(-R_2(0,t) + \int_0^x n_1(s,t)ds\right)\\
&\times[u_y(0,t) - b(0,t)u(0,t)]dt\\
&+ \int_0^x \int_0^y h(s,t)\exp\left(-R_2(s,t) + \int_s^x n_1(\sigma,t)d\sigma\right) dtds.
\end{aligned}
$$

Thus

$$z_2(x,y) = u(x,0)\exp(-R_2(x,0)) \tag{4.2.45}$$
$$+u(0,y)\exp\left(-R_2(0,y) + \int_y^x n_1(s,y)ds\right)$$
$$-u(0,0) - \int_0^y u(0,t)\exp\left(-R_2(0,t) + \int_0^x n_1(s,t)ds\right)$$
$$\times[\int_0^x n_{1y}(s,t)ds]dt$$
$$+\int_0^x \int_0^y h(s,t)\exp\left(-R_2(s,t) + \int_s^x n_1(\sigma,t)d\sigma\right)dtds.$$

So, we obtain:

Corollary 4.2.1 *Under the conditions of Theorem 4.2.2, the unique solution u of (4.2.6) on R_{ab} is given by the formula*

$$u(x,y) = \exp(R_1(x,y))\{z_1(x,y) + \int_0^x \int_0^y K(x,y;s,t)z_1(s,t)dtds\}, \tag{4.2.46}$$

or, by the formula

$$u(x,y) = \exp(R_2(x,y))\{z_2(x,y) + \int_0^x \int_0^y M(x,y;s,t)z_2(s,t)dtds\}. \tag{4.2.47}$$

Next, suppose that

$$b_x(x,y) = c_y(x,y), \quad (x,y) \in R_{ab}, \tag{4.2.48}$$

that is, the two Laplace invariants are equal. This happens when, for example, the coefficients b and c are constants, or more generally, when the ordinary differential equation $c(x,y)dx + b(x,y)dy = 0$ is exact. In this case, $m \equiv m_1 \equiv n \equiv n_1 \equiv 0$ on R_{ab} and moreover, $R_1 \equiv R_2$, $z_1 \equiv z_2$ and $K \equiv M$ on R_{ab}, so that the two formulas coincide on R_{ab} and we have:

Corollary 4.2.2 *Suppose that (4.2.48) holds. Then, under the conditions of Corollary 4.2.1, the unique solution u of (4.2.6) is given by the formula*

$$u(x,y) = \exp(R(x,y))\{w(x,y) + \int_0^x \int_0^y N(x,y;s,t)w(s,t)dtds\},$$

where R, w and N are defined by

$$R(x,y) = \int_0^y b(0,t)dt + \int_0^x c(s,y)ds = \int_0^x c(s,0)ds + \int_0^y b(x,t)dt,$$

$$
\begin{aligned}
w(x,y) &= u(x,0)\exp(-R(x,0)) + u(0,y)\exp(-R(0,y)) \\
&\quad -u(0,0) + \int_0^x \int_0^y h(s,t)\exp(-R(s,t))dtds,
\end{aligned}
$$

and,

$$
N(x,y;s,t) = \sum_{i=1}^{\infty} N_i(x,y;s,t),
$$

with

$$
N_1(x,y;s,t) = L_1(s,t) = L_2(s,t), \ \ and \ for \ i \geq 2,
$$

$$
N_i(x,y;s,t) = \int_s^x \int_t^y N_1(x,y;\sigma,\tau)N_{i-1}(\sigma,\tau;s,t)d\tau d\sigma.
$$

If both the Laplace invariants are zero, that is,

$$
a(x,y) + b(x,y)c(x,y) \equiv b_x(x,y) \equiv c_y(x,y), \ (x,y) \in R_{ab}, \tag{4.2.49}
$$

then both the gradient terms v_x on v_y, as well as the v terms in equations (4.2.10) and (4.2.15) are eliminated so that $N \equiv 0$ on R_{ab}, and we obtain

Corollary 4.2.3 *If* (4.2.49) *holds, then the unique solution* u *of* (4.2.42) *on* R_{ab} *is given by the formula*

$$
u(x,y) = \exp(R(x,y))w(x,y). \tag{4.2.50}
$$

Finally, we consider the important case, that of constant coefficients. Suppose that $a(x,y) \equiv \alpha$, $b(x,y) \equiv \beta$ and $c(x,y) \equiv \gamma$. Corollary 4.2.2 obviously applies in this case. But more than that, the resolvent kernel N is a constant multiple of the Bessel function J_0 defined in (4.2.5). Indeed, we have

Corollary 4.2.4 *The unique solution* u *on* R_{ab} *of*

$$
u_{xy} = \alpha u + \beta u_x + \gamma u_y + h(x,y), \tag{4.2.51}
$$

where α, β, γ *are constants and* $h \in C[R_{ab}, R]$, *is given by*

$$
\begin{aligned}
u(x,y) &= \exp(\gamma x + \beta y)\{W(x,y) \\
&\quad + \int_0^x \int_0^y (\alpha + \beta\gamma)J_0(2[-(\alpha + \beta\gamma)(x-s)(y-t)]^{1/2})W(s,t)dtds\},
\end{aligned}
$$

where

$$
\begin{aligned}
W(x,y) &= u(x,0)\exp(-\gamma x) + u(0,y)\exp(-\beta y) - u(0,0) \\
&\quad + \int_0^x \int_0^y h(s,t)\exp(-\gamma s - \beta t)dtds.
\end{aligned}
$$

In some applications, such as the monotone iterative techniques, α, β, and γ may be chosen so as to satisfy

$$\alpha + \beta\gamma = 0. \qquad (4.2.52)$$

In this case, we have

Corollary 4.2.5 *If (4.2.52) holds, then the unique solution u of (4.2.51) on R_{ab} is given by*

$$
\begin{aligned}
u(x,y) &= u(x,0)\exp(\beta y) + u(0,y)\exp(\gamma x) - u(0,0)\exp(\gamma x + \beta y) \\
&\quad + \int_0^x \int_0^y h(s,t)\exp[\gamma(x-s) + \beta(y-t)]dtds.
\end{aligned}
$$

As an illustration of our results, consider

Example 4.2.1 *The unique solution of the initial boundary value problem*

$$u_{xy} = 2y(1 - xy^2)u + 2xyu_x + y^2 u_y + \exp(xy^2),$$

$$u(x,0) = x, \quad u(0,y) = y$$

is, by Corollary 4.2.3, given by $u(x,y) = (x + y + xy)\exp(xy^2)$.

Utilizing the formulas developed so far, we now derive some maximum principles of interest. Next, we deduce a comparison theorem (a Wendroff-type differential inequality), useful in establishing quadratic convergence of sequences in the generalized quasilinearization technique.

A straightforward result in this direction is:

Theorem 4.2.3 *Suppose that $u(0,0) = 0$, and $L[u]$, given by (4.2.1) satisfies $L[u] \le 0$ on R_{ab}, that is, $h(x,y) \le 0$ on R_{ab}. Further let*

(i) *$u(x,0) \le 0$, $x \in I_a$, $u(0,y) \le 0$, $y \in I_b$; and*

(ii) *either $\lambda > 0$ on I_b and $L_2 \ge L_1 \ge 0$ on R_{ab} or $\mu > 0$ on I_a, and $L_1 \ge L_2 \ge 0$ on R_{ab}.*

Then $u(x,y) \le 0$ everywhere in R_{ab}.

Theorem 4.2.3 does not yield maximum principles for the gradient terms u_x and u_y, even when the coefficients a, b, and c are constants, unless certain sign conditions are satisfied. However, we have the following general result.

Theorem 4.2.4 *Suppose that* $L[u] \leq 0$ *on* R_{ab}. *Further, let*

(i) $u(x,0) \leq 0$ *for* $x \in I_a$, $u(0,y) \leq 0$ *for* $y \in I_b$;

(ii) $L_1(x,y) \geq 0$, $L_2(x,y) \geq 0$ *for* $(x,y) \in R_{ab}$;

(iii) $u_x(x,0) - c(x,0)\, u(x,0) \leq 0$ *for* $y \in I_a$, $u_y(0,y) - b(0,y)\, u(0,y) \leq 0$
    for $y \in I_b$;

(iv) $b(x,y) \geq 0$, $c(x,y) \geq 0$, *for* $(x,y) \in R_{ab}$.

   Then,
$$(u, u_x, u_y) \leq (0,0,0) \text{ everywhere on } R_{ab}. \tag{4.2.53}$$

Proof   We apply Corollary 4.2.1 along with the substitutions

$$\begin{aligned} u(x,y) &= v(x,y)\exp(R_1(x,y)) \tag{4.2.54} \\ &= v(x,y)\exp(R_2(x,y)), \tag{4.2.55} \end{aligned}$$

and note that the functions m in (4.2.12) and n in (4.2.17) respectively
become m_1 in (4.2.38) and n_1 in (4.2.39). From (i), (iii), (4.2.42), and
(4.2.44), it follows that $z_1 \leq 0$ and $z_2 \leq 0$ on R_{ab}. Also (ii), via (4.2.21),
(4.2.25), (4.2.26), and (4.2.29), (4.2.33), (4.2.34) yield $K \geq 0$ and $M \geq 0$,
on R_{ab}. It therefore follows from (4.2.24) or (4.2.32) that $v \leq 0$ on R_{ab}.
This obviously implies $u \leq 0$ on R_{ab}. Differentiating (4.2.54) with respect
to x, (4.2.55) with respect to y, and using (iii), we obtain $v_x(x,0) \leq 0$ on
I_a and $v_y(0,y) \leq 0$ on I_b. Therefore, (4.2.19), (4.2.27), along with the facts
that $v \leq 0$, $h_a \leq 0$, $h_2 \leq 0$, $L_1 \geq 0$ and $L_2 \geq 0$ imply $v_x \leq 0$ and $v_y \leq 0$
everywhere on R_{ab}. This, together with (iv) establishes the three inequalities
in (4.2.53), completing the proof.
   Conclusion (4.2.53) in Theorem 4.2.4 obviously remains valid if condition
(ii) is replaced by

(ii)′ $b(x,y)$ is nondecreasing in x, $c(x,y)$ is nondecreasing in y and (4.2.48)
    holds.

   However, if we replace condition (ii) by

(ii)″ $u(0,0) = 0$ and (4.2.49) holds,

which is stronger than (ii)′, then we can relax condition (iii) somewhat in
Theorem 4.2.4 and still obtain the conclusion (4.2.53). This is the content
of

Corollary 4.2.6 *Suppose that $L[u] \leq 0$ on R_{ab}. If*

(i) $u_x(x, 0) \leq 0$ *for* $x \in I_a$, *and* $u_y(0, y) \leq 0$ *for* $y \in I_b$;

(ii) *condition* (ii)″ *holds*;

(iii) $b(x, y) \geq 0$, $c(x, y) \geq 0$ *on* R_{ab}.

Then, the conclusion (4.2.53) *holds.*

Proof   Since $u(0, 0) = 0$, the fact that $u \leq 0$ on R_{ab} follows from formula (4.2.50) in Corollary 4.2.3. Also, differentiating (4.2.50) with respect to x and with respect to y, we obtain $u_x(x, y) \leq 0$ and $u_y(x, y) \leq 0$ everywhere on R_{ab}, completing the proof.

For the constant coefficients case, we have, as a consequence of Theorem 4.2.4 and Corollary 4.2.1, the following:

Corollary 4.2.7 *Suppose that*

$$u_{xy} \leq \alpha u + \beta u_x + \gamma u_y, \quad (x, y) \in R_{ab}$$

where α, β, γ *are constants and* $\beta \geq 0$, $\gamma \geq 0$. *If either*

(a)   (i) $u(x, 0) \leq 0$, $x \in I_a$, $u(0, y) \leq 0$, $y \in I_b$;

     (ii) $u_x(x, 0) - \gamma u(x, 0) \leq 0$, $x \in I_a$, $u_y(0, y) - \beta u(0, y) \leq 0$, $y \in I_b$;

     (iii) $\alpha + \beta\gamma \geq 0$,

or

(b)   (i)′ $u_x(, x0) \leq 0$, $x \in I_a$, $u_y(0, y) \leq 0$, $y \in I_b$ *and* $u(0, 0) = 0$;

     (ii)′ $\alpha + \beta\gamma = 0$,

then, the conclusion of (4.2.53) *holds.*

The linear comparison theorem is an immediate consequence of Theorem 4.2.2 and the fundamental result on hyperbolic partial differential inequalities in Theorem A.5.2

Corollary 4.2.8 *Suppose that*

$$u_{xy} \leq a(x, y)u + b(x, y)u_x + c(x, y)u_y + h(x, y), \quad (x, y) \in R_{ab},$$

where $u \in C^2[R_{ab}, \mathbb{R}]$, $a, b, c \in C^1[R_{ab}, \mathbb{R}]$, $h \in C[R_{ab}, \mathbb{R}]$ and $a, b, c \geq 0$ on R_{ab}. Then, for $(x, y) \in R_{ab}$, we have

$$u(x, y) \leq \lambda(y) \exp\left(\int_0^x c(s, y) ds\right) \left\{w_1(x, y) + \int_0^x \int_0^y K(x, y; s, t) w_1(s, t) dt ds\right\}$$

or

$$u(x, y) \leq \mu(x) \exp\left(\int_0^y b(x, t) dt\right) \left\{w_2(x, y) + \int_0^x \int_0^y M(x, y; s, t) w_2(s, t) dt ds\right\}$$

$$
\begin{aligned}
u_x(x, y) \leq{} & \lambda(y) \exp\left(\int_0^x c(s, y) ds\right) \left\{c(x, y) w_1(x, y) + w_{1x}(x, y)\right. \\
& + \int_0^x \int_0^y \left(c(x, y) K(x, y; s, t) + K_x(x, y; s, t)\right) w_1(s, t) dt ds \\
& \left. + \int_0^y K(x, y; x, t) w_1(x, t) dt\right\},
\end{aligned}
$$

and

$$
\begin{aligned}
u_x(x, y) \leq{} & \mu(x) \exp\left(\int_0^y b(x, t) dt\right) \left\{b(x, y) w_2(x, y) + w_{2y}(x, y)\right. \\
& + \int_0^x \int_0^y \left(b(x, y) M(x, y; s, t) + M_y(x, y; s, t)\right) w_2(s, t) dt ds \\
& \left. + \int_0^x M(x, y; s, y) w_2(s, y) ds\right\},
\end{aligned}
$$

where $\lambda \in C^1[I_b, \mathbb{R}]$, $\mu \in C^1[I_a, \mathbb{R}]$ are any never vanishing functions, and w_1, K, w_2, M are respectively given by (4.2.23), (4.2.25), (4.2.31) and (4.2.33).

4.3 Monotone Iterative Technique

Having obtained the variation of parameters formulas and the consequent comparison results in Section 4.2, relative to the linear nonhomogeneous hyperbolic equation, we shall, in this section, embark on the study of the nonlinear hyperbolic IBVP of the following type

$$
\left[
\begin{array}{ll}
u_{xy} & = f(x, y, u, u_x, u_y) + g(x, y, u, u_x, u_y) \\
u(x, 0) & = \sigma(x), x \in I_a, u(0, y) = \tau(y), y \in I_b, \sigma(0) = u_0 = \tau(0)
\end{array}
\right.
\tag{4.3.1}
$$

and develop the monotone iterative technique for (4.3.1). We shall use the following notation for convenience. For $a, b > 0$, let I_a, I_b and I_{ab} denote the intervals $[0, a]$, $[a, b]$ and the rectangle $I_a \times I_b$ respectively. By $v \in C^{1,2}[I_{ab}, R]$, we mean that v is continuous and its partial derivatives v_x, v_y and v_{xy} exist and are continuous on I_{ab}. For any $v \in C^{1,2}[I_{ab}, R]$, the triple (v, v_x, v_y) is denoted by $\langle v \rangle$. For any two functions $v, w \in C^{1,2}[I_{ab}, R]$, the sector is defined by $(v, w) = [z \in C^{1,2}[I_{ab}, R] : v \leq z \leq w$ on $I_{ab}]$. For any smooth function f (x, y, u, v, w) defined on $I_{ab} \times R^3$, $f_i(x, y, u, v, w)$, for $3 \leq i \leq 5$, denote respectively the first-order partial derivatives of f with respect to u, v and w. Similarly, $f_{i,j}(x, y, \langle v \rangle)$, for $3 \leq i, j \leq 5$, would mean the second-order partial derivatives of f relative to the respective arguments. We shall suppress the dependent variables x, y, when there is no danger of ambiguity in the function f, g. For example, $f_{3,5}(\langle v \rangle)$ means

$$\frac{\partial^2 f(x, y, v, v_x, v_y)}{\partial v \partial v_y}.$$

We shall use the notation $[f]$ for the triple (f_3, f_4, f_5). Thus the expression $f_3 v + f_4 v_x + f_5 v_y$ would imply the usual inner product $[f] \cdot \langle v \rangle$. For $v, w \in C^{1,2}[I_{ab}, R]$, the inequality $\langle v \rangle \leq \langle w \rangle$ means

$$v(x, y) \leq w(x, y), v_x(x, y) \leq w_x(x, y), v_y(x, y) \leq w_y(x, y) \text{ on } I_{ab}.$$

For $v^0, w^0 \in C^{1,2}[I_{ab}, R]$ such that $\langle v^0 \rangle \leq \langle w_0 \rangle$ on I_{ab}, we define the closed set $D = [(x, y, z, p, q) : \langle v^0 \rangle \leq (z, p, q) \leq \langle w^0 \rangle$ on $I_{ab}]$.

We shall suppose that $f, g \in C[I_{ab} \times R^3, R]$, $\sigma \in C^1[I_a, R]$ and $\tau \in C^1[I_b, R]$ and the IBVP (4.3.1). Relative to IBVP (4.3.1), we shall prove the following result on the monotone iterative technique.

Theorem 4.3.1 *Assume that*

(A_1) $v^0, w^0 \in C^{1,2}[I_{ab}, R]$, $\langle v^0 \rangle \leq \langle w^0 \rangle$ *on* I_{ab} *and* v^0, w^0 *are coupled lower and upper solutions of* (4.3.1), *that is,*

$$\left[\begin{array}{l} v_{xy}^0 \leq f(\langle v^0 \rangle) + g(\langle w^0 \rangle), \\ v_x^0(x, 0) \leq \sigma'; (x), v_y^0(0, y) \leq \tau'(y), v^0(0, 0) = u_0, \end{array} \right. \tag{4.3.2}$$

$$\left[\begin{array}{l} w_{xy}^0 \geq f(\langle w^0 \rangle) + g(\langle v^0 \rangle), \\ w_x^0(x, 0) \geq \sigma'(x), w_y^0(0, y) \geq \tau'(y), w^0(0, 0) = u_0; \end{array} \right. \tag{4.3.3}$$

(A_2) $f(\langle u \rangle)$ *is nondecreasing in* $\langle u \rangle$ *for* $(x, u) \in I_{ab}$, *and* $g(\langle u \rangle)$ *is nonincreasing in* $\langle u \rangle$ *in* D.

Then there exist monotone sequences $\{\langle v^n(x,y)\rangle\}$, $\{\langle w^n(x,y)\rangle\}$ $\in$ $C^{1,2}[I_{ab}, R]$ *such that* $\langle v^n \rangle \to \langle \rho \rangle$, $\langle w^n \rangle \to \langle r \rangle$ *in* $C^{1,2}[I_{ab}, R]$ *and* $\langle \rho \rangle$, $\langle r \rangle$ *are the coupled minimal and maximal solutions of IBVP* (4.3.1), *that is,* $\langle p \rangle$, $\langle r \rangle$ *satisfy*

$$\rho_{xy} = f(\langle \rho \rangle) + g(\langle r \rangle), \rho(x,0) = \sigma(x), \rho(0,y) = \tau(y), \rho(0,0) = u_0,$$

$$r_{xy} = f(\langle r \rangle) + g(\langle \rho \rangle), r(x,0) = \sigma(x), r(0,y) = \tau(y), r(0,0) = u_0,$$

and $\langle v^0 \rangle \le \langle \rho \rangle \le \langle r \rangle \le \langle w^0 \rangle$ *on* I_{ab}.

Proof   We consider the simple linear IBVPs for $n = 0, 1, 2, \ldots$,

$$\left[\begin{array}{l} v_{xy}^{n+1} = f(\langle v^n \rangle) + g(\langle w^n \rangle), \\ v^{n+1}(x,0) = \sigma(x), v^{n+1}(0,y) = \tau(y), v^{n+1}(0,0) = u_0, \end{array} \right. \tag{4.3.4}$$

and

$$\left[\begin{array}{l} w_{xy}^{n+1} = f(\langle w^n \rangle) + g(\langle v^n \rangle), \\ w^{n+1}(x,0) = \sigma(x), w^{n+1}(0,y) = \tau(y), w^{n+1}(0,0) = u_0. \end{array} \right. \tag{4.3.5}$$

It is clear that for all n, v^n, $w^n \in C^{1,2}[I_{ab}, R]$ are the unique solutions of (4.3.4), (4.3.5) respectively. We shall show that

$$\langle v^0 \rangle \le \langle v^1 \rangle \le \langle w^1 \rangle \le \langle w^0 \rangle \text{ on } I_{ab}. \tag{4.3.6}$$

Setting $p = v^0 - v^1$, it follows that $p_x(x,0) \le 0$, $p_y(0,y) \le 0$, $p(0,0) = 0$ and $p_{xy} \le f(\langle v^0 \rangle) + g(\langle w^0 \rangle) - f(\langle v^0 \rangle) - g(\langle w^0 \rangle) = 0$, and therefore, Corollary 4.2.7 gives $\langle p \rangle \le 0$ on I_{ab} and this implies $\langle v^0 \rangle \le \langle v^1 \rangle$ on I_{ab}. Similarly we get $\langle w^1 \rangle \le \langle w^0 \rangle$ on I_{ab}. To prove $\langle v^1 \rangle \le \langle w^1 \rangle$, we let $p = v^1 - w^1$ and note that $p_x(x,0) = 0$, $p_y(0,y) = 0$, $p(0,0) = 0$ and

$$p_{xy} = f(\langle v^0 \rangle) + g(\langle w^0 \rangle) - f(\langle w^0 \rangle) = g(\langle v^0 \rangle) \le 0,$$

in view of the monotone character of $f(\langle u \rangle)$, $g(\langle u \rangle)$ and the assumption $\langle v^0 \rangle \le \langle w^0 \rangle$. Hence Corollary 4.2.7 yields $\langle v^1 \rangle \le \langle w^1 \rangle$ and (4.3.6) holds.

Assume that for some $n > 1$,

$$\langle v^{n-1} \rangle \le \langle v^n \rangle \le \langle w^n \rangle \le \langle w^{n-1} \rangle \text{ on } I_{ab}. \tag{4.3.7}$$

Then we shall show that

$$\langle v^n \rangle \le \langle v^{n+1} \rangle \le \langle w^{n+1} \rangle \le \langle w^n \rangle \text{ on } I_{ab}. \tag{4.3.8}$$

To do this, let $p = v^n - v^{n+1}$ so that $p_x(x,0) = 0$, $p_y(0,y) = 0$, $p(0,0) = 0$
and
$$p_{xy} = f(\langle v^{n-1}\rangle) + g(\langle w^{n-1}\rangle) - f(\langle v^{n-1}\rangle) - g(\langle w^{n-1}\rangle) \le 0,$$
because of the monotonicity of $f(\langle u\rangle)$, $g(\langle u\rangle)$ and (4.3.7). Thus we get from
Corollary 4.2.7, $\langle v^n\rangle \le \langle v^{n+1}\rangle$ on I_{ab}. A similar argument yields $\langle w^{n+1}\rangle \le$
$\langle w^n\rangle$ on I_{ab}. Now letting $p = v^{n+1} - w^{n+1}$, we find $p_x(x,0) = 0$, $p_y(0,y) = 0$,
$p(0,0) = 0$ and
$$p_{xy} = f(\langle v^n\rangle) + g(\langle w^n\rangle) - f(\langle w^n\rangle) - g(\langle v^n\rangle) \le 0,$$
in view of (4.3.7) and the monotone nature of $f(\langle u\rangle)$, $g(\langle u\rangle)$. Hence, we
obtain by Corollary 4.2.7, $\langle v^{n+1}\rangle \le \langle w^{n+1}\rangle$ on I_{ab}, proving (4.3.8). Hence
by induction, we arrive at, for all $n = 1, 2, \ldots$

$$\langle v^0\rangle \le \langle v^1\rangle \le \ldots \le \langle v^n\rangle \le \langle w^n\rangle \le \ldots \le \langle w^1\rangle \le \langle w^0\rangle \text{ on } I_{ab}. \quad (4.3.9)$$

Using the monotone character of the sequences $\{\langle v^n\rangle\}$, $\{\langle w^n\rangle\}$, together
with the Ascoli–Arzela theorem, it can be shown using standard arguments
that $\langle v^n\rangle \to \langle \rho\rangle$, $\langle w^n\rangle \to \langle r\rangle$ uniformly on I_{ab}, $\langle \rho\rangle, \langle r\rangle \in C^{1,2}[I_{ab}, R]$ are the
coupled solutions of the IBVP (4.3.1).

To show that $\langle \rho\rangle$, $\langle r\rangle$ are the coupled extremal solutions of (4.3.1), let u
be any solution of (4.3.1) such that $\langle v^0\rangle \le \langle u\rangle \le \langle w^0\rangle$ on I_{ab}. Assume that
for some $n > 1$, we have

$$\langle v^n\rangle \le \langle u\rangle \le \langle w^n\rangle \text{ on } I_{ab}. \quad (4.3.10)$$

Then letting $p = v^{n+1} - u$, we note $p_x(x,0) =)$, $p_y(0,y) = 0$, $p(0,0) = 0$ and

$$p_{xy} = f(\langle v^n\rangle) + g(\langle w^n\rangle) - f(\langle u\rangle) - g(\langle u\rangle) \le 0,$$

because of (4.3.10) and the monotonicity assumption of f, g. Corollary 4.2.7
now shows that $\langle v^{n+1}\rangle \le \langle u\rangle$ on I_{ab}. Similarly, $\langle u\rangle \le \langle w^{n+1}\rangle$ on I_{ab}. It
then follows by induction that $\langle v^n\rangle \le \langle u\rangle \le \langle w^n\rangle$ on I_{ab} for all $n = 1, 2, \ldots$
and this implies, in turn, that $\langle \rho\rangle \le \langle u\rangle \le \langle r\rangle$ on I_{ab}, proving $\langle \rho\rangle, \langle r\rangle$ are
coupled minimal and maximal solutions of IBVP (4.3.1).

In order to prove uniqueness, we have the following result.

Lemma 4.3.1 *In addition to the assumptions of Theorem 4.3.1, suppose
that f, g satisfy for $\langle u_1\rangle \le \langle u_2\rangle$, $\langle v^0\rangle \le \langle u_1\rangle$, $\langle u_2\rangle \le \langle w^0\rangle$,*

$$f(\langle u_1\rangle) - f(\langle u_2\rangle) \le [N] \cdot \langle u_1 - u_2\rangle,$$

$$g(\langle u_1\rangle) - g(\langle u_2\rangle) \ge -[M] \cdot \langle u_1 - u_2\rangle, [N], [M] > 0,$$

*where $[N] \cdot \langle u\rangle = N_1 u + N_2 u_x + N_3 u_y$ with a similar meaning for $[M]$. Then
$\rho = r = u$ is the unique solution of IBVP (4.3.1).*

Proof Since $\langle \rho \rangle \le \langle r \rangle$, we need to show $\langle r \rangle \le \langle \rho \rangle$ on I_{ab}. Hence setting $p = r - \rho$, we find $p_x(x,0) = 0$, $p_y(0,y) = 0$, $p(0,0) = 0$ and

$$\begin{aligned} p_{xy} &= f(\langle r \rangle) + g(\langle \rho \rangle) - f(\langle \rho \rangle) - g(\langle r \rangle) \\ &\le [M + N] \cdot p. \end{aligned}$$

Theorem A.5.2 now implies that $\langle p \rangle \le 0$ on I_{ab} which proves Lemma 4.3.1.

Instead of u_{xy} in the IBVP (4.3.1), one could use the more general $Lu = u_{xy} - (\alpha u + \beta u_x + \gamma u_y)$, where α, β, γ are constants satisfying $\alpha + \beta\gamma = 0$. With this change, Theorem 4.3.1 remains true when we utilize Corollary 4.2.5 and 4.2.7 suitably.

As before, Theorem 4.3.1 includes several special cases, which we list below.

Remarks 4.3.1

(i) If $g(\langle u \rangle) \equiv 0$ in IBVP (4.3.1), then we get a result for a nondecreasing right-hand side function f.

(ii) If $f(\langle u \rangle) \equiv 0$ in IBVP (4.3.1), we get a dual result for the nonincreasing case.

(iii) If $g(\langle u \rangle) \equiv 0$ and $f(\langle u \rangle)$ is not nondecreasing in $\langle u \rangle$ but $\tilde{f}(\langle u \rangle) = f(\langle u \rangle) + (\alpha u + \beta u_x + \gamma u_y)$, α, β, γ are constants such that $\alpha + \beta\gamma = 0$, is nondecreasing in $\langle u \rangle$. Then we can consider the IBVP

$$Lu = u_{xy} - (\alpha u + \beta u_x + \gamma u_y) = \tilde{f}(\langle u \rangle), u(x,0) = \sigma(x), \quad (4.3.11)$$

$$u(0,0) = u_0.$$

It is clear that v^0, w^0 are lower and upper solutions of (4.3.11) as well. As a result, Theorem 4.3.1 applied to (4.3.11) yields the same conclusion as in (i).

(iv) Suppose that $g(\langle u \rangle)$ is nonincreasing in $\langle u \rangle$ and $f(\langle u \rangle)$ is not nondecreasing in $\langle u \rangle$ but $\tilde{f}$ as defined in (iii) is nondecreasing in $\langle u \rangle$. Then we consider

$$\left[\begin{array}{l} u_{wy} - (\alpha u + \beta u_x + \gamma u_y) = \tilde{f}(\langle u \rangle) + g(\langle u \rangle), \\ u(x,0) = \sigma(x), \; u(0,y) = \tau(y), \; u(0,0) = u_0. \end{array} \right. \quad (4.3.12)$$

Then the conclusion of Theorem 4.3.1 applied to (4.3.12) holds, since v^0 w^0 are also coupled lower and upper solutions of (4.3.12).

(v) If $f(\langle u \rangle) \equiv 0$ and $g(\langle u \rangle)$ is not nonincreasing in $\langle u \rangle$ but $\tilde{g}(\langle u \rangle) = g(\langle u \rangle) - (\alpha u + \beta u_x + \gamma u_y)$, α, β, γ are constants such that $\alpha + \beta \gamma = 0$, is nonincreasing in $\langle u \rangle$. We then consider the IBVP

$$\left[\begin{array}{l} u_{xy} + (\alpha u + \beta u_x + \gamma u_y) \;\; = \tilde{g}(\langle u \rangle), \\ u(x,0) = \sigma(x), \;\; u(0,y) = \tau(y), \;\; u(0,0) = u_0. \end{array} \right. \tag{4.3.13}$$

Assuming that v^0, w^0 are coupled lower and upper solutions of (4.3.13), an application of Theorem 4.3.1 to (4.3.13) yields the same result as in (ii) for (4.3.13).

(vi) Suppose that $f(\langle u \rangle)$ is nondecreasing in $\langle u \rangle$, $g(\langle u \rangle)$ is not nonincreasing in $\langle u \rangle$ but $\tilde{g}(\langle u \rangle)$ is nonincreasing in $\langle u \rangle$, where $\tilde{g}$ is the same function as in (v). Then we consider the IBVP

$$\left[\begin{array}{l} u_{xy} + (\alpha + \beta u_x + \gamma u_y) \;\; = f(\langle u \rangle) + \tilde{g}(\langle u \rangle), \\ u(x,0) = \sigma(x), \;\; u(0,y) = \tau(y), \;\; u(0,0) = u_0, \end{array} \right. \tag{4.3.14}$$

and assume that v^0, w^0 are also coupled lower and upper solutions of (4.3.14). Then the conclusion of Theorem 4.3.1 applied to IBVP (4.3.14) holds for the IBVP (4.3.14).

(vii) Suppose that $f(\langle u \rangle)$, $g(\langle u \rangle)$ are both not monotone but $\tilde{f}(\langle u \rangle) = f(\langle u \rangle) + (\alpha u + \beta u_x + \gamma u_y)$ is nondecreasing in $\langle u \rangle$ and $\tilde{g}(\langle u \rangle) = g(\langle u \rangle) - (\alpha u + \beta u_x + \gamma u_y)$ is nonincreasing in $\langle u \rangle$, α, β, γ satisfying $\alpha + \beta \gamma = 0$. Then we consider the IBVP

$$\left[\begin{array}{l} u_{xy} = \tilde{f}(\langle u \rangle) + \tilde{g}(\langle u \rangle), \\ u(x,0) = \sigma(x), \;\; u(0,y) = \tau(y), u(0,0) = u_0, \end{array} \right. \tag{4.3.15}$$

and suppose that v^0, w^0 are coupled lower and upper solutions of IBVP (4.3.15). Then we get the same conclusion of Theorem 4.3.1 for IBVP (4.3.15).

Note If, in addition to the conditions given in (v), (vi), and (vii), the uniqueness condition as in Lemma 4.3.1 holds, then Theorem 4.3.1 applied to the corresponding problems in (v), (vi), and (vii) yields the same conclusion for IBVP (4.3.1).

4.4   The Method of Generalized Quasilinearization

We shall continue to consider the IBVP (4.3.1) and the notation described in Section 4.3. We shall extend, in this section, the method of generalized quasilinearization to IBVP (4.3.1) under suitable conditions. Specifically, we shall prove the following theorem.

Theorem 4.4.1 *Assume that*

(A_1) $v^0, w^0 \in C^{1,2}[I_{ab}, R]$, $\langle v^0 \rangle \leq \langle w^0 \rangle$ *on I_{ab} and v^0, w^0 are lower and upper solutions of (4.3.1), that is,*

$$v_{xy}^0 \leq f(\langle v^0 \rangle) + g(\langle v^0 \rangle), v_x^0(x, 0) \leq \sigma'(x), v_y^0(0, y) \leq \tau'(y), v^0(0, 0) \leq u_0,$$

$$w_{xy}^0 \geq f(\langle w^0 \rangle) + g(\langle w^0 \rangle), w_x^0(x, 0) \geq \sigma'(x), w_y^0(0, y) \geq \tau'(y), w^0(0, 0) \geq u_0;$$

(A_2) $f, g \in C[D, R]$, *for $3 \leq i \leq 5$, f_i, g_i, $f_{i,j}$, $g_{i,j}$ exist and satisfy $f_{i,j}(\langle u \rangle) \geq 0$ and $g_{i,j}(\langle u \rangle) \leq 0$ on D;*

(A_3) $f(\langle u \rangle)$, $g(\langle u \rangle)$ *are nondecreasing in $\langle u \rangle$ on D.*

Then there exist monotone sequences $\{\langle v^n \rangle\}$, $\{\langle w^n \rangle\} \in C^{1,2}[I_{ab}, R]$ such that $\langle v^n \rangle \to \langle \rho \rangle$, $\langle w^n \rangle \to \langle r \rangle$ in $C^{1,2}[I_{ab}, R]$ and $\langle \rho \rangle = \langle r \rangle = \langle u \rangle$ is the unique solution of IBVP (4.3.1) and the convergence is quadratic.

Proof   In view of (A_2), we have on D,

$$f(\langle u \rangle) \geq f(\langle v \rangle) + [f(\langle v \rangle)] \cdot \langle u - v \rangle, \tag{4.4.1}$$

$$g(\langle u \rangle) \geq g(\langle v \rangle) + [g(\langle u \rangle)] \cdot \langle u - v \rangle, \tag{4.4.2}$$

for $\langle u \rangle \geq \langle v \rangle$. Also f, g satisfy a Lipschitz condition

$$f(\langle u \rangle) - f(\langle v \rangle) \leq [N_1] \cdot \langle u - v \rangle, \tag{4.4.3}$$

$$g(\langle u \rangle) - g(\langle v \rangle) \leq [N_2] \cdot \langle u - v \rangle, \text{ for } \langle u \rangle \geq \langle v \rangle, [N_1], [N_2] > 0. \tag{4.4.4}$$

We consider the following linear IBVPs for $n = 0, 1, 2, \ldots$,

$$v_{xy}^{n+1} = F(\langle v^{n+1} \rangle; \langle v^n \rangle, \langle w^n \rangle), v^{n+1}(x, 0) = \sigma(x), \tag{4.4.5}$$

$$v^{n+1}(0, y) = \tau(y), v^{n+1}(0, 0) = u_0,$$

$$w_{xy}^{n+1} = G(\langle w^{n+1} \rangle; \langle v^n \rangle, \langle w^n \rangle), w^{n+1}(x, 0) = \sigma(x), w^{n+1}(0, y) = \tau(y), \tag{4.4.6}$$

$$w^{n+1}(0, 0) = u_0,$$

where for each $n = 0, 1, 2, \ldots$,

$$
\begin{aligned}
F(\langle u \rangle; \langle v^n \rangle, \langle w^n \rangle) &= f(\langle v^n \rangle) + g(\langle v^n \rangle) + [f(\langle v^n \rangle)] \cdot \langle u - v^n \rangle \quad (4.4.7) \\
&\quad + [g(\langle w^n \rangle)] \cdot \langle u - v^n \rangle, \\
G(\langle u \rangle; \langle v^n \rangle, \langle w^n \rangle) &= f(\langle w^n \rangle) + g(\langle w^n \rangle) + [f(\langle v^n \rangle)] \cdot \langle u - w^n \rangle \quad (4.4.8) \\
&\quad + [g(\langle w^n \rangle)] \cdot \langle u - w^n \rangle.
\end{aligned}
$$

We find

$$
v_{xy}^0 \leq f(\langle v_0 \rangle) + g(\langle v_0 \rangle) \equiv F(\langle v^0 \rangle; \langle v^0 \rangle, \langle w \rangle),
$$

and

$$
\begin{aligned}
w_{x,y}^0 &\geq f(\langle w^0 \rangle) + g(\langle w^0 \rangle) \geq f(\langle v^0 \rangle) + [f(\langle v_0 \rangle)] \cdot \langle w^0 - v^0 \rangle \\
&\quad + g\langle v^0 \rangle + [g(\langle w^0 \rangle)] \cdot \langle w^0 - v^0 \rangle \\
&\equiv F(\langle w^0 \rangle; \langle v^0 \rangle; \langle w^0 \rangle),
\end{aligned}
$$

using (4.4.1), (4.4.2) and the fact that $\langle v^0 \rangle \leq \langle w^0 \rangle$. By (A_2) and (A_3), $F(\langle u \rangle;$ $\langle v^0 \rangle, \langle w^0 \rangle)$ is nondecreasing in $\langle u \rangle$ and is linear. Consequently, Theorem A.5.4 gives the existence of the unique solution $v^1 \in C^{1,2}[I_{ab}, R]$ of

$$
v_{xy}^1 = F(\langle v' \rangle; \langle v^0 \rangle; \langle w^0 \rangle), v^1(x,0) = \sigma(x), v^1(0,y) = \tau(y), v^1(0,0) = u_0,
$$
$$
(4.4.9)
$$

such that $\langle v^0 \rangle \leq \langle v^1 \rangle \leq \langle w^0 \rangle$ on I_{ab}. A similar argument shows that $w^1 \in C^{1,2}[I_{ab}, R]$ is the unique solution of

$$
w_{xy}^1 = G(\langle w^1 \rangle; \langle v^0 \rangle, \langle w^0 \rangle), w^1(x,0) = \sigma(x), w^1(0,y) = \tau(y), w^1(0,0) = u_0,
$$
$$
(4.4.10)
$$

such that $\langle v^0 \rangle \leq \langle w^1 \rangle \leq \langle w^0 \rangle$ on I_{ab}.

We wish to show that

$$
\langle v^0 \rangle \leq \langle v^1 \rangle \leq \langle w^1 \rangle \leq \langle w^0 \rangle \text{ on } I_{ab}. \qquad (4.4.11)
$$

Hence we need to show $\langle v^1 \rangle \leq \langle w^1 \rangle$ on I_{ab}. The relations (4.4.9), (4.4.10) together with (4.4.1), (4.4.2), in view of the definition of F, G yields

$$
v_{xy}^1 \leq f(\langle v^1 \rangle) + g(\langle v^1 \rangle),
$$

$$
w_{xy}^1 \geq f(\langle w^1 \rangle) + g(\langle w^1 \rangle),
$$

showing that v^1, w^1 are also lower and upper solutions of IBVP (4.3.1). Clearly, $v_x^1(x,0) = \sigma'(x) = w_x^1(x,0)$, $v_y^1(0,y) = w_y^1(0,y) = \tau'(y)$, $v^1(0,0) =$

$w^1(0,0)$. We therefore obtain by Theorem A.5.2 that $\langle v^1 \rangle \leq \langle w^1 \rangle$ as I_{ab}, proving (4.4.11).

We shall next prove that if

$$\langle v^{n-1} \rangle \leq \langle v^n \rangle \leq \langle w^n \rangle \leq \langle w^{n+1} \rangle, \text{ on } I_{ab}, \qquad (4.4.12)$$

for some $n > 1$, then it follows that

$$\langle v^n \rangle \leq \langle v^{n+1} \rangle \leq \langle w^{n+1} \rangle \leq \langle w^n \rangle \text{ on } I_{ab}. \qquad (4.4.13)$$

Since v^n satisfies

$$v^n_{xy} = F(\langle v^n \rangle; \langle v^{n-1} \rangle, \langle w^{n-1} \rangle), v^n(x,0) = \sigma(x), v^n(0,y) = \tau(y), v^n(0,0) = u_0,$$

we obtain using (4.4.1), (4.4.2) and (4.4.12)

$$v^n_{xy} \leq F(\langle v^n \rangle; \langle v^n \rangle, \langle w^n \rangle),$$

and similarly, $w^n_{xy} \geq G(\langle w^n \rangle; \langle v^n \rangle, \langle w^n \rangle)$, with the same initial boundary conditions.

As before, by Theorem A.5.4, we can conclude that there exists a unique solution $v^{n+1} \in C^{1,2}[I_{ab}, R]$ of (4.4.5) satisfying $\langle v^n \rangle \leq \langle v^{n+1} \rangle \leq \langle w^n \rangle$ on I_{ab}. Similar reasoning gives the existence of $w^{n+1} \in C^{1,2}[I_{ab}, R]$ of (4.4.6) satisfying $\langle v^n \rangle \leq \langle w^{n+1} \rangle \leq \langle w^n \rangle$ on I_{ab}. To show that $\langle v^{n+1} \rangle \leq \langle w^{n+1} \rangle$, we see that (4.4.5), (4.4.6) together with (4.4.1), (4.4.2), give

$$v^{n+1}_{xy} \leq f(\langle v^{n+1} \rangle) + g(\langle v^{n+1} \rangle),$$

$$w^{n+1}_{xy} \geq f(\langle w^{n+1} \rangle) + g(\langle w^{n+1} \rangle),$$

which shows that v^{n+1}, w^{n+1} are lower and upper solutions of (4.3.1). Obviously, the initial and boundary conditions hold with equality. As a result, Theorem A.5.2 shows that $\langle v^{n+1} \rangle \leq \langle w^{n+1} \rangle$ on I_{ab}, proving (4.4.13). Hence, we have by induction

$$\langle v^0 \rangle \leq \langle v^1 \rangle \leq \ldots \leq \langle v^n \rangle \leq \langle w^n \rangle \leq \ldots \leq \langle w^1 \rangle \leq \langle w^0 \rangle \text{ on } I_{ab}, \qquad (4.4.14)$$

for all $n = 0, 1, 2, \ldots$.

Now the monotone nature of sequences $\{\langle v^n \rangle\}$, $\{\langle w^n \rangle\}$, coupled with Ascoli–Arzela theorem, can be used with standard estimates to show that $\langle v^n \rangle \to \langle \rho \rangle$, $\langle w^n \rangle \to \langle r \rangle$ uniformly on I_{ab}, $\langle \rho \rangle$, $\langle r \rangle \in C^{1,2}[I_{ab}, R]$ and ρ, r are solutions of IBVP (4.3.1).

Since $\langle \rho \rangle \leq \langle r \rangle$, taking $\langle v \rangle = \langle r \rangle$, $\langle w \rangle = \langle \rho \rangle$ and using Theorem A.5.2, we obtain $\langle r \rangle \leq \langle \rho \rangle$ on I_{ab}, proving $\langle r \rangle = \langle \rho \rangle = \langle u \rangle$ is the unique solution of IBVP (4.3.1).

To prove quadratic convergence of $\{\langle v^n \rangle\}$, $\{\langle w^n \rangle\}$, to the unique solution u of IBVP (4.3.1), we set $p^{n+1} = u - v^{n+1}$, $q^{n+1} = w^{n+1} - u$ so that we have $p_x^{n+1}(x,0) = p_y^{n+1}(0,y) = p^{n+1}(0,0) = 0$ and

$$
\begin{aligned}
p_{xy}^{n+1} &= f(\langle u \rangle) + g(\langle u \rangle) - F(\langle v^{n+1} \rangle; \langle v^n \rangle, \langle w^n \rangle) \\
&= [f(\langle z \rangle)] \cdot \langle p^n \rangle + [f(\langle v^n \rangle)] \cdot \langle p^{n+1} - p^n \rangle, \\
&\quad + [g(\langle \sigma \rangle)] \cdot \langle p^n \rangle + [g(\langle w^n \rangle)] \cdot \langle p^{n+1} - p^n \rangle,
\end{aligned}
$$

where $\langle v^n \rangle \leq \langle z \rangle$, $\langle \sigma \rangle \leq \langle u \rangle$. Using the monotone nature of $[f(\langle u \rangle)]$, $[g(\langle w \rangle)]$ and rearranging, we get

$$
p_{xy}^{n+1} \leq [f(\langle u \rangle) - f(\langle v^n \rangle)] \cdot \langle p^n \rangle + [g(\langle v^n \rangle) - g(\langle w^n \rangle)] \cdot \langle p^n \rangle
$$

$$
+ [f(\langle v^n \rangle)] \cdot \langle p^{n+1} \rangle + [g(\langle w^n \rangle)] \cdot p^{n+1}.
$$

Now $[f(\langle v^n \rangle)] \cdot \langle p^{n+1} \rangle \leq N_1 \parallel \langle p^{n+1} \rangle \parallel$, $[g(\langle w^n \rangle)] \cdot \langle p^{n+1} \rangle \leq N_2 \parallel \langle p^{n+1} \rangle \parallel$, where $\parallel \langle u \rangle \parallel = \mid u \mid + \mid u_x \mid + \mid u_y \mid$, $N_1 = \max_{3 \leq i \leq 5} [\max_D \mid f_i(\langle u \rangle) \mid]$, and $N_2 = \max_{3 \leq i \leq 5} [\max_D \mid g_i(\langle u \rangle) \mid]$. Also, after some computation, we arrive at

$$
[f(\langle u \rangle) - f(\langle v^n \rangle)] \cdot \langle p^n \rangle \leq M_1 \parallel \langle p^n \rangle \parallel^2,
$$

$$
\begin{aligned}
[g(\langle v^n \rangle) - g(\langle w^n \rangle)] \cdot \langle p^n \rangle &\leq M_2 [\parallel \langle p^n \rangle \parallel^0 (\parallel \langle p^n \rangle \parallel + \parallel \langle q^n \rangle \parallel)] \\
&\leq 2M_2 \parallel \langle p^n \rangle \parallel^2 + M_2 \parallel \langle q^n \rangle \parallel^2,
\end{aligned}
$$

where

$$
M_1 = \max_{3 \leq i,j \leq 5} [\max_D \mid f_{i,j}(\langle u \rangle) \mid]
$$

and

$$
M_2 = \max_{3 \leq i,j \leq 5} [\max_D \mid g_{i,j}(\langle u \rangle) \mid].
$$

It then follows that

$$
p_{xy}^{n+1} \leq N \parallel \langle p^{n+1} \rangle \parallel + M(\parallel \langle p^n \rangle \parallel^2 + \parallel \langle q^n \rangle \parallel^2 \quad \text{on } I_{ab},
$$

where $N = N_1 + N_2$ and $M = 2M_1 + M_2$; or equivalently,

$$
p_{xy}^{n+1} \leq N \parallel \langle p^{n+1} \rangle \parallel + Q, \tag{4.4.15}
$$

where $Q = M(\| \langle p^n \rangle \|_0^2 + \| \langle q^n \rangle \|_0)$ and $\| \langle z \rangle \|_0 = \max_{I_{ab}} \| \langle z \rangle \|$.

Choose $w = Q \exp[N(x + y + xy) + N^2 xy]$ and $v = p^{n+1}$. Then, an easy computation shows that $w_{xy} \geq N \| \langle w \rangle \| + Q$ and

$$p_x^{n+1}(x, 0) \leq w_x(x, 0), p_y^{n+1}(0, y) \leq w_y(0, y), p^{n+1}(0, 0) \leq w(0, 0).$$

Consequently, an application of Theorem A.5.2 to (4.4.15) shows that

$$\langle p^{n+1} \rangle \leq \langle w \rangle \text{ on } I_{ab}.$$

This, in turn, yields

$$\max_{I_{ab}} \| \langle p^{n+1} \rangle \|_0 \leq C(\| \langle p^n \rangle \|_0^2 + \| \langle q^n \rangle \|_0^2)$$

where $C = \max[Me^{N(a+b+ab)+N^2 ab}(1, N + Nb + N^2 b, N + Na + N^2 a)]$, which proves the quadratic convergence of $\{\langle v^n \rangle\}$ to $\langle u \rangle$. A similar estimate is valid for $\| \langle q^{n+1} \rangle \|_0$ arguing in a similar fashion.

Theorem 4.4.1 covers many special cases which we list below.

Remarks 4.4.1

(i) If $g(\langle u \rangle) \equiv 0$ in (4.3.1), we get a result when $f(\langle u \rangle)$ is convex.

(ii) If $f(\langle u \rangle) \equiv 0$ in (4.3.1), one obtains a dual result when $g(\langle u \rangle)$ is concave.

(iii) Consider the case when $g(\langle u \rangle) \equiv 0$ and $f(\langle u \rangle)$ is not convex but $\tilde{f}(\langle u \rangle) = f(\langle u \rangle) + G(\langle u \rangle)$ is convex with $G(\langle u \rangle)$ convex. Then IBVP (4.3.1) can be rewritten as

$$u_{xy} = \tilde{f}(\langle u \rangle) + \tilde{g}(\langle u \rangle), u(x, 0) = \sigma(x), u(0, y) = \tau(y), u(0, 0) = u_0,$$

where $\tilde{g}(\langle u \rangle) = -G(\langle u \rangle)$. If $\tilde{f}, \tilde{g}$ are nondecreasing in $\langle u \rangle$, then we get the same conclusion even when $f(\langle u \rangle)$ is not convex.

(iv) A dual situation arises when $f(\langle u \rangle) = 0$ and $g(\langle u \rangle)$ is not concave but $\tilde{g}(\langle u \rangle) = g(\langle u \rangle) + F(\langle u \rangle)$ is concave with $F(\langle u \rangle)$ concave. Then IBVP (4.3.1) appears as

$$u_{xy} = \tilde{f}(\langle u \rangle) + \tilde{g}(\langle u \rangle), u(x, 0) = \sigma(x), u(0, y) = \tau(y), u(0, 0) = u_0,$$

where $\tilde{f}(\langle u \rangle) = -F(\langle u \rangle)$. If $\tilde{f}, \tilde{g}$, are nondecreasing in $\langle u \rangle$, then we get the same conclusion even when $g(\langle u \rangle)$ is not concave.

(v) Suppose that $f(\langle u \rangle)$ is not convex but $g(\langle u \rangle)$ is concave. As in (iii), assume $\tilde{f}(\langle u \rangle)$ is convex and let $\tilde{g}(\langle u \rangle) = g(\langle u \rangle) - F(\langle u \rangle)$ which is concave as well. If $\tilde{f}, \tilde{g}$ are nondecreasing in $\langle u \rangle$, then we consider the modified IBVP (4.3.1) with $\tilde{f}$ and $\tilde{g}$ and obtain the same conclusion from Theorem 4.4.1 for this case as well.

(vi) Suppose that $f(\langle u \rangle)$ is convex and $g(\langle u \rangle)$ is not concave. As in (iv), assume $\tilde{g}(\langle u \rangle)$ is concave and set $\tilde{f}(\langle u \rangle) = f(\langle u \rangle) - G(\langle u \rangle)$ which is also convex. If $\tilde{f}, \tilde{g}$ are nondecreasing in $\langle u \rangle$, Theorem 4.4.1 gives the same conclusion for this situation also.

(vii) Assume that $f(\langle u \rangle)$, $g(\langle u \rangle)$ do not satisfy the required assumptions in IBVP (4.3.1) but $\tilde{f}(\langle u \rangle) = f(\langle u \rangle) + G(\langle u \rangle) - F(\langle u \rangle)$ is convex and $\tilde{g}(\langle u \rangle) = g(\langle u \rangle) + F(\langle u \rangle) - G(\langle u \rangle)$ is concave as in (iii), (iv). If $\tilde{f}, \tilde{g}$ are also nondecreasing in $\langle u \rangle$, we get from Theorem 4.4.1, the same conclusion for this case also.

4.5   Notes and Comments

For the discussion of the variation of parameters formula and the derived comparison theorems for the linear nonhomogeneous hyperbolic problem, given in Section 4.2, see Pandit [73]. The general results in a unified setting for the monotone iterative technique described in Section 4.3 and the method of generalized quasilinearization investigated in Section 4.4 are new in the presented framework. Refer also for special cases to Blakley and Pandit [9], Deo and Pandit [23], Marshall and Pandit [67] and Ladde et al. [47]. For the results on the monotone iterative technique relative to nonlocal hyperbolic differential equations, see Byszewski and Lakshmikantham [12].

Part B

Chapter 5

Elliptic Equations

5.1 Introduction

We shall devote our attention, from this chapter on, to the study of partial differential equations relative to the monotone iterative technique and the method of generalized quasilinearization via variational techniques and consequently, we shall be concerned with generalized or weak solutions. As a result, we need to consider second-order partial differential operators in divergence form for elliptic, parabolic and hyperbolic IBVPs.

We begin, in Section 5.2, with elliptic BVPs of second-order partial differential equations in divergence form, prepare the framework, define the notion of weak solutions, and weak lower and upper solutions, and indicate the typical results that can be proven in the monotone iterative technique and generalized quasilinearization. We then recall an existence and uniqueness theorem for linear elliptic BVPs utilizing the Lax–Milgram lemma, in a suitable form that is helpful in our investigation. Finally, we prove a comparison theorem for semilinear BVPs via a variational technique and state a corollary that is most useful. Section 5.3 describes the monotone iterative technique in a unified framework and proves more general results that cover several special cases including those stated in Section 5.2. We then extend the monotone iterative technique to quasilinear elliptic BVPs with p-Laplacian, proving first an appropriate comparison theorem. Section 5.5 develops the monotone iterative technique for degenerate elliptic BVPs in the framework of weighted Sobolev spaces together with needed comparison results.

The method of generalized quasilinearization is investigated in Section 5.6 for semilinear elliptic BVPs employing the required results of Section 5.2.

Quasilinear elliptic BVPs are considered in Section 5.7 for proving generalized quasilinearization following the development of Section 5.4. In Section 5.8, we extend the method of generalized quasilinearization to degenerate elliptic BVPs on the basis of the results of Section 5.5.

5.2 Comparison Result

We shall consider the BVP

$$
\left[
\begin{array}{ll}
Lu = F(x, u) & \text{in } \Omega, \\
u = 0 & \text{on } \partial\Omega \text{ (in the sense of trace)};
\end{array}
\right.
\tag{5.2.1}
$$

where L denotes the second-order partial differential operator in the divergence form

$$
Lu = -\sum_{i,j}^{n} (a_{ij}(x)u_{x_i})_{x_j} + c(x)u
\tag{5.2.2}
$$

and Ω is an open, bounded subset of R^n. We assume that $a_{ij}, c \in L^\infty(\Omega)$, $i, j = 1, 2, \ldots, n$, $a_{ij} = a_{ji}$,

$$
\sum_{i,j=j1}^{n} a_{ij}(x)\xi_i\xi_j \geq \theta \mid \xi \mid^2 \quad \text{for } x \in \Omega, \text{ a.e. } \xi \in R^n
$$

with $\theta > 0$ (uniform elliptic condition), and $c(x) \geq 0$. We shall always mean that the boundary condition is in the sense of trace and hence we shall not repeat it to avoid monotony. Also, $F : \bar{\Omega} \times R \to R$, $F(x, u)$ is a Caratheodory function, that is $F(\cdot, u)$ is measurable for all $u \in R$ and $F(x, \cdot)$ is continuous a.e. $x \in \Omega$. The bilinear form $B[\ ,\ ]$ associated with the operator L is

$$
B[u, v] = \int_\Omega \left[\sum_{i,j=1}^{n} a_{ij}(x)u_{x_i}v_{x_j} + c(x)uv \right] dx,
\tag{5.2.3}
$$

for $u, v \in H_0^1(\Omega)$.

Definition 5.2.1 *The function $u \in H_0^1(\Omega)$ is said to be a weak solution of (5.2.1) if $F(x, u) \in L^1(\Omega)$, $F(x, u)u \in L^1(\Omega)$, and*

$$
B[u, v] = (F, v)
\tag{5.2.4}
$$

for all $v \in H_0^1(\Omega)$ where $(\ ,\ )$ denotes the inner product in $L_2(\Omega)$.

Definition 5.2.2 *The function $\alpha_0 \in H^1(\Omega)$ is said to be a weak lower solution of (5.2.1) if $\alpha_0 \leq 0$ on $\partial\Omega$ and*

$$\int_\Omega \left[\sum_{i,j=1}^n a_{ij}(x)(\alpha_0)_{x_i} v_{x_j} + c(x)\alpha_0 v \right] dx \qquad (5.2.5)$$

$$\leq \int_\Omega F(x, \alpha_0) v \, dx,$$

for each $v \in H_0^1(\Omega)$, $v \geq 0$. If the inequalities are reversed, then α_0 is said to be a weak upper solution of (5.2.1).

In this set-up, one can prove the following result on the monotone iterative technique.

Theorem 5.2.1 *Assume that*

(i) $\alpha_0, \beta_0 \in H^1(\Omega)$ *are weak lower and upper solutions of (5.2.1) such that $\alpha_0 \leq \beta_0$ in Ω a.e.;*

(ii) $F : \bar{\Omega} \times R \to R$ *is a Caratheodory function satisfying*

$$F(x, u) - F(x, v) \geq -M(x)(u - v), u \geq v, \ a.e. \ in \ x \in \Omega,$$

where $M(x) \geq 0$, $M \in L^\infty(\Omega)$;

(iii) $0 < N \leq c(x) + M(x)$ *a.e., $x \in \Omega$ and for any $\eta \in H_0^1(\Omega)$,*

$$F(x, \eta) + M(x)\eta \in L^2(\Omega) \ where \ \alpha_0 \leq \eta \leq \beta_0 \ in \ \Omega, \ a.e.$$

Then there exist monotone sequences $\{\alpha_n(x)\}$, $\{\beta_n(x)\} \in H_0^1(\Omega)$ such that $\alpha_n \to \rho$, $\beta_n \to r$ weakly in $H_0^1(\Omega)$ with $\rho, r \in H_0^1(\Omega)$, (ρ, r) are minimal and maximal weak solutions of (5.2.1) respectively satisfying $\alpha_0 \leq \rho \leq r \leq \beta_0$ in Ω, a.e.

The special case when $F(x, u)$ is monotone nondecreasing in u is included in Theorem 5.2.1 with $M(x) \equiv 0$. We shall prove in the next section, a more general result, when $F(x, u)$ admits a splitting of a difference of two monotone functions, which includes several cases of interest.

An extension of the method of generalized quasilinearization is also valid in the present framework. We shall merely state a theorem corresponding to Theorem 1.2.4 leaving the proof of the more general case to a later section.

Theorem 5.2.2 *Assume that*

  (i) *condition* (i) *of Theorem* 5.2.1 *holds; and suppose further that*

 (ii) $F : \bar{\Omega} \times R \to R$ *is a Caratheodory function and* $F_u(x, u)$, $F_{uu}(x, u)$ *exist and are Caratheodory functions such that* $F_{uu}(x, u) \geq 0$ *a.e.,* $x \in \Omega$;

(iii) $0 < N \leq c(x) - F_u(x, \beta_0)$ *a.e.,* $x \in \Omega$ *and for any* $\eta \in H_0^1(\Omega)$ *with* $\alpha_0 \leq \eta \leq \beta_0$, $F(x, \eta) - F_u(x, \eta)\eta \in L^2(\Omega)$.

Then there exist monotone sequences $\{\alpha_n(x)\}$, $\{\beta_n(x)\} \in H_0^1(\Omega)$ *such that* $\alpha_n \to \rho$, $\beta_n \to r$ *weakly in* $H_0^1(\Omega)$ *with* $\rho, r \in H_0^1(\Omega)$, $\rho(x) = r(x)$ $= u(x)$ *is the unique weak solution of* (5.2.1) *satisfying* $\alpha_0 \leq u \leq \beta_0$ *in* Ω, *a.e. and the convergence is quadratic.*

A dual result when $F(x, u)$ is concave in u is also true. We shall show later that when $F(x, u)$ admits a splitting of a difference of two convex or concave functions, we can construct monotone sequences that converge to the unique weak solution and the convergence is quadratic. This would include the foregoing results as well as several interesting cases.

In order to discuss the results on the monotone iterative technique and generalized quasilinearization in a unified setting, we need to consider the existence and uniqueness of weak solutions of linear boundary value problems. For this purpose, let H be a real Hilbert space, with norm $\| \cdot \|$ and inner product (,). We let $\langle \ , \ \rangle$ denote the pairing of H with its dual space. The result on the existence of weak solutions for the linear BVP can be obtained from the Lax–Milgram theorem which is a generalization of the Riez representation theorem.

Theorem 5.2.3 *Assume that* $B : H \times H \to R$ *is a bilinear mapping, for which there exist constants* $\alpha, \beta > 0$ *such that*

  (i) $|B[u, v]| \leq \alpha \parallel u \parallel \parallel v \parallel$, $u, v \in H$, *and*

 (ii) $\beta \parallel u \parallel^2 \leq B[u, u]$, $u \in H$.

Assume also that $f : H \to R$ *is a bounded linear functional on* H. *Then there exists a unique element* $u \in H$ *such that*

$$B[u, v] = \langle f, v \rangle$$

for all $v \in H$.

We can prove the following.

Theorem 5.2.4 *Consider the linear BVP*

$$\begin{bmatrix} Lu = h(x) & in\ \Omega, \\ u = 0 & on\ \Omega\ (in\ the\ sense\ of\ trace). \end{bmatrix} \tag{5.2.6}$$

Then there exists a unique solution $u \in H_0^1(\Omega)$ for the linear BVP (5.2.6) provided $0 < c^* \le c(x)$ a.e. in Ω and $h \in L^2(\Omega)$.

Proof   Let us verify the hypothesis of Theorem 5.2.3 for the specific bilinear form $B[u, v]$ defined by (5.2.3). We readily check

$$|B[u,v]| \le \sum_{i,j=1}^{n} \|a_{ij}\|_{L^\infty} \int_\Omega |Du|\,|Dv|\,dx + \|c\|_{L^\infty} \int_\Omega |u|\,|v|\,dx$$

$$\le \alpha \|u\|_{H_0^1(\Omega)} \|v\|_{H_0^1(\Omega)},$$

for some suitable constant α. Moreover, in view of the ellipticity condition and the assumption of the theorem on $c(x)$, we get

$$\theta \int_\Omega |Du|^2\,dx + c^* \int_\Omega |u|^2\,dx \le B[u,u],$$

which implies that

$$\beta \|u\|_{H_0^1(\Omega)}^2 \le B[u,u],$$

for a suitable constant β. Now fix $h \in L^2(\Omega)$ and set $\langle h, v \rangle = (h,v)_{L^2(\Omega)}$, which is a bounded linear functional on $L^2(\Omega)$ and thus on $H_0^1(\Omega)$. Consequently, Theorem 5.2.3 gives a unique function $u \in H_0^1(\Omega)$ which is the desired solution of (5.2.6).

Let us next prove the basic comparison result in the variational set-up.

Theorem 5.2.5 *Let α_0, β_0 be weak lower and upper solutions of (5.2.1). Suppose further that F satisfies*

$$F(x, u_1) - F(x, u_2) \le K(u_1 - u_2) \tag{5.2.7}$$

whenever $u_1 \ge u_2$ a.e., $x \in \Omega$ and $K > 0$. Then, if $0 < c - K \in L^1(\Omega)$, we have

$$\alpha_0(x) \le \beta_0(x)\ in\ \Omega,\ a.e. \tag{5.2.8}$$

Proof   From the definition of lower and upper solutions, we get

$$\int_\Omega \left[\sum_{i,j=1}^{n} a_{ij}(x)(\alpha_{0,x_i} - \beta_{0,x_i})v_{x_j} + c(x)(\alpha_0 - \beta_0)v \right] dx \tag{5.2.9}$$

$$\leq \int_\Omega [f(x,\alpha_0) - f(x,\beta_0)]vdx,$$

for each $v \in H_0^1(\Omega)$, $v \geq 0$, a.e. choose $v = (\alpha_0 - \beta_0)^+ \in H_0^1(\Omega)$, $v \geq 0$, a.e. Then we have

$$\int_\Omega \left[\sum_{i,j=1}^n a_{ij}(x)(\alpha_{0,x_i} - \beta_{0,x_i})(\alpha_0 - \beta_0)_{x_j}^+ + c(x)(\alpha_0 - \beta_0)(\alpha_0 - \beta_0)^+ \right] dx$$

$$\leq \int_\Omega [f(x,\alpha_0) - f(x,\beta_0)](\alpha_0 - \beta_0)^+ dx.$$

Since

$$(\alpha_0 - \beta_0)_{x_j}^+ = \begin{bmatrix} (\alpha_{0,x_j} - \beta_{0,x_j}) & \text{a.e. on } \{\alpha_0 > \beta_0\} \\ 0 & \text{a.e. on } \{\alpha_0 \leq \beta_0\}, \end{bmatrix}$$

we obtain using the ellipticity condition and (5.2.7)

$$\int_{\alpha_0 > \beta_0} [\theta \mid (\alpha_{0,x_i} - \beta_{0,x_i}) \mid^2 + c(x) \mid \alpha_0 - \beta_0 \mid^2]dx \leq K \int_{\alpha_0 > \beta_0} \mid \alpha_0 - \beta_0 \mid^2 dx,$$

which implies $\alpha_0(x) \leq \beta_0(x)$ in Ω a.e. This completes the proof.

The following corollary is useful in our discussion.

Corollary 5.2.1 *For $p \in H^1(\Omega)$ satisfying*

$$\int_\Omega \left[\sum_{i,j=1}^n a_{ij}(x)p_{x_i}v_{x_j} + c_0(x)pv \right] dx \leq 0,$$

for each $v \in H_0^1(\Omega)$, $v \geq 0$, a.e. and $p \leq 0$ on $\partial\Omega$ we have $p(x) \leq 0$ in Ω, a.e., provided $c_0(x) > 0$.

5.3 Monotone Iterative Technique (MIT): Semilinear Problems

In this section, we shall describe the monotone iterative technique in a unified way proving more general theorems which contain, as special cases, several important results of interest.

Let us first consider the following semilinear elliptic boundary value problem in divergence form

$$\begin{bmatrix} Lu = f(x,u) + g(x,u) & \text{in } \Omega, \\ u = 0 & \text{on } \partial\Omega, \text{ (in the sense of trace)}, \end{bmatrix} \tag{5.3.1}$$

where L denotes the same second-order partial differential operator in the divergence form (5.2.2) with the bilinear form $B[u, v]$ described in (5.2.3). Here $f, g : \bar{\Omega} \times R \to R$ are Caratheodory functions, other assumptions being as before.

In order to develop the monotone iterative technique for the BVP (5.3.1), we need to utilize appropriate coupled lower and upper solutions of (5.3.1). We shall continue, in the following sections, to extend the monotone iterative technique to quasilinear elliptic BVPs and semilinear degenerate elliptic BVPs employing suitable coupled lower and upper solutions corresponding to each such problem.

In view of the discussion in Section 1.3, with respect to the coupled lower and upper solutions of various types and the conclusion, it is sufficient to define, in the present set-up, only the following.

Definition 5.3.1 *Relative to the BVP (5.3.1), the functions α_0, $\beta_0 \in H^1(\Omega)$ are said to be*

(i) *weakly coupled lower and upper solutions of type I if*

$$B[\alpha_0, v] \leq (f(x, \alpha_0) + g(x, \beta_0), v),$$

$$B[\beta_0, v] \geq (f(x, \beta_0) + g(x, \alpha_0), v),$$

for each $v \in H_0^1(\Omega)$, $v \geq 0$ a.e. in Ω;

(ii) *weakly coupled lower and upper solutions of type II if*

$$B[\alpha_0, v] \leq (f(x, \beta_0) + g(x, \alpha_0), v),$$

$$B[\beta_0, v] \geq (f(x, \alpha_0) + g(x, \beta_0), v),$$

for each $v \in H_0^1(\Omega)$, $v \geq 0$, a.e. in Ω.

We are now in a position to prove the first result parallel to Theorem 1.3.2.

Theorem 5.3.1 *Assume that*

(A_1) *$\alpha_0, \beta_0 \in H^1(\Omega)$ are the weak coupled lower and upper solutions of type I with $\alpha_0(x) \leq \beta_0(x)$ a.e. in Ω;*

(A_2) *$f, g : \bar{\Omega} \times R \to R$ are Caratheodory functions such that $f(x, u)$ is nondecreasing in u, $g(x, u)$ is nonincreasing in u for $x \in \Omega$, a.e.;*

(A_3) $c(x) \geq N > 0$ in Ω, a.e. and for any $\eta, \mu \in H^1(\Omega)$ with $\alpha_0 \leq \eta$, $\mu \leq \beta_0$, the function $h(x) = f(x, \eta) + g(x, \mu) \in L^2(\Omega)$.

Then there exist monotone sequences $\{\alpha_n(x)\}$, $\{\beta_n(x)\} \in H_0^1(\Omega)$ such that $\alpha_n \to \rho$, $\beta_n \to r$ weakly in $H_0^1(\Omega)$ as $n \to \infty$ and (ρ, r) are weak coupled minimal and maximal solutions of (5.3.1) respectively, that is,

$$L\rho = f(x, \rho) + g(x, r) \text{ in } \Omega, \ \rho = 0 \text{ on } \partial\Omega,$$

$$Lr = f(x, r) + g(x, \rho) \text{ in } \Omega, \ r = 0 \text{ on } \partial\Omega.$$

Proof   Consider the linear BVPs

$$L\alpha_{n+1} = f(x, \alpha_n) + g(x, \beta_n) \text{ in } \ \Omega, \alpha_{n+1} = 0 \text{ on } \partial\Omega, \qquad (5.3.2)$$

and

$$L\beta_{n+1} = f(x, \beta_n) + g(x, \alpha_n) \text{ in } \ \Omega, \ \beta_{n+1} = 0 \text{ on } \partial\Omega. \qquad (5.3.3)$$

The variational forms associated with (5.3.2) and (5.3.3) are

$$B[\alpha_{n+1}, v] = \int_\Omega [f(x, \alpha_n) + g(x, \beta_n)]v dx, \qquad (5.3.4)$$

$$B[\beta_{n+1}, v] = \int_\Omega [f(x, \beta_n) + g(x, \alpha_n)]v dx, \qquad (5.3.5)$$

for all $v \in H_0^1(\Omega)$, $v \geq 0$ a.e. in Ω.

We shall show that the weak solutions α_n, β_n of (5.3.2) and (5.3.3) are uniquely defined and satisfy

$$\alpha_0 \leq \alpha_1 \leq \alpha_2 \leq \ldots \leq \alpha_n \leq \beta_n \leq \ldots \leq \beta_2 \leq \beta_1 \leq \beta_0 \text{ a.e. in } \Omega. \quad (5.3.6)$$

For each $n \geq 1$, if we have $\alpha_0 \leq \alpha_n \leq \beta_n \leq \beta_0$, then by assumption (A_3), $h_1(x) \equiv f(x, \alpha_n) + g(x, \beta_n) \in L_2(\Omega)$ and $h_2(x) \equiv f(x, \beta_n) + g(x, \alpha_n) \in L_2(\Omega)$. Therefore, Theorem 5.2.4 implies that BVPs (5.3.2) and (5.3.3) have unique solutions α_n and β_n in view of $c(x) \geq N > 0$. Hence, we need to show (5.3.6) holds.

For this, we first claim that $\alpha_1 \geq \alpha_0$ a.e. in Ω. Now, let $p = \alpha_0 - \alpha_1$ so that $p \leq 0$ on $\partial\Omega$ and for $v \in H_0^1(\Omega)$, $v \geq 0$ a.e. in Ω,

$$B[p, v] = \int_\Omega \left[\sum_{i,j=1}^n a_{ij}(x)p_{x_i}v_{x_j} + c(x)pv \right] dx$$

$$\leq \int_\Omega [f(x, \alpha_0) + g(x, \beta_0)]v dx - \int_\Omega [f(x, \alpha_0) + g(x, \beta_0)]v dx = 0.$$

Hence by Corollary 5.2.1 $p(x) \leq 0$, that is, $\alpha_0 \leq \alpha_1$ a.e. in Ω. Similarly, we can show that $\beta_1 \leq \beta_0$ a.e. in Ω.

Assume that, for some fixed $n > 1$, $\alpha_n \leq \alpha_{n+1}$ and $\beta_n \geq \beta_{n+1}$ a.e. in Ω. Now consider $p = \alpha_{n+1} - \alpha_{n+2}$. Note that $p = 0$ on $\partial\Omega$, and using the monotone nature of f, g, we get

$$B[p, v] = \int_\Omega [f(x, \alpha_n) - f(x, \alpha_{n+1}) + g(x, \beta_n) - g(x, \beta_{n+1})]v dx \leq 0.$$

This implies by Corollary 5.2.1 that $\alpha_{n+1} \leq \alpha_{n+2}$ a.e. in Ω.

Also, by setting $p = \beta_{n+2} - \beta_{n+1}$, one can easily show in the same way, that $\beta_{n+2} \leq \beta_{n+1}$ a.e. in Ω. Hence, using an induction argument, we have $\alpha_n \geq \alpha_{n+1}$ and $\beta_n \geq \beta_{n+1}$ a.e. in Ω for all $n \geq 1$.

We now show that $\alpha_1 \leq \beta_1$ a.e. in Ω. Consider $p = \alpha_1 - \beta_1$ and note that $p = 0$ on $\partial\Omega$, and

$$B[p, v] = \int_\Omega [f(x, \alpha_0) + g(x, \beta_0) - f(x, \beta_0) - g(x, \alpha_0)]v dx \leq 0,$$

because of the fact that f, g are monotone by assumption.

Hence Corollary 5.2.1 yields $\alpha_1 \leq \beta_1$ a.e. in Ω. Employing similar arguments, it is easy to show that if we assume, for some fixed $n > 1$, we have $\alpha_n \leq \beta_n$ a.e. in Ω, then it follows that $\alpha_{n+1} \leq \beta_{n+1}$ a.e. in Ω. Consequently, using the induction argument, it is clear that (5.3.6) holds for all $n \geq 1$.

By the monotone character of $\{\alpha_n\}$, $\{\beta_n\}$ there exist pointwise limits

$$\lim_{n \to \infty} \alpha_n(x) = \rho(x) \text{ a.e. in } \Omega \text{ and } \lim_{n \to \infty} \beta_n(x) = r(x) \text{ a.e. in } \Omega.$$

Moreover, since $\alpha_0 \leq \alpha_n \leq \beta_n \leq \beta_0$ a.e. in Ω, it follows by Lebesgue's dominated convergence theorem that

$$\alpha_n \to \rho \text{ and } \beta_n \to r \text{ in } L^2(\Omega).$$

For each $n \geq 1$, we note that α_n satisfies for each $v \in H_0^1(\Omega)$, $v \geq 0$, a.e. in Ω

$$\int_\Omega \left[\sum_{i,j=1}^n a_{ij}(x)(\alpha_n)_{x_i} v_{x_j} + c(x)\alpha_n v \right] dx = \int_\Omega h_{n-1}(x)v dx,$$

where $h_{n-1}(x) = f(x, \alpha_{n-1} + g(x, \beta_{n-1})$.

We now use the ellipticity condition and the fact that $c(x) \geq N > 0$ with $v = \alpha_n$ to get

$$\int_\Omega [\theta \mid (\alpha_n)_x \mid^2 + N \mid \alpha_n \mid^2] dx \leq \int_\Omega h_{n-1}(x) v dx.$$

Since the integrand on the right-hand side belongs to $L^2(\Omega)$, we obtain the estimate

$$\sup_n \| \alpha_n \|_{H_0^1(\Omega)} < \infty.$$

Hence there exists a subsequence $\{\alpha_{n_k}\}$ which converges weakly to $\rho(x)$ in $H_0^1(\Omega)$.

A similar argument implies that $\sup_n \| \beta_n \|_{H_0^1(\Omega)} < \infty$. Hence there exist subsequences $\{\alpha_{n_k}\}$, $\{\beta_{n_k}\}$ which converge weakly in $H_0^1(\Omega)$ to (ρ, r) $\in H_0^1(\Omega)$ respectively.

Since α_n and β_n satisfy (5.3.4) and (5.3.5) for a fixed $v \in H_0^1(\Omega)$, $v \geq 0$ a.e. in Ω by taking the limit as $n \to \infty$, we see that

$$B[\rho, v] = \int_\Omega [f(x, \rho) + g(x, r)] v dx,$$

and

$$B[r, v] = \int_\Omega [f(x, r) + g(x, \rho)] v dx.$$

Finally, we claim that ρ and r are the weak coupled minimal and maximal solutions of (5.3.1) that is, if u is any weak solution of (5.3.1) such that $\alpha_0(x) \leq u(x) \leq \beta_0(x)$ a.e. in Ω, then

$$\alpha_0(x) \leq \rho(x) \leq u(x) \leq r(x) \leq \beta_0 \text{ a.e. in } \Omega. \tag{5.3.7}$$

Suppose that for some fixed $n \geq 1$, $\alpha_n \leq u \leq \beta_n$ a.e. in Ω. Setting $p = \alpha_{n+1} - u$, we have $p = 0$ on $\partial\Omega$ and employing the monotone character of f and g, it follows that

$$B[p, v] = \int_\Omega [f(x, \alpha_n) + g(x, \beta_n) - f(x, u) - g(x, u)] v dx \leq 0,$$

which implies by Corollary 5.2.1 $\alpha_{n+1} \leq u$ a.e. in Ω. In a similar way, we obtain $u \leq \beta_{n+1}$ a.e. in Ω so that $\alpha_{n+1} \leq u \leq \beta_{n+1}$ a.e. in Ω. By induction, $\alpha_n \leq u \leq \beta_n$ a.e. in Ω for all $n \geq 1$. Now taking the limit as $n \to \infty$, we get (5.3.7), completing the proof.

To show uniqueness, we have the following corollary.

Corollary 5.3.1 *Assume, in addition to the conditions of Theorem 5.3.1, that f and g satisfy*

$$f(x, u_1) - f(x, u_2) \leq N_1(u_1 - u_2)$$

and

$$g(x, u_1) - g(x, u_2) \geq -N_2(u_1 - u_2)$$

where $u_1 \geq u_2$, $N_1 > 0$, $N_2 > 0$ and $c(x) - (N_1 + N_2) > 0$ a.e. in Ω. Then $\rho = u = r$, is the unique weak solution of (5.3.1).

Proof   Since we have $\rho \leq r$, it is enough to show that $r \leq \rho$. Setting $p = r - \rho$, we have $p = 0$ on $\partial\Omega$ and using the assumptions of f, g, it follows that

$$
\begin{aligned}
B[p, v] &= \int_\Omega [f(x, r) + g(x, \rho) - f(x, \rho) - g(x, r)]v dx \\
&\leq \int_\Omega [N_1(r - \rho) + N_2(r - \rho)]v dx = ((N_1 + N_2)p, v).
\end{aligned}
$$

This implies by Corollary 5.2.1, that $p \leq 0$ a.e. in Ω. Hence $u = \rho = r$ is the weak solution of (5.3.1).

Now consider the linear iteration scheme

$$L\alpha_{n+1} = f(x, \beta_n) + g(x, \alpha_n) \text{ in } \Omega, \quad \alpha_{n+1} = 0 \text{ on } \partial\Omega \qquad (5.3.8)$$

$$L\beta_{n+1} = f(x, \alpha_n) + g(x, \beta_n) \text{ in } \Omega, \quad \beta_{n+1} = 0 \text{ on } \partial\Omega \qquad (5.3.9)$$

and the associated variational forms

$$B[\alpha_{n+1}, v] = \int_\Omega [f(x, \beta_n) + g(x, \alpha_n)]v dx \qquad (5.3.10)$$

$$B[\beta_{n+1}, v] = \int_\Omega [f(x, \alpha_n) + g(x, \beta_n)]v dx \qquad (5.3.11)$$

for all $v \in H_0^1(\Omega)$, $v \geq 0$ a.e. in Ω. Then, for the weak solutions of (5.3.1), we have the following theorem which is analogous to Theorem 1.3.3.

Theorem 5.3.2 *Assume that the conditions (A_2) and (A_3) of Theorem 5.3.1 hold. Then for any weak solution $u(x)$ of (5.3.1) such that $\alpha_0(x) \leq u(x) \leq \beta_0(x)$ a.e. in Ω, we have the iterates $\{\alpha_n(x)\}, \{\beta_n(x)\}$ satisfying*

$$\alpha_0 \leq \alpha_2 \leq \ldots \leq \alpha_{2n} \leq u \leq \alpha_{2n+1} \leq \ldots \leq \alpha_3 \leq \alpha_1, \text{ a.e. in } \Omega, \quad (5.3.12)$$

$$\beta_1 \le \beta_3 \le \ldots \le \beta_{2n+1} \le u \le \beta_{2n} \le \ldots \le \beta_2 \le \beta_0, \text{ a.e. on } \Omega, \qquad (5.3.13)$$

provided $\alpha_0 \le \alpha_2$ and $\beta_2 \le \beta_0$ a.e. in Ω, where the iterative schemes are given by (5.3.8) and (5.3.9) respectively. Moreover, the monotone sequences $\{\alpha_{2n}\}$, $\{\alpha_{2n+1}\}$, $\{\beta_{2n}\}$, $\{\beta_{2n+1}\} \in H_0^1(\Omega)$ converge weakly in $H_0^1(\Omega)$ to ρ, r, ρ^*, r^* respectively and they satisfy the relations

$$Lr = f(x, \rho^*) + g(x, \rho) \text{ in } \Omega, r = 0 \text{ on } \partial\Omega,$$

$$L\rho = f(x, r^*) + g(x, r) \text{ in } \Omega, \rho = 0 \text{ on } \partial\Omega,$$

$$Lr^* = f(x, \rho) + g(x, \rho^*) \text{ in } \Omega, r^* = 0 \text{ on } \partial\Omega,$$

$$L\rho^* = f(x, r) + g(x, r^*) \text{ in } \Omega, \rho^* = 0 \text{ on } \partial\Omega.$$

Also $\rho \le u \le r$ and $r^* \le u \le \rho^*$, a.e. in Ω.

Proof   Using an argument similar to that employed in Theorem 1.3.3 with suitable modifications, it is easy to show that there exist α_0, $\beta_0 \in H_0^1(\Omega)$ with $\alpha_0 \le \beta_0$ a.e. in Ω satisfying the weak coupled lower and upper solutions of (5.3.1) of type II. Also, as in Theorem 5.3.1 using the condition (A_3), we have the existence of unique weak solutions $\alpha_n(x)$, $\beta_n(x)$ of (5.3.8), (5.3.9) respectively. We shall therefore proceed to establish the inequalities (5.3.12) and (5.3.13).

In view of the assumptions $\alpha_0 \le \alpha_2$ and $\beta_2 \le \beta_0$ a.e. in Ω, we shall first show that

$$\alpha_0 \le \alpha_2 \le u \le \alpha_3 \le \alpha_1 \text{ and } \beta_1 \le \beta_3 \le u \le \beta_2 \le \beta_0 \text{ a.e. in } \Omega. \qquad (5.3.14)$$

Setting $p = u - \alpha_1$, we get, because of the monotone character of f and g,

$$B[p, v] = \int_\Omega [f(x, u) + g(x, u) - f(x, \beta_0) - g(x, \alpha_0)] v dx \le 0,$$

for each $v \in H_0^1(\Omega)$, $v \ge 0$ a.e. in Ω. Since $p = 0$ on $\partial\Omega$, this implies, by Corollary 5.2.1, that $u \le \alpha_1$ a.e. in Ω. Hence we have $\alpha_0 \le u \le \alpha_1$ a.e. in $\bar{\Omega}$. Similarly, one can obtain $\beta_1 \le u \le \beta_0$ a.e. in Ω. Now we set $p = \alpha_3 - \alpha_1$ and get, by using similar reasoning,

$$B[p, v] = \int_\Omega [f(x, \beta_2) + g(x, \alpha_2) - f(x, \beta_0) - g(x, \alpha_0)] v dx \le 0,$$

and $p = 0$ on $\partial\Omega$ so that by Corollary 5.2.1, it follows that $\alpha_3 \le \alpha_1$ a.e. in Ω. Also setting $p = \beta_2 - u$ and $p = u - \alpha_2$ yields $\beta_2 \ge u$ and $u \ge \alpha_2$ a.e. in

Ω. In addition, if $p = u - \alpha_3$ and $p = \beta_3 - u$, we obtain $\alpha_3 \geq u$ and $\beta_3 \leq u$ a.e. in Ω.

Thus, the inequalities in (5.3.14) hold a.e. in Ω. Now, for some fixed $n > 1$, assuming

$$\alpha_{2n-4} \leq \alpha_{2n-2} \leq u \leq \alpha_{2n-1} \leq \alpha_{2n-3},$$

and

$$\beta_{2n-3} \leq \beta_{2n-1} \leq u \leq \beta_{2n-2} \leq \beta_{2n-4}, \text{ a.e. in } \Omega,$$

hold a.e. in Ω if we let p be $\alpha_{2n-2} - \alpha_{2n}$, $\alpha_{2n} - u$, $\beta_{2n-1} - \beta_{2n+1}$, $\beta_{2n+1} - u$, $u - \alpha_{2n+1}$, $\alpha_{2n+1} - \alpha_{2n-1}$, $u - \beta_{2n}$, and $\beta_{2n} - \beta_{2n-1}$, and use the monotone nature of f and g, and Corollary 5.2.1, respectively, we obtain

$$\alpha_{2n-1} \leq \alpha_{2n} \leq u \leq u_{2n+1} \leq \alpha_{2n-1},$$

and

$$\beta_{2n-1} \leq \beta_{2n+1} \leq u \leq \beta_{2n} \leq \beta_{2n-2} \text{ a.e. in } \Omega.$$

Hence, by induction, we obtain the inequalities (5.3.12) and (5.3.13) a.e. in Ω.

We note that α_{2n}, $\alpha_{2n+1} \in H_0^1(\Omega)$ and β_{2n}, $\beta_{2n+1} \in H_0^1(\Omega)$, and hence arguing as in the proof of Theorem 5.3.1 with appropriate modifications, we see that

$$\alpha_{2n} \to \rho \text{ weakly in } H_0^1(\Omega),$$

$$\alpha_{2n+1} \to r \text{ weakly in } H_0^1(\Omega),$$

$$\beta_{2n+1} \to r^* \text{ weakly in } H_0^1(\Omega),$$

$$\beta_{2n} \to \rho^* \text{ weakly in } H_0^1(\Omega).$$

Now passing to the limit in the variational forms (5.3.10) and (5.3.11), we see that (ρ, r) and (r^*, ρ^*) satisfy

$$B[\rho, r] = \int_\Omega [f(r^*) + g(r)]v dx,$$

$$B[r, v] = \int_\Omega [f(\rho^*) + g(\rho)]v dx,$$

$$B[\rho^*, v] = \int_\Omega [f(r) + g(r^*)]v dx,$$

$$B[r^*, v] = \int_\Omega [f(\rho) + g(\rho^*)]v dx,$$

for $v \in H_0^1(\Omega)$, $v \geq 0$ a.e. in Ω.

Also, from (5.3.12) and (5.3.13), we get $\rho \leq u \leq r$ and $r^* \leq u \leq \rho^*$ a.e. in Ω. This completes the proof.

Corollary 5.3.2 *In addition to Theorem 5.3.2 assume that f and g are as in Corollary 5.3.1. Then, $u = \rho = r = \rho^* = r^*$ a.e. in Ω is the unique solution of (5.3.1).*

Proof   Let $p_1 = r - \rho$ and $p_2 = \rho^* - r^*$. By Theorem 5.3.1, $p_1 \geq 0$ and $p_2 \geq 0$ a.e. in Ω. Then, utilizing the conditions and the monotone character of f, g it follows that

$$B[p_1, v] \leq \int_\Omega [N_1(\rho^* - r^*) + N_2(r - \rho)]v dx$$

$$B[p_2, v] \leq \int_\Omega [N_1(r - \rho) + N_2(\rho^* - r^*)]v dx,$$

and

$$B[p_1 + p_2, v] \leq \int_\Omega [(N_1 + N_2)(p_1 + p_2)]v dx,$$

for $v \in H_0^1(\Omega)$, $v \geq 0$ a.e. in Ω.

Consequently, we have

$$\int_\Omega \left[\sum_{i,j=1}^n a_{ij}(x)(p_1 + p_2)_{x_i} v_{x_j} + (c(x) - (N_1 + N_2))(p_1 + p_2)v \right] dx \leq 0$$

which yields, by Corollary 5.2.1, $(p_1 + p_2) \leq 0$ a.e. in Ω. This implies $p_1 \leq 0$ and $p_2 \leq 0$ a.e. in Ω, that is, $\rho = r$, $\rho^* = r^*$ a.e. in Ω. This, in turn, implies $u = \rho = r = \rho^* = r^*$ a.e. in Ω is the unique solution of (5.3.1). This completes the proof.

5.4   Monotone Iterative Technique (MIT): Quasi-linear Problems

In this section we shall extend the monotone iterative technique to quasi-linear elliptic boundary value problems. As before, we shall investigate the situation where the nonlinear function admits a splitting of a difference of two monotone functions so that the results provide a unified setting that covers several interesting special cases. We begin with necessary preliminaries and the comparison results.

Let $\Omega \subset R^n$ be a smooth bounded domain and $1 < p < n$. We let $W^{1,p}(\Omega)$, $W_0^{1,p}(\Omega)$ be Sobolev spaces. Let Δ_p denote the p-Laplacian, namely, $\Delta_p u = div(| Du |^{p-2} Du)$, where Du denotes the gradient of u.

Suppose that $f, g : \Omega \times R \to R$ are Caratheodory functions. Assume that $c : \Omega \to R$ is a measurable function with $c(x) > 0$ a.e. $x \in \Omega$ and $c \in L^1(\Omega)$. We consider the quasilinear boundary value problem (BVP)

$$\left[\begin{array}{ll} -\Delta_p u + c(x)u = f(x, u) + g(x, u) \text{ in } \Omega \\ u = 0 \text{ on } \partial\Omega \quad \text{in the sense of trace.} \end{array} \right. \tag{5.4.1}$$

We shall always mean that the boundary condition is in the sense of trace and therefore we shall not repeat it to avoid monotony.

Definition 5.4.1 *The function* $u \in W_0^{1,p}(\Omega)$ *is said to be a weak solution of (5.4.1) if* $f(x, u) + g(x, u) \in L^1(\Omega)$, $f(x, u_1)u_1 \in L^1(\Omega)$ *and* $g(x, u_2)u_2 \in L^1(\Omega)$ *and for any* $v \in W_0^{1,p}(\Omega) \cap L^\infty(\Omega)$, *we have*

$$\int_\Omega | Du |^{p-2} Du Dv + c(x)uv = \int_\Omega \{f(x, u) + g(x, u)\}v.$$

Definition 5.4.2 *The functions* $u_1, u_2 \in W_0^{1,p}(\Omega)$ *are said to be coupled weak solutions of (5.4.1) if* $f(x, u_1) + g(x, u_2) \in L^1(\Omega)$, $[f(x, u_1) + g(x, u_2)]u_1 \in L^1(\Omega)$ *and for any* $v \in W_0^{1,p}(\Omega) \cap L^\infty(\Omega)$, *it follows that*

$$\int_\Omega | Du_1 |^{p-2} Du_1 Dv + c(x)u_1 v = \int_\Omega [f(x, u_1) + g(x, u_2)]v,$$

and

$$\int_\Omega | Du_2 |^{p-2} Du_2 Dv + c(x)u_2 v = \int_\Omega [f(x, u_2) + g(x, u_1)]v.$$

Definition 5.4.3 *The functions* $\alpha_0, \beta_0 \in W^{1,p}(\Omega) \cap L^\infty(\Omega)$ *are said to be coupled weak lower and upper solutions of (5.4.1) if* $\alpha_0(x) \le 0 \le \beta_0(x)$ *on* $\partial\Omega$ *and*

$$\int_\Omega | D\alpha_0 |^{p-2} D\alpha_0 Dv + c(x)\alpha_0 v \le \int_\Omega [f(x, \alpha_0) + g(x, \beta_0)]v,$$

$$\int_\Omega | D\beta_0 |^{p-2} D\beta_0 Dv + c(x)\beta_0 v \le \int_\Omega [f(x, \beta_0) + g(x, \alpha_0)]v,$$

are satisfied for each $v \in W^{1,p}(\Omega)$ *with* $v \ge 0$ *a.e. in* Ω.

We need the following inequality which holds for $p > 1$ and $u, v \in W_0^{1,p}(\Omega)$, namely,

$$\int_\Omega [| Du |^{p-2} Du - | Dv |^{p-2} Dv](Du - Dv) \tag{5.4.2}$$

$$\geq (\parallel u \parallel^{p-1} - \parallel v \parallel^{p-1})(\parallel u \parallel - \parallel v \parallel) \geq 0,$$

where $\parallel u \parallel = (\int_\Omega \mid Du \mid^p)^{1/p}$ is the equivalent norm for $W_0^{1,p}(\Omega)$ which is linked in a natural way to the p-Laplacian, in view of Friedrich's inequality. If $p \geq 2$, we have the following stronger inequality,

$$\int_\Omega [\mid Du \mid^{p-2} Du - \mid Dv \mid^{p-2} Dv][Du - Dv] \geq c(p) \parallel u - v \parallel^p, \qquad (5.4.3)$$

where $c(p) > 0$.

We are now in a position to prove the following comparison result.

Theorem 5.4.1 *Let α_0, β_0 be coupled lower and upper solutions of (5.4.1). Suppose further, a.e. in Ω,*

$$\begin{bmatrix} f(x, u_1) - f(x, u_2) & \leq d_1(x)(u_1 - u_2), \\ g(x, u_1) - g(x, u_2) & \geq -d_2(x)(u_1 - u_2) \end{bmatrix} \qquad (5.4.4)$$

whenever $u_1 \geq u_2$ and $d_i(x) : \Omega \to R$, $i = 1, 2$ are measurable, $d_i \in L^1(\Omega)$ and $c(x) - (d_1(x) + d_2(x)) > 0$ a.e. in Ω. Then $\alpha_0(x) \leq \beta_0(x)$ a.e. in Ω.

Proof   From the definition of coupled weak lower and upper solutions of (5.4.1), we get

$$\int_\Omega [\mid D\alpha_0 \mid^{p-2} D\alpha_0 - \mid D\beta_0 \mid^{p-2} D\beta_0]Dv + c(x)(\alpha_0 - \beta_0)v$$

$$\leq \int_\Omega [f(x, \alpha_0) - f(x, \beta_0) + g(x, \beta_0) - g(x, \alpha_0)]v$$

for each $v \in W_0^{1,p}(\Omega)$, $v \geq 0$ a.e. in Ω. Choose $v = (\alpha_0 - \beta_0)^+ \in W_0^{1,p}(\Omega)$, $v \geq 0$ a.e. in Ω. Then, using (5.4.4), we obtain

$$\int_{\alpha_0 > \beta_0} [\mid D\alpha_0 \mid^{p-2} D\alpha_0 - \mid D\beta_0 \mid^{p-2} D\beta_0]D(\alpha_0 - \beta_0) + c(x)(\alpha_0 - \beta_0)^2$$

$$\leq \int_{\alpha_0 > \beta_0} [d_1(x) + d_2(x)] \mid \alpha_0 - \beta_0 \mid^2,$$

which, because of the estimate (5.4.2) and the assumption $c - (d_1 + d_2) > 0$, implies

$$0 < (\parallel \alpha_0 \parallel^{p-1} - \parallel \beta_0 \parallel^{p-1})(\parallel \alpha_0 \parallel - \parallel \beta_0 \parallel)(\parallel \alpha_0 \parallel - \parallel \beta_0 \parallel)$$

$$+ \int_{\alpha_0 > \beta_0} [c(x) - d_1(x) - d_2(x)] \mid \alpha_0 - \beta_0 \mid^2$$

$$\leq 0.$$

This contradiction proves the claim $\alpha_0(x) \leq \beta_0(x)$ a.e. in Ω.

We require the following existence theorem for our discussion.

Theorem 5.4.2 *Let $g : \Omega \times R \to R$ be a Caratheodory function such that $g(x,0) = 0$ a.e. in Ω, $\sup_{|u| \leq \alpha} | g(x,u) | \leq \phi_\alpha(x)$ with $\phi_\alpha \in L^1(\Omega)$ for all $\alpha > 0$ and $g(x,u)$ is increasing in u, a.e. in Ω. Then, for any $h \in W^{-1,p'}(\Omega)$ ($\frac{1}{p} + \frac{1}{p'} = 1$), there exists a unique weak solution $u \in W_0^{1,p}(\Omega)$ for the BVP*

$$\left[\begin{array}{c} -\Delta_p u + g(x,u) = h(x) \ \ in \ \Omega, \\ u = 0 \ \ on \ \partial\Omega. \end{array} \right. \tag{5.4.5}$$

If, in addition, we assume that $h \in L^\delta(\Omega)$ where $\delta > \frac{n}{p}$, then the unique weak solution of (5.4.5) satisfies $u \in W_0^{1,p}(\Omega) \cap L^\infty(\Omega)$.

See [10], [27] for the statement of Theorem 5.4.2 and [27], [26] for $L^\infty(\Omega)$ bounds and other details.

We are now in a position to prove our main result on the unified monotone iterative technique.

Theorem 5.4.3 *Assume that*

(A_1) *$\alpha_0, \beta_0 \in W^{1,p}(\Omega) \cap L^\infty(\Omega)$ are coupled lower and upper solutions of (5.4.1) such that $\alpha_0(x) \leq \beta_0(x)$ a.e. in Ω;*

(A_2) *$f, g : \bar{\Omega} \times R \to R$ are Caratheodory functions such that $f(x,u)$ is nondecreasing in u and $g(x,u)$ is nonincreasing in u for $x \in \Omega$, a.e.;*

(A_3) *for any $\eta, \mu \in W^{1,p}(\Omega) \cap L^\infty(\Omega)$ such that $\alpha_0 \leq \eta$, $\mu \leq \beta_0$, $h \in L^\delta(\Omega)$, where $h(x) = f(x,\eta) + g(x,\mu)$, $\delta > \frac{n}{p}$ and $0 < N \leq c(x)$ a.e. in Ω.*

Then there exist monotone sequences $\{\alpha_k(x)\}$, $\{\beta_k(x)\} \in W_0^{1,p}(\Omega) \cap L^\infty(\Omega)$, $k = 1, 2, \ldots$, with $\alpha_k \to \rho$, $\beta_k \to r$ weakly in $W_0^{1,p}(\Omega)$ and ρ, r are the coupled minimal and maximal solutions of (5.4.1).

Proof   Consider the following BVPs for each $k = 1, 2, \ldots$,

$$-\Delta_p \alpha_k + c(x)\alpha_k = f(x,\alpha_{k-1}) + g(x,\beta_{k-1}) \ in \ \Omega, \alpha_k = 0 \ on \ \partial\Omega, \tag{5.4.6}$$

$$-\Delta_p \beta_k + c(x)\beta_k = f(x,\beta_{k-1}) + g(x,\alpha_{k-1}) \ in \ \Omega, \beta_k = 0 \ on \ \partial\Omega, \tag{5.4.7}$$

whose variational forms are given by

$$\int_\Omega | D\alpha_k |^{p-2} D\alpha_k Dv + c(x)\alpha_k v = \int_\Omega [f(x,\alpha_{k-1}) + g(x,\beta_{k-1})]v, \tag{5.4.8}$$

$$\int_\Omega \mid D\beta_k \mid^{p-2} D\beta_k Dv + c(x)\beta_k v = \int_\Omega [f(x,\beta_{k-1}) + g(x,\alpha_{k-1})]v, \quad (5.4.9)$$

for each $v \in W_0^{1,p}(\Omega)$, $v \geq 0$ a.e. in Ω.

By (A_3), $h \in L^\delta(\Omega)$, where $h(x) = f(x,\alpha_0)+g(x,\beta_0)$. Hence by Theorem 5.4.2 there exists a unique weak solution $\alpha_1 \in W_0^{1,p}(\Omega)\cap L^\infty(\Omega)$ of the BVP (5.4.6). Also, by a similar argument $\beta_1 \in W_0^{1,p}(\Omega)\cap L^\infty(\Omega)$ is a unique weak solution of (5.4.7). We shall show first that

$$\alpha_0 \leq \alpha_1 \leq \beta_1 \leq \beta_0 \text{ a.e. in } \Omega. \quad (5.4.10)$$

To prove $\alpha_0 \leq \alpha_1$, we find that α_1 satisfies the relation

$$\int_\Omega \mid D\alpha_1 \mid^{p-2} D\alpha_1 Dv + c(x)\alpha_1 v = \int_\Omega [f(x,\alpha_0) + g(x,\beta_0)]v, \quad (5.4.11)$$

and note that α_0 satisfies

$$\int_\Omega \mid D\alpha_0 \mid^{p-2} D\alpha_0 Dv + c(x)\alpha_0 v \leq \int_\Omega [f(x,\alpha_0) + g(x,\beta_0)]v, \quad (5.4.12)$$

for each $v \in W_0^{1,p}(\Omega)$, $v \geq 0$ a.e. in Ω. Then by Theorem 5.4.1, it follows that $\alpha_0 \leq \alpha_1$ a.e. in Ω. Similar reasoning proves that $\beta_1 \leq \beta_0$ a.e. in Ω. To show that $\alpha_1 \leq \beta_1$, we see from (5.4.9) that β_1 satisfies

$$\int_\Omega \mid D\beta_1 \mid^{p-2} D\beta_1 Dv + c(x)\beta_1 v \quad (5.4.13)$$
$$= \int_\Omega [f(x,\beta_0) + g(x,\alpha_0)]v$$
$$\geq \int_\Omega [f(x,\alpha_0) + g(x,\beta_0)]v$$

for each $v \in W_0^{1,p}(\Omega)$, $v \geq 0$ a.e. in Ω, using the monotone character of f and g. By Theorem 5.4.1, we readily obtain, in view of (5.4.11) and (5.4.13), $\alpha_1 \leq \beta_1$ a.e. in Ω, proving (5.4.10).

Next we shall prove that if for some $j > 1$,

$$\alpha_0 \leq \alpha_{j-1} \leq \alpha_j \leq \beta_j \leq \beta_{j-1} \leq \beta_0 \text{ a.e. in } \Omega, \quad (5.4.14)$$

then

$$\alpha_j \leq \alpha_{j+1} \leq \beta_{j+1} \leq \beta_j \text{ a.e. in } \Omega. \quad (5.4.15)$$

Because of (A_3) and the assumed relation (5.4.14), Theorem 5.4.2 guarantees the existence of unique weak solutions α_j, α_{j+1}, β_j, $\beta_{j+1} \in W_0^{1,p}(\Omega) \cap L^\infty(\Omega)$ of (5.4.6) and (5.4.7) with $k = j$, $k = j+1$. To show that $\alpha_j \leq \alpha_{j+1}$, consider

$$\int_\Omega |D\alpha_{j+1}|^{p-2} D\alpha_{j+1}Dv + c(x)\alpha_{j+1}v = \int_\Omega [f(x,\alpha_j) + g(x,\beta_j)]v, \quad (5.4.16)$$

and

$$\int_\Omega |D\alpha_j|^{p-2} D\alpha_j Dv + c(x)d_j v \qquad\qquad (5.4.17)$$

$$= \int_\Omega [f(x,\alpha_{j-1}) + g(x,\beta_{j-1})]v$$

$$\leq \int_\Omega [f(x,\alpha_j) + g(x,\beta_j)]v,$$

for each $v \in W_0^{1,p}(\Omega)$, $v \geq 0$ a.e. in Ω. In (5.4.17), we have utilized the monotone nature of f and g and (5.4.14). Theorem 5.4.1, then yields that $\alpha_j \leq \alpha_{j+1}$ a.e. in Ω. A similar argument proves that $\beta_{j+1} \leq \beta_j$ a.e. in Ω. To prove $\alpha_{j+1} \leq \beta_{j+1}$, we find, using the monotone character of f and g and (5.4.14)

$$\int_\Omega |D\beta_{j+1}|^{p-2} D\beta_{j+1}Dv + c(x)\beta_{j+1}v \qquad\qquad (5.4.18)$$

$$= \int_\Omega [f(x,\beta_j) + g(x,\alpha_j)]v$$

$$\geq \int_\Omega [f(x,\alpha_j) + g(x,\beta_j)]v$$

for each $v \in W_0^{1,p}(\Omega)$, $v \geq 0$ a.e. in Ω. This, together with (5.4.16), proves by Theorem 5.4.1 that $\alpha_{j+1} \leq \beta_{j+1}$ a.e. in Ω, showing that (5.4.15) is valid. Hence by induction, we arrive at

$$\alpha_0 \leq \alpha_1 \leq \alpha_2 \leq \ldots \alpha_k \leq \beta_k \leq \ldots \leq \beta_2 \leq \beta_1 \leq \beta_0 \text{ a.e. in } \Omega, \quad (5.4.19)$$

for all k. As a result of (5.4.19), the pointwise limits

$$\rho(x) = \lim_{k\to\infty} \alpha_k(x), \quad r(x) = \lim_{k\to\infty} \beta_k(x), \qquad\qquad (5.4.20)$$

exist a.e. in Ω. Moreover, $\alpha_k(x) \to \rho(x)$, $\beta_k(x) \to r(x)$ in $L^p(\Omega)$ which is guaranteed by the dominated convergence theorem and (5.4.19). Setting $v = \alpha_k$ in (5.4.8) and using (5.4.2), it follows that

$$\| \alpha_k \|^p + N \int_\Omega | \alpha_k |^2 \leq \int_\Omega [f(x,\alpha_{k-1}) + g(x,\beta_{k-1})]\alpha_k,$$

and similarly,

$$\| \beta_k \|^p + N \int_\Omega | \beta_k |^2 \leq \int_\Omega [f(x,\beta_{k-1}) + g(x,\alpha_{k-1})]\alpha_k.$$

Since by (A_3), the right-hand sides of the foregoing inequalities belong to $L^\delta(\Omega)$, this implies that

$$\sup_k \| \alpha_k \|_{W_0^{1,p}(\Omega)}, \sup_k \| \beta_k \|_{W_0^{1,p}(\Omega)} < \infty.$$

Hence, there exist subsequences $\{\alpha_{k_j}\}$, $\{\beta_{k_j}\}$ that converge weakly in $W_0^{1,p}(\Omega)$ to $\rho, r \in W_0^{1,p}(\Omega)$ respectively.

We need to show that ρ, r are coupled weak minimal and maximal solutions of (5.4.1). We shall first show that they are weak solutions of (5.4.1). We set $Q(Du) = | Du |^{p-2} Du$ and note that Q satisfies, because of (5.4.2),

$$\left[\begin{array}{l} \int_\Omega [Q(D\alpha_k) - Q(Dw)][D\alpha_k - Dw] \geq 0, \\ \int_\Omega [Q(D\beta_k) - Q(dw)][D\beta_k - Dw] \geq 0, \end{array} \right. \tag{5.4.21}$$

for each $w \in W^{1,p}(\Omega)$, $\alpha_k, \beta_k \neq w$. Since $Q(D\alpha_k), Q(D\beta_k)$ are bounded in $L^p(\Omega)$, we have

$$Q(D\alpha_k) \to \xi, \ Q(D\beta_k) \to \eta \text{ weakly in } L^p(\Omega),$$

for some $\xi, \eta \in L^p(\Omega)$. This implies, taking the limit as $k \to \infty$ in (5.4.8), (5.4.9),

$$\left[\begin{array}{l} \int_\Omega \xi Dv + c(x)\rho v = \int_\Omega [f(x,\rho) + g(x,r)]v, \\ \int_\Omega \eta Dv + c(x)rv = \int_\Omega [f(x,r) + g(x,\rho)]v. \end{array} \right. \tag{5.4.22}$$

Setting $v = \alpha_k$, $v = \beta_k$ in (5.4.8), (5.4.9) respectively and substituting in (5.4.21), we arrive at

$$\int_\Omega \{[f(x,\alpha_{k-1}) + g(x,\beta_{k-1})]\alpha_k - c(x)\alpha_k^2 - Q(D\alpha_k)$$
$$\cdot Dw - Q(Dw)[D\alpha_k - Dw]\} \geq 0,$$

$$\int_\Omega \{[f(x,\beta_{k-1}) + g(x,\alpha_{k-1})]\beta_k - c(x)\beta_k^2 - Q(D\beta_k)$$
$$\cdot Dw - Q(Dw)[D\beta_k - Dw] \geq 0.$$

As $k \to \infty$, we get

$$\left[\begin{array}{l} \int_\Omega \{[f(x,\rho) + g(x,r)]\rho - c(x)\rho^2 - \xi \cdot D(w) - Q(Dw)[D\rho - Dw]\} \geq 0, \\ \int_\Omega \{[f(x,r) + g(x,\rho)]r - c(x)r^2 - \eta \cdot D(w) - Q(Dw)[Dr - Dw]\} \geq 0. \end{array} \right.$$
$$\tag{5.4.23}$$

Setting $v = \rho$, $v = r$ respectively in (5.4.22) and using (5.4.22) in (5.4.23), it follows that

$$\int_\Omega \xi \cdot D\rho - \xi \cdot D(w) - Q(Dw)[D(\rho - w)] \geq 0,$$

$$\int_\Omega \eta Dr - \eta \cdot D(w) - Q(Dw)[D(r - w)] \geq 0;$$

which leads to

$$\int_\Omega [\xi - Q(Dw)][D(\rho - w)] \geq 0, \quad \int_\Omega [\eta - Q(Dw)][D(r - w)] \geq 0.$$

Choosing $w = \rho - \lambda v$, $w = r - \lambda v$, $\lambda > 0$, we get

$$\int_\Omega [\xi - Q(D(\rho - \lambda v))]Dv \geq 0, \quad \int_\Omega [\eta - Q(D(r - \lambda v))]Dv \geq 0.$$

As $\lambda \to 0$, this results in

$$\int_\Omega [\xi - Q(D\rho)]Dv \geq 0, \quad \int_\Omega [\eta - Q(Dr)]Dv \geq 0.$$

Setting $v = -v$, we find

$$\int_\Omega [\xi - Q(D\rho)]Dv \leq 0, \quad \int_\Omega [\eta - Q(Dr)]Dv \leq 0.$$

Hence,

$$\int_\Omega \xi Dv = \int Q(D\rho)Dv = \int_\Omega [f(x, \rho) + g(x, r) - c(x)\rho]v,$$

$$\int_\Omega \eta Dv = \int Q(Dr)Dv = \int_\Omega [f(x, r) + g(x, \rho) - c(x)r]v,$$

or equivalently,

$$\int_\Omega Q(D\rho)Dv + c(x)\rho v = \int_\Omega [f(x, \rho) + g(x, r)]v,$$

$$\int_\Omega Q(Dr)Dv + c(x)rv = \int_\Omega [f(x, r) + g(x, \rho)]v,$$

for $v \in W_0^{1,p}(\Omega)$, $v \geq 0$ a.e. in Ω. This shows that ρ, r are coupled weak solutions of (5.4.1) in view of Definition 5.4.2.

Let u be any weak solution of (5.4.1) such that $\alpha_0 \le u \le \beta_0$ a.e. in Ω. Then we show that $\alpha_1 \le u \le \beta_1$ a.e. in Ω. Since α_1 satisfies (5.4.11), we get using the monotone nature of f, g,

$$\int_\Omega | D\alpha_1 |^{p-2} D\alpha_1 Dv + c(x)\alpha_1 v \le \int_\Omega [f(x,u) + g(x,u)]v$$

for each $v \in W_0^{1,p}(\Omega)$, $v \ge 0$ a.e. in Ω. Since u is a weak solution of (5.4.1), we have

$$\int_\Omega | Du |^{p-2} DuDv + c(x)uv = \int_\Omega [f(x,u) + g(x,u)]v,$$

for each $v \in W_0^{1,p}(\Omega)$, $v \ge 0$ a.e. in Ω. Therefore, by Theorem 5.4.1, we conclude that $\alpha_1 \le u$ a.e. in Ω. Similarly, we show that $u \le \beta_1$ a.e. in Ω. Next suppose that for some $j > 1$, $\alpha_0 \le \alpha_j \le u \le \beta_j \le \beta_0$ a.e. in Ω. Then using the monotone character of f and g, we have

$$\int_\Omega | D\alpha_{j+1} |^{p-2} D\alpha_{j+1} Dv + c(x)\alpha_{j+1}v \; = \; \int_\Omega [f(x,\alpha_j) + g(x,\beta_j)]v$$

$$\le \; \int_\Omega [f(x,u) + g(x,u)]v,$$

for each $v \in W_0^{1,p}(\Omega)$, $v \ge 0$ a.e. in Ω. It then follows by Theorem 5.4.1 that $\alpha_{j+1} \le u$ a.e. in Ω. Similarly, one can show that $u \le \beta_{j+1}$ a.e. in Ω. Thus we arrive at

$$\alpha_0 \le \alpha_j \le \alpha_{j+1} \le u \le \beta j + 1 \le \beta_j \le \beta_0 \text{ a.e. in } \Omega.$$

Hence by induction, $\alpha_k \le u \le \beta_k$ a.e. in Ω for all k and consequently, taking limits as $k \to \infty$, we get $\rho \le u \le r$ a.e. in Ω, proving that ρ, r are the coupled weak minimal and maximal solutions of (5.4.1). The proof is therefore complete.

Corollary 5.4.1 *In addition to the assumptions of Theorem 5.4.3 suppose that f, g satisfy the conditions (5.4.4) of Theorem 5.4.1. Then $u = \rho = r$ is the unique weak solution of (5.4.1).*

Proof   Since $\rho \le r$ a.e. in Ω, it is sufficient to show that $r \le \rho$ a.e. in Ω. Setting $\alpha_0 = r$, $\beta_0 = \rho$ in Theorem 5.4.1, we find immediately that $r \le \rho$ a.e. in Ω, which proves uniqueness.

Theorem 5.4.3 includes several results and thus offers a unified approach to the monotone iterative technique. This is made clear from the following remarks.

Remarks 5.4.1

(i) If, in Theorem 5.4.3, we suppose that $g(x, u) \equiv 0$, then we get a result where ρ, r are weak minimal and maximal solutions of (5.4.1).

(ii) If $f(x, u) \equiv 0$ in Theorem 5.4.3, then we have a result corresponding to the nonincreasing case, which is new.

(iii) If $g(x, u) \equiv 0$ and $f(x, u)$ is not nondecreasing in u but $\tilde{f}(x, u) = f(x, u) + d(x)u$ is nondecreasing in u, where $d : \Omega \to R_+$ is measurable and $d \in L^1(\Omega)$, then one can consider the BVP

$$-\Delta_p u + [c(x) + d(x)]u = \tilde{f}(x, u) \text{ in } \Omega, u = 0 \text{ on } \partial\Omega. \qquad (5.4.24)$$

It is clear that α_0, β_0 are weak lower and upper solutions of (5.4.24), $\tilde{f}$ is nondecreasing in u and $c(x) + d(x) > 0$ a.e. in Ω. Then Theorem 5.4.3 yields the same conclusion as in (1) for the BVP (5.4.1).

(iv) If $f(x, u) \equiv 0$ and $g(x, u)$ is not nonincreasing in u but $\tilde{g}(x, u) = g(x, u) - d(x)u$ is nonincreasing in u where d is the same function as in (3), then we consider the BVP

$$-\Delta_p u + c(x)u - d(x)y = \tilde{g}(x, u) \text{ in } \Omega, u = 0 \text{ on } \partial\Omega. \qquad (5.4.25)$$

If we assume α_0, β_0 are coupled lower and upper solutions of (5.4.25), Theorem 5.4.3 gives the same conclusion as in (2) for BVP (5.4.25). In addition, if $c(x) - d(x) > 0$ a.e. in Ω, then Theorem 5.4.3 applied to the BVP (5.4.25) guarantees the same conclusion for the BVP

$$-\Delta_p u + c(x)u = f(x, u) + g(x, u) \text{ in } \Omega, \quad u = 0 \text{ on } \partial\Omega$$

where $g(x, u)$ does not satisfy the monotonicity requirement and $f(x, u) \equiv 0$.

(v) If in Theorem 5.4.3, $g(x, u)$ is nonincreasing in u and $f(x, u)$ is not nondecreasing in u but $\tilde{f}(x, u) = f(x, u) + d(x)$ is nondecreasing in u, where d is the same function as in (3), then we consider the BVP

$$-\Delta_p u + [c(x) + d(x)]u = \tilde{f}(x, u) + g(x, u) \text{ in } \Omega, u = 0 \text{ on } \partial\Omega. \quad (5.4.26)$$

It is easy to verify that α_0, β_0 are also coupled lower and upper solutions of (5.4.26). Hence Theorem 5.4.3 applied to (5.4.26) provides the same conclusion of Theorem 5.4.3 for BVP (5.4.1).

(vi) Assume that $f(x, u)$ is nondecreasing in u and $g(x, u)$ is not nonincreasing in u but $\tilde{g}(x, u) = g(x, u) - d(x)u$ is nonincreasing in u with
d as in (3). Then consider the BVP

$$-\Delta_p u + c(x)u - d(x)u = f(x, u) + \tilde{g}(x, u) \text{ in } \Omega, u = 0 \text{ on } \partial\Omega \quad (5.4.27)$$

If we assume α_0, β_0 are coupled lower and upper solutions of (5.4.27),
Theorem 5.4.3 applied to (5.4.27) yields the conclusion for the BVP
(5.4.27). Moreover, if $c(x) - (d_1(x) + d(x)) > 0$ a.e. in Ω where $d_1(x)$ as
in Theorem 5.4.1, then Theorem 5.4.3 applied to BVP (5.4.27) implies
the same conclusion for the BVP

$$-\Delta_p u + c(x)u = f(x, u) + g(x, u) \text{ in } \Omega, \quad u = 0 \text{ on } \partial\Omega$$

where $g(x, u)$ is not nonincreasing in u.

(vii) Assume in Theorem 5.4.3 that both f and g are not monotone but
$\tilde{f}(x, u) = f(x, u) + \tilde{d}(x)u$ is nondecreasing in u and $\tilde{g}(x, u) = g(x, u) -
d_1(x)u$ is nonincreasing in u where $d(x)$ and $\tilde{d}(x)$ are similar functions
as in Theorem 5.4.1. We then consider the BVP

$$-\Delta_p u + [c(x) + \tilde{d}(x) - d(x)]u = \tilde{f}(x, u) + \tilde{g}(x, u) \text{ in } \Omega, u = 0 \text{ on } \partial\Omega. \quad (5.4.28)$$

The conclusion of Theorem 5.4.3 applied to BVP (5.4.28) holds for the BVP
(5.4.28), if again we assume that α_0, β_0 are coupled lower and upper solutions
of (5.4.28). Moreover, if $c(x) - (d_1(x) + d_2(x) + d(x) - \tilde{d}(x)) > 0$ where $d_1(x)$
and $d_2(x)$ as in Theorem 5.4.1, then the conclusion of Theorem 5.4.3 applied
to the BVP (5.4.27) holds for the BVP

$$\Delta_p u + c(x)u = f(x, u) + g(x, u) \text{ in } \Omega, \quad u = 0 \text{ on } \partial\Omega$$

where $f(x, u)$ and $g(x, u)$ do not satisfy the required monotonicity conditions.

5.5 Monotone Iterative Technique (MIT): Degenerate Problems

This section extends the monotone iterative technique to degenerate elliptic
boundary value problems providing a very general result which includes
several important cases. We begin with the necessary preliminaries.

Let $\Omega \subset R^n$ be a smooth bounded domain in R^n. We let $H^1(\Omega, w)$, $H_0^1(\Omega, w)$ be the usual weighted Sobolev (Hilbert) spaces with the weight function $w : \Omega \to R_+$ such that

$$w \in L_{\text{loc}}^1(\Omega), \quad w^{-1} \in L_{\text{loc}}^1(\Omega). \tag{5.5.1}$$

The norm of $H^1(\Omega, w)$ is given by

$$\| u \|_{H^1(\Omega, w)} = (\int_\Omega | u |^2 + \sum_{i=1}^n w \, | \frac{\partial u}{\partial x_i} |^2 \, dx)^{1/2}.$$

Let L denote the second-order degenerate partial differential operator in divergence form

$$Lu = - \sum_{i,j=1}^n (a_{ij}(x)u_{x_i})_{x_j} + c(x)u \tag{5.5.2}$$

where we assume that $\frac{a_{ij}}{w} \in L^\infty(\Omega)$, $a_{ij} = a_{ji}$, $c \in L^\infty(\Omega)$, $c(x) \geq c^* > 0$ and $\sum_{ij}^n a_{ij}(x)\xi_i\xi_j \geq w(x) \, | \xi |^2$ for $x \in \Omega$, a.e. The bilinear form associated with the operator L is

$$B[u, v] = \int_\Omega \left[\sum_{i,j} a_{ij}(x)u_{x_i}v_{x_j} + c(x)uv \right] dx \tag{5.5.3}$$

for $u, v \in H_0^1(\Omega, w)$. We consider the linear BVP

$$\left[\begin{array}{ll} Lu = h(x) & \text{in } \Omega, \\ u = 0 \text{ in } \partial\Omega & \text{in the sense of trace.} \end{array} \right. \tag{5.5.4}$$

We shall always mean that the boundary condition is in the sense of trace and therefore we shall not repeat it to avoid monotony.

A function $u \in H_0^1(\Omega, w)$ is said to be a weak solution of (5.5.4) if $h \in L^2(\Omega)$ and

$$B[u, v] = (h, v) \tag{5.5.5}$$

for all $v \in H_0^1(\Omega, w)$, where $(,)$ denotes the inner product in $L^2(\Omega)$.

We also consider the semilinear degenerate elliptic BVP

$$\left[\begin{array}{ll} Lu = f(x, u) & \text{in } \Omega, \\ u = 0 & \text{on } \partial\Omega \end{array} \right. \tag{5.5.6}$$

where $f : \bar{\Omega} \times R \to R$, $f(x, u)$ is a Caratheodory function.

Definition 5.5.1 *A function $u \in H_0^1(\Omega, w)$ is said to be a weak solution of* (5.5.6) *if $f(x, u) \in L^1(\Omega)$, $f(x, u)u \in L^1(\Omega)$ and*

$$B[u, v] = (f(x, u), v) \tag{5.5.7}$$

for all $v \in H_0^1(w, \Omega)$, where (,) is the inner product in $L^2(\Omega)$.

Definition 5.5.2 *The function α_0 is said to be a weak lower solution of* (5.5.6) *if $\alpha_0 \leq 0$ on $\partial\Omega$ and*

$$\int_\Omega \left[\sum_{i,j}^n a_{ij}(x)\alpha_{0,x_i} v_{x_j} + c(x)\alpha_0 v \right] dx \leq \int_\Omega f(x, \alpha_0)v dx \tag{5.5.8}$$

for each $v \in H_0^1(\Omega, w)$, $v \geq 0$ a.e. in Ω. If the inequalities are reversed, then α_0 is said to be a weak upper solution of (5.5.6).

In this set-up one can prove the following result.

Theorem 5.5.1 *Assume that*

(i) *$\alpha_0, \beta_0 \in H_0^1(\Omega, w)$ are weak lower and upper solutions of* (5.5.6) *such that $\alpha_0(x) \leq \beta_0(x)$, a.e. in Ω;*

(ii) *$f : \bar{\Omega} \times R \to R$ is a Caratheodory function satisfying*

$$f(x, u) - f(x, v) \geq -M(x)(u - v), u \geq v, \text{ a.e. in } \Omega,$$

where $M(x) \geq 0$, $M \in L^\infty(\Omega)$;

(iii) *$0 < N \leq c(x) + M(x)$, a.e. in Ω and for every $\eta \in H_0^1(\Omega, w)$ with $\alpha_0 \leq \eta \leq \beta_0$ in Ω, a.e. in $h(x) \equiv f(x, \eta) + (x)\eta \in L^2(\Omega)$.*

Then there exist monotone sequences $\{\alpha_n(x)\}$, $\{\beta_n(x)\} \in H_0^1(\Omega, w)$ such that $\alpha_n \to \rho$, $\beta_n \to r$ weakly in $H_0^1(\Omega, w)$ with $\rho, r \in H_0^1(\Omega, w)$, (ρ, r) are minimal and maximal weak solutions of (5.5.6), *respectively.*

The case when $f(x, u)$ is nondecreasing in u is covered in Theorem 5.5.1 when $M(x) \equiv 0$. However, the special case when $f(x, u)$ is nonincreasing in u is not included in Theorem 5.5.1 and is of particular interest. We shall prove, in the next section, when $f(x, u)$ admits a splitting of two monotone functions, which includes the case when $f(x, u)$ is nonincreasing in u as well as a few other important generalizations.

We shall now prove the following comparison result.

Theorem 5.5.2 *Let α_0, β_0 be weak lower and upper solutions of (5.5.6) respectively. Suppose further than $f(x, u)$ satisfies*

$$f(x, u_1) - f(x, u_2) \leq K(x)(u_1 - u_2) \qquad (5.5.9)$$

whenever $u_1 \geq u_2$ a.e. in Ω and $K(x) > 0$ a.e. in Ω. Then, if $c^* - K(x) > 0$ a.e. in Ω, we have

$$\alpha_0(x) \leq \beta_0(x) \text{ in } \Omega, \text{ a.e.}$$

Proof The definition of lower and upper solutions of (5.5.6) yields

$$\int_\Omega \left[\sum_{i,j}^n a_{ij}(x)(\alpha_{0,x_i} - \beta_{0,x_i})v_{x_j} + c(x)(\alpha_0 - \beta_0)v \right] dx \qquad (5.5.10)$$

$$\leq \int_\Omega [f(x, \alpha_0) - f(x, \beta_0)]v dx,$$

for each $v \in H_0^1(\Omega, w)$, $v \geq 0$ a.e. in Ω. Take $v = (\alpha_0 - \beta_0)^+ \in H_0^1(\Omega, w)$, $v \geq 0$ a.e. in Ω. Then we get

$$\int_\Omega \left[\sum_{i,j=1}^n a_{ij}(x)(\alpha_{0,x_i} - \beta_{0,x_i})(\alpha_0 - \beta_0)_{x_j}^+ + c(x)(\alpha_0 - \beta_0)(\alpha_0 - \beta_0)^+ \right] dx$$

$$\leq \int_\Omega [f(x, \alpha_0) - f(x, \beta_0)](\alpha_0 - \beta_0)^+ dx.$$

Since

$$(\alpha_0 - \beta_0)_{x_j}^+ = \begin{bmatrix} (\alpha_{0,x_j} - \beta_{0,x_j}) & \text{a.e. on } \alpha_0 > \beta_0, \\ 0 & \text{a.e. on } \alpha_0 \leq \beta_0, \end{bmatrix}$$

we get, using the ellipticity condition and (5.5.9)

$$\int_{\alpha_0 > \beta_0} [w(x) \mid (\alpha_{0,x} - \beta_{0,x} \mid^2 + c^* \mid \alpha_0 - \beta_0 \mid^2] dx \leq \int_{\alpha_0 > \beta_0} L(x) \mid \alpha_0 - \beta_0 \mid^2 dx,$$

which yields the claimed conclusion $\alpha_0(x) \leq \beta_0(x)$ a.e. in Ω, completing the proof.

The following corollary is useful in our discussions.

Corollary 5.5.1 *For p in $H^1(\Omega, w)$ satisfying*

$$\int_\Omega \left[\sum_{i,j}^n a_{ij}(x)p_{x_j}v_{x_j} + c_0(x)pv \right] dx \leq 0$$

for each $v \in H_0^1(\Omega, w)$, $v \geq 0$, a.e. in Ω and $p \leq 0$ on $\partial\Omega$, we have $p \leq 0$ a.e. in Ω, provided $c(x) > 0$ a.e. in Ω.

Theorem 5.5.3 *There exists a unique weak solution $u \in H_0^1(\Omega, w)$ for the linear BVP (5.5.4) provided $0 < c^* \le c(x)$ a.e. in Ω and $h \in L^2(\Omega)$.*

See [36] for details of degenerate elliptic problems.

We will now consider more general BVPs which include (5.5.4) as well as other cases, namely,

$$\left[\begin{array}{ll} Lu = f(x, u) + g(x, u) & \text{in } \Omega, \\ u = 0 & \text{on } \partial\Omega \end{array} \right. \tag{5.5.11}$$

where $f, g : \bar{\Omega} \times R \to R$ are Caratheodory functions. For any $u \in H_0^1(\Omega, w)$, we assume that $f, g \in L^1(\Omega)$. We shall prove the following result relative to (5.5.11).

Theorem 5.5.4 *Assume that*

(A_1) $\alpha_0, \beta_0 \in H^1(\Omega, w)$ *are coupled lower and upper solutions of (5.5.11), that is,*

$$B[\alpha_0, v] \le (f(x, \alpha_0) + g(x, \beta_0), v),$$

$$B[\beta_0, v] \ge (f(x, \beta_0) + g(x, \alpha_0), v)$$

for all $v \in H_0^1(\Omega, w)$, $v \ge 0$ a.e. in Ω such that $\alpha_0(x) \le \beta_0(x)$ a.e. in Ω;

(A_2) f, g *satisfy, in addition, that $f(x, u)$ is nondecreasing and $g(x, u)$ is nonincreasing in u, a.e. in Ω;*

(A_3) *for any $\eta \in H^1(\Omega, w)$ with $\alpha_0 \le \eta$, $\mu \le \beta_0$, the function $h(x) = f(x, \eta) + g(x, \mu) \in L^2(\Omega)$.*

Then there exist monotone sequences $\{\alpha_n(x)\}$, $\{\beta_n(x)\} \in H_0^1(\Omega, w)$ such that $\alpha_n \to \rho$, $\beta_n \to r$ weakly in $H_0^1(\Omega, w)$ with $\rho, r \in H_0^1(\Omega, w)$ and (ρ, r) are the coupled minimal and maximal weak solutions of (5.5.11), that is, (ρ, r) satisfy

$$B[\rho, v] = (f(x, \rho) + g(x, r), v)$$

$$B[r, v] = (f(x, r) + g(x, \rho), v)$$

for all $v \in H_0^1(\Omega, w)$, $v \ge 0$ a.e. in Ω and $\rho \le u \le r$ a.e. in Ω, for any weak solution u, of (5.5.11) such that $\alpha_0 \le u \le \beta_0$ a.e. in Ω.

Proof   Consider the linear BVPs for each $k \geq 1$

$$\left[\begin{array}{ll} L\alpha_{k+1} = f(x, \alpha_k) + g(x, \beta_k) & \text{in } \Omega, \\ \alpha_{k+1} = 0 & \text{on } \partial\Omega, \end{array} \right. \tag{5.5.12}$$

and

$$\left[\begin{array}{ll} L\beta_{k+1} = f(x, \beta_k) + g(x, \alpha_k) & \text{in } \Omega, \\ \beta_{k+1} = 0 & \text{on } \partial\Omega, \end{array} \right. \tag{5.5.13}$$

whose bilinear form associated with the operator L are given by

$$B[\alpha_{k+1}, v] = (f(x, \alpha_k) + g(x, \beta_k, v) \tag{5.5.14}$$

and

$$B[\beta_{k+1}, v] = (f(x, \beta_k) + g(x, \alpha_k), v) \tag{5.5.15}$$

for all $v \in H_0^1(\Omega, w)$, $v \geq 0$ a.e. in Ω, $k = 1, 2, \ldots$. By (A_3), $h(x) \in L^2(\Omega)$ where $h(x) = f(x, \eta) + g(x, \mu)$ for any $\eta, \mu \in H^1(\Omega, w)$ such that $\alpha_0 \leq \eta$, $\mu \leq \beta_0$. As a result, by Theorem 5.5.3, there exist weak solutions of (5.5.12) and (5.5.13) for each $k = 1, 2, \ldots$. Our aim is to show that the weak solutions $\alpha_k, \beta_k \in H_0^1(\Omega, w)$ satisfy

$$\alpha_0 \leq \alpha_1 \leq \ldots \leq \alpha_k \leq \beta_k \leq \ldots \leq \beta_1 \leq \beta_0 \text{ a.e. in } \Omega. \tag{5.5.16}$$

We shall first show

$$\alpha_0 \leq \alpha_1 \leq \beta_1 \leq \beta_0 \text{ a.e. in } \Omega. \tag{5.5.17}$$

Since α_0 is a lower solution of (5.5.11), we have

$$B[\alpha_0, v] \leq (f(x, \alpha_0) + g(x, \beta_0), v)$$

and

$$B[\alpha_1, v] = (f(x, \alpha_0) + g(x, \beta_0), v)$$

for each $v \in H_0^1(\Omega, w)$, $v \geq 0$ a.e. in Ω. Therefore,

$$B[\alpha_0 - \alpha_1, v] \leq 0 \text{ for each } v \in H_0^1(\Omega, w), \ v \geq 0 \text{ a.e. in } \Omega.$$

Corollary 5.5.1 then yields $\alpha_0 \leq \alpha_1$ a.e. in Ω since $\alpha_0 - \alpha_1 \leq 0$ on $\partial\Omega$. Similarly, we get $\beta_1 \leq \beta_0$ a.e. in Ω. Now to prove that $\alpha_1 \leq \beta_1$ a.e. in Ω, we see from (5.5.12) and (5.5.13),

$$B[\alpha_1 - \beta_1, v] = (f(x, \alpha_0) + g(x, \beta_0) - f(x, \beta_0) - g(x, \alpha_0), v) \leq 0$$

for all $v \in H_0^1(\Omega, w), v \geq 0$ a.e. in Ω

because of the fact that $f(x, u)$ in nondecreasing in u, $g(x, u)$ is nonincreasing in u a.e. in Ω, and $\alpha_0 \leq \beta_0$. This again implies by Corollary 5.5.1 that $\alpha_1 \leq \beta_1$ a.e. in Ω since $\alpha_1 - \beta_1 = 0$ on $\partial\Omega$ proving (5.5.17).

Assuming now that for some $k > 1$,

$$\alpha_{k-1} \leq \alpha_k \leq \beta_k \leq \beta_{k-1} \text{ a.e. in } \Omega. \qquad (5.5.18)$$

Then, in view of (5.5.12), we obtain using (5.5.18) and the monotone character of f and g in u,

$$B[\alpha_{k+1} - \alpha_k, v] = (f(x, \alpha_k) + g(x, \beta_k) - f(x, \alpha_{k-1}) - g(x, \beta_{k-1}), v) \leq 0$$

and

$$\alpha_{k+1} - \alpha_k = 0 \text{ on } \partial\Omega,$$

which proves by Corollary 5.5.1 that $\alpha_k \leq \alpha_{k-1}$ a.e in Ω. A similar argument shows that $\beta_{k+1} \leq \beta_k$ a.e. in Ω. To show $\alpha_{k+1} \leq \beta_{k+1}$, it follows by using (5.5.12), (5.5.13) and the monotone nature of f, g, for each $v \in H_0^1(\Omega, w)$, $v \geq 0$, a.e. in Ω

$$B[\alpha_{k+1} - \beta_{k+1}, v] = (f(x, \alpha_k) + g(x, \beta_k) - f(x, \beta_k) - g(x, \alpha_k), v) \leq 0,$$

proving $\alpha_{k+1} \leq \beta_{k+1}$ by Corollary 5.5.1 since $\alpha_{k+1} - \beta_{k+1} = 0$ on $\partial\Omega$. Thus (5.5.18) is valid.

Hence by induction (5.5.16) is also true. Consequently, the pointwise limits

$$\rho(x) = \lim_{k \to \infty} \alpha_k(x), \quad r(x) = \lim_{k \to \infty} \beta_k(x)$$

exist a.e. in Ω. Furthermore, $\alpha_k(x) \to \rho(x)$, $\beta_k(x) \to r(x)$ in $L^2(\Omega)$ by the dominated convergence theorem and (5.5.16). We note that α_k satisfies for each $v \in H_0^1(\Omega, w)$, $v \geq 0$ a.e. in Ω,

$$\int_\Omega \left[\sum_{ij} a_{ij}(x)\alpha_{k,x_i} v_{x_j} + c(x)\alpha_k v \right] dx = \int_\Omega [f(x, \alpha_{k-1}) + g(x, \beta_{k-1})]v dx.$$

Using the ellipticity condition with $v = \alpha_k$, we get

$$\int_\Omega [w(x) \mid \alpha_{k,x} \mid^2 + c^* \mid \alpha_k \mid^2] dx \leq \int_\Omega [f(x, \alpha_{k-1}) + g(x, \beta_{k-1})]\alpha_k dx.$$

Since by (A_3), the integrand on the right-hand side belongs to $L^2(\Omega)$, in view of the fact that $\alpha_{k-1}, \beta_{k-1} \in H_0^1(\Omega, w)$, we get the estimate

$\sup_k \| \alpha_k \|_{H_0^1(\Omega,w)} < \infty$. Similarly, $\sup_k \| \beta_k \|_{H_0^1(\Omega,w)} < \infty$. Hence there exist subsequences $\{\alpha_{k_j}(x)\}$, $\{\beta_{k_j}(x)\}$ which converge weakly in $H_0^1(\Omega, w)$ to $\rho, r \in H_0^1(\Omega, w)$, respectively.

To verify that ρ, r are coupled weak solutions of (5.5.11), we fix $v \in H_0^1(\Omega, w)$ and find that $\alpha_{k+1}, \beta_{k+1}$ satisfy (5.5.14) and (5.5.15), respectively. Taking the limit as $k \to \infty$, we obtain

$$B[\rho, v] = (f(x, \rho) + g(x, r), v),$$

$$B[r, v] = (f(x, r) + g(x, \rho), v),$$

and this shows (ρ, r) are coupled weak solutions of (5.5.11). Now, let $u \in H_0^1(\Omega, w)$ be any weak solution of (5.5.1) such that $\alpha_0 \leq u \leq \beta_0$ a.e. in Ω. Then

$$B[u, v] = (f(x, u) + g(x, u), v)$$

and

$$B[\alpha_1, v] = (f(x, \alpha_0) + g(x, \beta_0)v)$$

for each $v \in H_0^1(\Omega, w)$, $v \geq 0$ a.e. in Ω. Hence

$$B[\alpha_1 - u, v] = (f(x, \alpha_0) - f(x, u) + g(x, \beta_0) - g(x, u), v) \leq 0$$

for each $v \in H_0^1(\Omega, w)$, $v \geq 0$ a.e. in Ω and $\alpha_1 - u = 0$ on $\partial\Omega$. Then Corollary 5.5.1 shows that $\alpha_1 \leq u$ a.e. in Ω. A similar argument proves that $u \leq \beta_1$ a.e. in Ω. Now assume for some $k > 1$, $\alpha_k \leq u \leq \beta_k$ a.e. in Ω. Then, we get by our assumptions

$$B[\alpha_{k+1} - u, v] = (f(x, \alpha_k) + g(x, \beta_k) - f(x, u) - g(x, u), v) \leq 0$$

for all $v \in H_0^1(\Omega, w)$, $v \geq 0$ a.e. in Ω. Since $\alpha_{k+1} - u = 0$ on $\partial\Omega$, Corollary 5.5.1 again yields $\alpha_{k+1} \leq u$ a.e. in Ω. Similarly, $u \leq \beta_{k+1}$ a.e. in Ω. Thus, by induction, we see that

$$\alpha_0 \leq \alpha_k \leq u \leq \beta_k \leq \beta_0 \quad \text{a.e. in } \Omega \text{ for all } k.$$

Consequently, $\alpha_0 \leq \rho \leq u \leq r \leq \beta_0$ a.e. in Ω, proving (ρ, r) are the coupled extremal solutions of (5.5.11).

Corollary 5.5.2 *Assume in addition to the conditions of Theorem 5.5.4, that f and g satisfy*

$$f(x, u_1) - f(x, u_2) \leq N_1(u_1 - u_2)$$

and

$$g(x, u_1) - g(x, u_2) \geq -N_2(u_1 - u_2)$$

where $u_1 \geq u_2$, $N_1 > 0$, $N_2 > 0$ and $c(x) - (N_1 + N_2) > 0$ a.e. in Ω. Then $\rho = u = r$ is the unique solution of (5.5.11).

Proof   Since we have $\rho \leq r$, it is enough to show that $r \leq \rho$. Setting $p = r - \rho$, we have $p = 0$ on $\partial\Omega$ and using the assumptions on f and g, we obtain

$$B[p, v] \leq ((N_1 + N_2)p, v).$$

This implies by Corollary 5.5.1, $p \leq 0$ a.e. in Ω proving that $u = \rho = r$ is the unique weak solution of (5.5.11).

Theorem 5.5.4 includes several special cases of interest which we list below as remarks.

Remarks 5.5.1

(i) In Theorem 5.5.4, assume that $g(x, u) \equiv 0$. Then we get Theorem 5.5.1 with $M(x) \equiv 0$.

(ii) If $g(x, u) \equiv 0$ in Theorem 5.5.4 and $f(x, u)$ is not nondecreasing in u but satisfies the condition that $f(x, u) + Mu$ is nondecreasing in u for $M > 0$, then one can consider the BVP

$$\tilde{L}u = Lu + Mu = f(x, u) + Mu = \tilde{f}(x, u) \text{ in } \Omega, u = 0 \text{ on } \partial\Omega.$$
$$(5.5.19)$$

It is clear that α_0, β_0 are the weak lower and upper solutions of (5.5.19) with $g(x, u) \equiv 0$, $\tilde{f}(x, u)$ is nondecreasing in u and $c^* + M > 0$ in Ω. Then Theorem 5.5.4 yields the same conclusion as Theorem 5.5.1.

(iii) If $f(x, u) \equiv 0$ in Theorem 5.5.4, then we get a result corresponding to the case when $g(x, u)$ is nonincreasing in u.

(iv) If $f(x, u) \equiv 0$ and $g(x, u)$ is not nonincreasing in u but $g(x, u) - Nu$, $N > 0$, is nonincreasing in u, then we consider the BVP

$$\tilde{L}u = Lu - Nu = \tilde{f}(x, u) + \tilde{g}(x, u) \text{ in } \Omega, u = 0 \text{ on } \partial\Omega \qquad (5.5.20)$$

where $\tilde{f}(x, u) = Nu$ and $\tilde{g}(x, u) = g(x, u) - Nu$ and assume that the coupled weak lower and upper solutions as given in Theorem 5.5.4 of (5.5.20) exist. Then we see that Theorem 5.5.4 yields the same conclusion as in Theorem 5.5.1 for the BVP (5.5.20). In addition, if

$c(x) - (N_2 + N) > 0$ a.e. in Ω, with N_2 as in Corollary 5.5.2, then applying Theorem 5.5.4 to the BVP (5.5.20) gives the same result for

$$Lu = f(x,u) + g(x,u) \text{ in } \Omega, \quad u = 0 \text{ on } \partial\Omega$$

when $g(x,u)$ is not nonincreasing and $f(x,u) \equiv 0$.

(v) If $g(x,u)$ is nonincreasing in u and $f(x,u)$ is not nondecreasing in u but $f(x,u) + Mu$ is nondecreasing in u for $M > 0$, then we consider the BVP

$$\tilde{L}u = Lu + Mu = \tilde{f}(x,u) + g(x,u) \text{ in } \Omega, u = 0 \text{ on } \partial\Omega, \quad (5.5.21)$$

where $\tilde{f}(x,u) = f(x,u) + Mu$ is nondecreasing in u. It is easy to verify that α_0, β_0 are the coupled lower and upper solutions of (5.5.21) and therefore we get the conclusion of Theorem 5.5.4 for the BVP (5.5.11).

(vi) If $f(x,u)$ is nondecreasing in u and $g(x,u)$ is not nonincreasing in u but $g(x,u) - Nu$, $N > 0$, is nonincreasing in u, we consider the BVP

$$\tilde{L}u = Lu - Nu = f(x,u) + \tilde{g}(x,u) \text{ in } \Omega, u = 0 \text{ on } \partial\Omega \quad (5.5.22)$$

where $\tilde{g}(x,u) = g(x,u) - Nu$ and assume that the coupled weak lower and upper solutions of (5.5.22) exist. Then Theorem 5.5.4 applied to (5.5.22) guarantees the same conclusion of Theorem 5.5.4.

(vii) If both functions $f(x,u)$, $g(x,u)$ do not satisfy the required monotone character in Theorem 5.5.4, but $f(x,u) + Mu$ and $g(x,u) - Nu$ for $M, N > 0$, are nondecreasing and nonincreasing in u, respectively, we consider the BVP

$$Lu = \tilde{f}(x,u) + \tilde{g}(x,u) \text{ in } \Omega, u = 0 \text{ on } \partial\Omega \quad (5.5.23)$$

where $\tilde{f}(x,u) = f(x,u) + Mu$, $\tilde{g}(x,u) = g(x,u) - Nu$. Clearly, the conditions of Theorem 5.5.4 are fulfilled by $\tilde{f}, \tilde{g}$, and hence assuming α_0 and β_0 are the coupled lower and upper weak solutions of (5.5.23), Theorem 5.5.4 shows that the same conclusion is valid for the BVP

$$Lu = f(x,u) + g(x,u) \text{ in } \Omega, u = 0 \text{ on } \partial\Omega,$$

where $f(x,u)$, $g(x,u)$ are not monotone functions provided $c(x) + M - N - (N_1 + N_2) > 0$, N_1, N_2 as in Corollary 5.5.2.

5.6   Generalized Quasilinearization (GQ): Semilinear Problems

This and the following two sections are devoted to the method of generalized
quasilinearization. In this section, we shall consider the semilinear elliptic
boundary value problem

$$\left[\begin{array}{ll} Lu = f(x,u) + g(x,u) & \text{in } \Omega, \\ u = 0 & \text{on } \partial\Omega, \end{array} \right. \tag{5.6.1}$$

where L denotes the same second-order partial differential operator in divergence form as given in (5.2.2) together with the bilinear form $B[u,v]$ defined
in (5.2.3) under the same assumptions. The functions $f, g : \bar{\Omega} \times R \to R$ are
Caratheodory functions as before.

   We shall discuss the method of generalized quasilinearization relative to
the BVP (5.6.1) using a unified approach so that several important cases
are included in the results proved.

   In the next two sections, we shall describe the method of generalized
quasilinearization to quasilinear elliptic BVPs and to degenerate semilinear
elliptic BVPs, respectively.

   We shall prove the following result for the BVP (5.6.1), employing Corollary 5.2.1 and Theorem 5.2.4.

Theorem 5.6.1 *Assume that*

$(B1)$ $\alpha_0, \beta_0 \in H^1(\Omega)$ *are lower and upper solutions of* (5.6.1) *such that*
$\alpha_0 \le \beta_0$ *in* Ω, *a.e.;*

$(B2)$ $f, g : \bar{\Omega} \times R \to R$ *are Caratheodory functions,* $f_u(x,u)$, $g_u(x,u)$,
$f_{uu}(x,u)$, $g_{uu}(x,u)$ *exist and are Caratheodory functions, and* $f_{uu}(x,u)$
≥ 0, $g_{uu}(x,u) \le 0$ *for* $\alpha_0 \le u \le \beta_0$ *a.e. in* Ω;

$(B3)$ $0 < N \le c(x) - f_u(x,\beta_0) - g_u(x,\alpha_0)$ *a.e. in* Ω, *and for any* $\mu, \eta \in$
$H^1(\Omega)$ *satisfying* $\alpha_0 \le \mu \le \eta \le \beta_0$, *the function* $h \in L^2(\Omega)$ *where*

$$h(x) = f(x,\eta) + g(x,\eta) - f_u(x,\mu)\eta - g_u(x,\eta)\eta.$$

   Then there exist monotone sequences $\{\alpha_k\}$, $\{\beta_k\} \in H_0^1(\Omega)$ *such that*
$\alpha_k \to \rho$, $\beta_k \to r$ *weakly in* $H_0^1(\Omega)$ *as* $k \to \infty$, *where* $\rho = r = u$ *is the*
unique weak solution of (5.6.1) *satisfying* $\alpha_0 \le u \le \beta_0$, *a.e. in* Ω *and the*
convergence is quadratic.

Proof We prove the conclusion in several steps.

(a) **Iterative schemes and generalized quasilinearization** Let us introduce the linearization of $f + g$ in the form

$$G(x, u; \alpha, \beta) = f(x, \beta) + g(x, \beta) + f_u(x, \alpha)(u - \beta) \quad (5.6.2)$$
$$+ g_u(x, \beta)(u - \beta),$$

and

$$F(x, u; \alpha, \beta) = f(x, \alpha) + g(x, \alpha) + f_u(x, \alpha)(u - \alpha) \quad (5.6.3)$$
$$+ g_u(x, \beta)(u - \alpha),$$

and consider the following related iterative schemes for $k = 0, 1, 2, \ldots$

$$L\beta_{k+1} = G(x, \beta_{k+1}; \alpha_k, \beta_k) \text{ in } \Omega, \quad \beta_{k+1} = 0 \text{ on } \partial\Omega, \quad (5.6.4)$$

and

$$L\alpha_{k+1} = F(x, \alpha_{k+1}; \alpha_k, \beta_k) \text{ in } \Omega, \quad \alpha_{k+1} = 0 \text{ on } \partial\Omega. \quad (5.6.5)$$

The variational forms associated with (5.6.4) and (5.6.5) are given by

$$B[\beta_{k+1}, v] = \int_\Omega G(x, \beta_{k+1}; \alpha_k, \beta_k) v \, dx, \quad (5.6.6)$$

and

$$B[\alpha_{k+1}, v] = \int_\Omega F(x, \alpha_{k+1}; \alpha_k, \beta_k) v \, dx, \quad (5.6.7)$$

for all $v \in H_0^1(\Omega)$, $v \geq 0$ a.e. We shall show that the weak solutions β_k, α_k of (5.6.4) and (5.6.5), respectively, are uniquely defined and satisfy

$$\alpha_0 \leq \alpha_1 \leq \alpha_2 \leq \ldots \leq \alpha_k \leq \beta_k \leq \ldots \leq \beta_2 \leq \beta_1 \leq \beta_0 \text{ a.e. in } \Omega \quad (5.6.8)$$

Let us consider $k = 0$. Then according to (5.6.2), $G(x, u; \alpha_0, \beta_0)$ is of the form $G(x, u; \alpha_0, \beta_0) = d(x)u + h(x)$, where $d(x) = f_u(x, \alpha_0) + g_u(x, \beta_0)$ and $h(x) = f(x, \beta_0) + g(x, \beta_0) - f_u(x, \alpha_0)\beta - g_u(x, \beta_0)\beta_0$. Since $f_u(x, u)$ is nondecreasing in u and $g_u(x, u)$ is nonincreasing in u for each $x \in \Omega$, a.e., we find that

$$[c(x) - d(x)] = c(x) - f_u(x, \alpha_0) - g_u(x, \beta_0)$$
$$\geq c(x) - f_u(x, \beta_0) - g_u(x, \alpha_0)$$
$$\geq N$$
$$> 0$$

and $h \in L^2(\Omega)$ by $(B3)$. Thus by Theorem 5.2.4 there is a unique weak solution $\beta_1 \in H_0^1 \in (\Omega)$ of (5.6.4). Similarly, one gets the existence of a weak solution $\alpha_1 \in H_0^1(\Omega)$ of (5.6.5). We shall next show that

$$\alpha_0 \le \alpha_1 \le \beta_1 \le \beta_0 \text{ a.e. in } \Omega. \qquad (5.6.9)$$

Since α_0 is a lower solution of (5.6.1), we see that

$$B[\alpha_0, v] \le \int_\Omega F(x, \alpha_0; \alpha_0, \beta_0) v dx$$

for all $v \in H_0^1(\Omega)$, $v \ge 0$ a.e. By the definition of the iterates, α_1 is the unique weak solution of (5.6.5) for $k = 0$ and thus α_1 satisfies the variational form (5.6.7) with $k = 0$. Let $p = \alpha_0 - \alpha_1$ so that $p(0) \le 0$ on $\partial\Omega$ and for $v \in H_0^1(\Omega)$, $v \ge 0$ a.e.

$$\int_\Omega \sum_{i,j=1}^n a_{ij}(x) p_{x_j} v_{x_j} + \int_\Omega c_0(x) p v dx \le 0$$

where $c_0(x) = c(x) - d(x) \ge 0$. Hence Corollary 5.2.1 yields $p \le 0$, that is, $\alpha_0 \le \alpha_1$ a.e. in Ω. Similarly, we can show that $\beta_1 \le \beta_0$ a.e. in Ω. Next we shall show that $\alpha_1 \le \beta_0$ and $\alpha_0 \le \beta_1$ a.e. in Ω. To show that $\alpha_1 \le \beta_0$ we employ the following inequalities which are consequences of $f_{uu}(x, u) \ge 0$ and $g_{uu}(x, u) \le 0$ imposed by $(B2)$,

$$f(x, u) \ge f(x, v) + f_u(x, v)(u - v), \qquad (5.6.10)$$

$$g(x, u) \ge g(x, v) + g_u(x, u)(u - v), \qquad (5.6.11)$$

for all $\alpha_0 \le v \le u \le \beta_0$. By (5.6.10) and (5.6.11), we obtain for all $v \in H_0^1(\Omega)$, $v \ge 0$ a.e.,

$$
\begin{aligned}
B[\alpha_1, v] &= \int_\Omega F(x, \alpha_1; \alpha_0, \beta_0) v dx \qquad (5.6.12)\\
&\le \int_\Omega [f(x, \beta_0) - f_u(x, \alpha_0)(\beta_0 - \alpha_0)\\
&\quad + g(x, \beta_0) - g_u(x, \beta_0)(\beta_0 - \alpha_0)\\
&\quad + f_u(x, \alpha_0)(\alpha_1 - \alpha_0) + g_u(x, \beta_0)(\alpha_1 - \alpha_0)] v dx\\
&\le \int_\Omega [f(x, \beta_0) + g(x, \beta_0)\\
&\quad + f_u(x, \alpha_0)(\alpha_1 - \alpha_0 + \alpha_0 - \beta_0)\\
&\quad + g_u(x, \beta_0)(\alpha_0 - \beta_0 + \alpha_1 - \alpha_0)] v dx\\
&= \int_\Omega G(x, \alpha_1; \alpha_0, \beta_0) v dx.
\end{aligned}
$$

Since β_0 is an upper solution of (5.6.1), it satisfies, for all $v \in H_0^1(\Omega)$, $v \geq 0$ a.e.,

$$B[\beta_0, v] \geq \int_\Omega [f(x, \beta_0) + g(x, \beta_0)]v dx = \int_\Omega G(x, \beta_0; \alpha_0, \beta_0)v dx.$$

Hence by Corollary 5.2.1, $\alpha_1 \leq \beta_0$ a.e. in Ω. A similar argument proves that $\alpha_0 \leq \beta_1$ a.e. in Ω. To prove $\alpha_1 \leq \beta_1$, we use (5.6.10) and (5.6.11) and the fact $g_u(x, u)$ is nonincreasing in u and $\alpha_1 \leq \beta_0$ to get

$$
\begin{aligned}
B[\alpha_1, v] &= \int_\Omega F(x, \alpha_1; \alpha_0, \beta_0)v dx && (5.6.13) \\
&\leq \int_\Omega [f(x, \alpha_1) + g(x, \alpha_1) + g_u(x, \alpha_1)(\alpha_0 - \alpha_1) \\
&\quad + g_u(x, \beta_0)(\alpha_1 - \alpha_0)]v dx \\
&\leq \int_\Omega [f(x, \alpha_1) + g(x, \alpha_1)]v dx \\
&= \int_\Omega F(x, \alpha_1; \alpha_1, \beta_1)v dx
\end{aligned}
$$

for all $v \in H_0^1(\Omega)$, $v \geq 0$ a.e. Similarly, for all $v \in H_0^1(\Omega)$, $v \geq 0$ a.e.,

$$
\begin{aligned}
B[\beta_1, v] &= \int_\Omega G(x, \beta_1; \alpha_0, \beta_0)v dx && (5.6.14) \\
&\geq \int_\Omega [f(x, \beta_1) + f_u(x, \beta_1)(\beta_0 - \beta_1) \\
&\quad + f_u(x, \alpha_0)(\beta_1 - \beta_0) + g(x, \beta_1)]v dx \\
&= \int_\Omega [f(x, \beta_1) + (-f_u(x, \beta_1) + f_u(x, \alpha_0))(\beta_1 - \beta_0) + g(x, \beta_1)]v dx \\
&\geq \int_\Omega [f(x, \beta_1) + g(x, \beta_1)]v dx \\
&= \int_\Omega F(x, \beta_1; \alpha_1, \beta_1)v dx,
\end{aligned}
$$

because of the fact that $f_u(x, u)$ is nondecreasing in u and $\alpha_0 \leq \beta_0$. The inequalities (5.6.13) and (5.6.14) imply that α_1 is a lower solution and β_1 is an upper solution of (5.6.1). Since $f + g$ is Lipschitz continuous in $\alpha_0 \leq u \leq \beta_0$, it follows from Theorem 5.2.5 that $\alpha_1 \leq \beta_1$ in Ω proving (5.6.9).

We shall next prove that if

$$\alpha_{k+1} \leq \alpha_k \leq \beta_k \leq \beta_{k+1} \text{ a.e. in } \Omega, \qquad (5.6.15)$$

for some $k > 1$, then it follows that

$$\alpha_k \leq \alpha_{k+1} \leq \beta_{k+1} \leq \beta_k \text{ a.e. in } \Omega. \tag{5.6.16}$$

Since α_k satisfies, for $v \in H_0^1(\Omega)$, $v \geq 0$ a.e.,

$$B[\alpha_k, v] = \int_\Omega F(x, \alpha_k; \alpha_{k-1}, \beta_{k-1}) v \, dx$$

by utilizing the arguments employed to obtain (5.6.13) and (5.6.14), we can show because of (5.6.15) that

$$B[\alpha_k, v] \leq \int_\Omega F(x, \alpha_k; \alpha_k, \beta_k) v \, dx, \tag{5.6.17}$$

and

$$B[\beta_k, v] \geq \int_\Omega G(x, \beta_k; \alpha_k, \beta_k) v \, dx, \tag{5.6.18}$$

for each $v \in H_0^1(\Omega)$, $v \geq 0$ a.e. Furthermore, $\alpha_{k+1}, \beta_{k+1}$ satisfy (5.6.5) and (5.6.4), respectively. Hence we conclude from Corollary 5.2.1 that $\alpha_k \leq \alpha_{k+1}$ and $\beta_{k+1} \leq \beta_k$ a.e. in Ω. Next we show that $\alpha_{k+1} \leq \beta_k$ and $\alpha_k \leq \beta_{k+1}$. We find using (5.6.10) and (5.6.11) that

$$
\begin{aligned}
B[\alpha_{k+1}, v] &= \int_\Omega F(x, \alpha_{k+1}; \alpha_k, \beta_k) v \, dx \\
&\leq \int_\Omega [f(x, \beta_k) - f_u(x, \alpha_k)(\beta_k - \alpha_k) \\
&\quad + g_u(x, \beta_k)(\beta_k - \alpha_k) + g(x, \beta_k) \\
&\quad + f_u(x, \alpha_k)(\alpha_{k+1} - \alpha_k) \\
&\quad + g_u(x, \beta_k)(\alpha_{k+1} - \alpha_k)] v \, dx \\
&\leq \int_\Omega [f(x, \beta_k) + g(x, \beta_k) \\
&\quad + f_u(x, \alpha_k)(-\beta_k + \alpha_k + \alpha_{k-1} - \alpha_k) \\
&\quad + g_u(x, \beta_k)(\alpha_k - \beta_k + \alpha_{k+1} - \alpha_k)] v \, dx \\
&= \int_\Omega G(x, \alpha_{k+1}; \alpha_k, \beta_k) v \, dx.
\end{aligned}
$$

This, together with (5.6.18), gives by Corollary 5.2.1, $\alpha_{k+1} \leq \beta_k$. Similarly, we can show that $\alpha_k \leq \beta_{k+1}$, using (5.6.17) and for each

$v \in H_0^1(\Omega)$, $v \geq 0$ a.e.

$$
\begin{aligned}
B[\beta_{k+1}, v] &= \int_\Omega G(x, \beta_{k+1}; \alpha_k, \beta_k) v dx \\
&\geq \int_\Omega [f(x, \alpha_k) + f_u(x, \alpha_k)(\beta_k - \alpha_k) + g(x, \alpha_k) \\
&\quad + f_u(x, \beta_k)(\beta_k - \alpha_k) \\
&\quad + f_u(x, \alpha_k)(\beta_{k+1} - \beta_k) \\
&\quad + g_u(x, \beta_k)(\beta_{k+1} - \beta_k)] v dx \\
&= \int_\Omega [f(x, \alpha_k) + g(x, \alpha_k) \\
&\quad + f_u(x, \alpha_k)(\beta_k - \alpha_k + \beta_{k+1} - \beta_k) \\
&\quad + g_u(x, \beta_k)(\beta_k - \alpha_k + \beta_{k+1} - \beta_k)] v dx \\
&= \int_\Omega F(x, \beta_{k+1}; \alpha_k, \beta_k) v dx.
\end{aligned}
$$

Finally, to prove $\alpha_{k+1} \leq \beta_{k+1}$ a.e. in Ω, we need to show that α_{k+1} and β_{k+1} satisfy

$$
B[\alpha_{k+1}, v] \leq \int_\Omega F(x, \alpha_{k+1}; \alpha_k, \beta_k) v dx
$$

and

$$
B[\beta_{k+1}, v] \geq \int_\Omega G(x, \beta_{k+1}; \alpha_k, \beta_k) v dx,
$$

for each $v \in H_0^1(\Omega)$, $v \geq 0$ a.e. This precisely employs the same arguments as we have utilized in proving (5.6.13) and (5.6.14) replacing α_0, β_0, α_a, β_1 by α_k, β_k, α_{k+1}, β_{k+1} and using (5.6.10) and (5.6.11) as well as the monotone character of f_u and g_u. Consequently, we get from Theorem 5.2.3 that $\alpha_{k+1} \leq \beta_{k+1}$, proving (5.6.16). Thus by induction, it follows that (5.6.8) is true for each $k = 1, 2, \ldots$.

(b) Convergence of $\{\alpha_k\}$, $\{\beta_k\}$ to the unique solution of (5.6.1)
By the monotone character of the iterates $\{\alpha_k\}$, $\{\beta_k\}$, according to (5.6.8) there exist pointwise limits

$$
\rho(x) = \lim_{k \to \infty} \alpha_k(x), \quad r(x) = \lim_{k \to \infty} \beta_k(x), \quad \text{a.e. in } \Omega.
$$

Moreover, since $\alpha_0 \leq \alpha_k \leq \beta_k \leq \beta_0$ a.e. in Ω, it follows by Lebesgue's dominated convergence theorem that

$$
\alpha_k \to \rho \text{ and } \beta_k \to r \text{ in } L^2(\Omega).
$$

We note that α_k satisfies, for each $v \in H_0^1(\Omega)$, $v \geq 0$ a.e. in Ω,

$$\int_\Omega \left[\sum_{i,j}^n a_{ij}(x)\alpha_{k,x_i}v_{x_j} + c_0(x)\alpha_k v \right] dx$$

$$= \int_\Omega [f(x,\alpha_{k-1}) - f_u(x,\alpha_{k-1})\alpha_{k-1} + g(x,\alpha_{k-1})$$
$$- g_u(x,\beta_{k-1})\alpha_{k-1}]v dx,$$

where $c_0(x) = c(x) - f_u(x,\alpha_{k-1}) - g_u(x,\beta_{k-1})$. We now use the ellipticity condition and $(B3)$ with $v = \alpha_k$ to get

$$\int_\Omega [\theta \mid \alpha_{k,x} \mid^2 + N \mid \alpha_k \mid^2] dx \;\leq\; \int_\Omega [f(x,\alpha_{k-1}) - f_u(x,\alpha_{k-1})\alpha_{k-1}$$
$$+ g(x,\alpha_{k-1}) - g_u(x,\beta_{k-1})\alpha_{k-1}]\alpha_k dx.$$

We then get, since by $(B3)$ the integrand on the right-hand side belongs to $L^2(\Omega)$, in view of the fact that α_{k-1}, $\beta_{k-1} \in H_0^1(\Omega)$, the estimate

$$\sup_k \parallel \alpha_k \parallel_{H_0^1(\Omega)} < \infty.$$

A similar argument implies that $\sup_k \parallel \beta_k \parallel_{H_0^1(\Omega)} < \infty$. Hence there exist subsequences $\{\alpha_{k_j}\}$, $\{\beta_{k_j}\}$ which converge weakly in $H_0^1(\Omega)$ to $\rho, r \in H_0^1(\Omega)$, respectively. To verify that ρ, r are weak solutions of (5.6.1), we fix $v \in H_0^1(\Omega)$, $v \geq 0$ a.e. and find that α_{k+1}, β_{k+1} satisfy (5.6.6) and (5.6.7) with F and G defined by (5.6.2) and (5.6.3). Taking limits as $k \to \infty$, we obtain

$$B[\rho,v] = \int_\Omega [f(x,\rho) + g(x,\rho)]v dx,$$

and

$$B[r,v] = \int_\Omega [f(x,r) + g(x,r)]v dx,$$

showing that ρ, r are weak solutions of (5.6.1).

To prove that $\rho = r = u$ is the unique solution of (5.6.1), it is enough to prove that $r \leq \rho$ a.e. in Ω, since we know that $\rho \leq r$ a.e. in Ω. Taking $\alpha = r$, $\beta = \rho$, and applying Theorem 5.2.5, we find that $r \leq \rho$ a.e. in Ω proving the claim.

(c) **Quadratic convergence of** $\{\alpha_k\}$, $\{\beta_k\}$ To prove the quadratic convergence of sequences $\{\alpha_k\}$, $\{\beta_k\}$ to the unique solution u respectively, we set

$$p_{k-1} = u - \alpha_{k+1}, \quad q_{k+1} = \beta_{k+1} - u,$$

so that $p_{k+1}(x) = 0$ on Ω and $q_{k+1}(x) = 0$ on Ω. We then have for $v \in H_0^1(\Omega)$, $v \geq 0$ a.e. in Ω, using the fact that f_u is nondecreasing in u and g_u is nonincreasing in u,

$$
\begin{aligned}
B[p_{k+1}, v] &= \int_\Omega [f(x, u) + g(x, u) - f(x, \alpha_k) \\
&\quad - f_u(x, \alpha_k)(\alpha_{k+1} - \alpha_k) \\
&\quad - g(x, \alpha_k) - g_u(x, \beta_k)(\alpha_{k+1} - \alpha_k)] v \, dx \\
&\leq \int_\Omega [(f_u(x, u) - f_u(x, \alpha_k)) p_k \\
&\quad - (g_u(x, \beta_k) - g_u(x, \alpha_k)) p_k \\
&\quad + (f_u(x, \alpha_k) + g_u(x, \beta_k) p_{k+1}] v \, dx \\
&= \int_\Omega [f_{uu}(x, \xi) p_k^2 - g_{uu}(x, \sigma)(\beta_k - \alpha_k) \\
&\quad + (f_u(x, \alpha_k) + g_u(x, \beta_k)) p_{k+1}] v \, dx,
\end{aligned}
$$

where $\alpha_k \leq \xi \leq u$, $\alpha_k \leq \sigma \leq \beta_k$. But

$$
\begin{aligned}
-g_{uu}(x, \sigma)(\beta_k - \alpha_k) p_k &\leq N_2(q_k + p_k) p_k \\
&\leq N_2(p_k^2 + p_k q_k) \\
&\leq \frac{3}{2} N_2 p_k^2 + \frac{1}{2} N_2 q_k^2,
\end{aligned}
$$

where $\mid g_{uu}(x, u) \mid_{L^\infty(\Omega)} \leq N_2$. Thus we get

$$\int_\Omega \left[\sum_{i,j=1}^n a_{ij}(x)(p_{k+1})_{x_i} v_{x_j} + c_0(x) p_{k+1} v \right] dx$$

$$\leq \int_\Omega [(N_1 + \frac{3}{2} N_2) p_k^2 + \frac{1}{2} N_2 q_k^2] v \, dx,$$

where $\mid f_{uu}(x, u) \mid_{L^\infty(\Omega)} \leq N_1$. Taking $v = p_{k+1}$ and using the ellipticity condition, we arrive at

$$\int_\Omega [\theta \mid (p_{k+1})_x \mid^2 + N \mid p_{k+1} \mid^2] dx$$

$$\leq N_0 \int_\Omega (\mid p_k \mid^2 + \mid q_k \mid^2) p_{k+1} dx,$$

where $N_0 = N_1 + \frac{3}{2} N_2$. Let $\theta_0 = \min(\theta, N)$. Then

$$
\begin{aligned}
\theta_0 \parallel p_{k+1} \parallel^2_{H_0^1(\Omega)} \leq{} & N_0 \left(\frac{2}{\theta_0}\right)^{1/2} [\parallel p_k^2 \parallel_{L^2(\Omega)} + \parallel q_k^2 \parallel_{L^2(\Omega)}] \left(\frac{\theta_0}{2}\right)^{1/2} \\
& \times \parallel p_{k+1} \parallel_{L^2(\Omega)} \\
\leq{} & \frac{1}{2}\left[\frac{2N_0^2}{\theta_0} \parallel p_k^2 \parallel^2_{L^2(\Omega)} + \frac{\theta_0}{2} \parallel p_{k+1} \parallel^2_{L^2(\Omega)}\right] \\
& + \frac{1}{2}\left[\frac{2N_0^2}{\theta_0} \parallel q_k^2 \parallel^2_{L^2(\Omega)} + \frac{\theta_0}{2} \parallel p_{k+1} \parallel^2_{L^2(\Omega)}\right] \\
={} & \frac{N_0^2}{\theta_0}[\parallel p_k^2 \parallel^2_{L^2(\Omega)} + \parallel q_k^2 \parallel^2_{L^2(\Omega)}] + \frac{\theta_0}{2} \parallel p_{k+1} \parallel^2_{L^2(\Omega)}
\end{aligned}
$$

We then get

$$\frac{\theta_0}{2} \parallel p_{k+1} \parallel^2_{H_0^1(\Omega)} \leq \frac{N_0^2}{\theta_0}[\parallel p_k^2 \parallel^2_{H_0^1(\Omega)} + \parallel q_k^2 \parallel^2_{H_0^1(\Omega)}]$$

or equivalently,

$$\parallel p_{k+1} \parallel^2_{H_0^1(\Omega)} \leq \frac{2N_0^2}{\theta_0^2}[\parallel p_k^2 \parallel^2_{H_0^1(\Omega)} + \parallel q_k^2 \parallel^2_{H_0^1(\Omega)}].$$

One can get a similar estimate for v_{k+1}. We omit the details.

5.7   Generalized Quasilinearization (GQ): Quasilinear Problem

This section deals with the method of generalized quasilinearization for quasilinear elliptic boundary value problems. We shall consider only a special case of (5.4.1) with $g(x, u) \equiv 0$. Accordingly, Definitions 5.4.1, 5.4.2 and 5.4.3 are to be modified for the special quasilinear elliptic BVP

$$\left[\begin{aligned} -\Delta_p u + c(x) u &= f(x, u) \quad \text{in } \Omega, \\ u &= 0 \text{ on } \partial\Omega \quad \text{in the sense of trace.} \end{aligned} \right. \tag{5.7.1}$$

We shall also utilize the comparison result, Theorem 5.4.1 with $g(x, u) \equiv 0$. With this specialization, we shall prove the method of generalized quasilinearization for the BVP (5.7.1) when $f(x, u)$ is convex in u.

We prove the following result in this direction.

Theorem 5.7.1 *Assume that*

(A_1) $\alpha_0, \beta_0 \in W^{1,p}(\Omega) \cap L^\infty(\Omega)$ *are lower and upper solutions of* (5.7.1) *such that* $\alpha_0(x) \leq \beta_0(x)$, *a.e. in* Ω;

(A_2) $f : \bar{\Omega} \times R \to R$, $f(x, u)$ *is a Caratheodory function such that* $f_u(x, u)$, $f_{uu}(x, u)$ *exist and* $f_{uu}(x, u) \geq 0$ *a.e. in* Ω;

(A_3) $0 < N \leq c(x) - f_u(x, \beta_0)$ *a.e. in* Ω, *for any* $\eta \in W^{1,p}(\Omega)$ *such that* $\alpha_0 \leq \eta \leq \beta_0$, $h(x) \in L^\delta(\Omega)$, *where* $h(x) = f(x, \eta) - f_u(x, \eta)\eta$, $\delta > \frac{n}{p}$, *and* $f_u(x, \eta) \leq d(x)$ *a.e. in* Ω *with* $d \in L^1(\Omega)$.

Then there exist monotone sequences $\{\alpha_k\}$, $\{\beta_k\} \in W_0^{1,p}(\Omega) \cap L^\infty(\Omega)$ *such that* $\alpha_n \to \rho$, $\beta_n \to r$ *weakly in* $W_0^{1,p}(\Omega)$, *and* $\rho = r = u$ *is the unique weak solution of* (5.7.1) *satisfying* $\alpha_0 \leq u \leq \beta_0$ *a.e. in* Ω. *Moreover, the convergence is quadratic for the case* $p \geq 2$.

Proof   Consider the following BVPs for each $k = 1, 2, \ldots$

$$\begin{bmatrix} -\Delta_p \alpha_{k+1} + c(x)\alpha_{k+1} &= f(x, \alpha_k) + f_u(x, \alpha_k)(\alpha_{k+1} - \alpha_k) \text{ in } \Omega, \\ \alpha_{k+1} &= 0 \text{ on } \partial\Omega \end{bmatrix}$$

(5.7.2)

and

$$\begin{bmatrix} -\Delta_p \beta_{k+1} + c(x)\beta_{k+1} &= f(x, \beta_k) + f_u(x, \alpha_k)(\beta_{k+1} - \beta_k) \text{ in } \Omega, \\ \beta_{k+1} &= 0 \text{ on } \partial\Omega \end{bmatrix}$$

(5.7.3)

whose variational forms are given by

$$\int_\Omega [| D\alpha_{k+1} |^{p-2} D\alpha_{k+1} Dv + c(x)\alpha_{k+1} v]$$

(5.7.4)

$$= \int_\Omega [f(x, \alpha_k) + f_u(x, \alpha_k)(\alpha_{k+1} - \alpha_k)]v,$$

and

$$\int_\Omega [| D\beta_{k+1} |^{p-2} D\beta_{k+1} Dv + c(x)\beta_{k+1} v]$$

(5.7.5)

$$= \int_\Omega [f(x, \beta_k) + f_u(x, \alpha_k)(\beta_{k+1} - \beta_k)]v,$$

for each $v \in W_0^{1,p}(\Omega)$, $v \geq 0$ a.e. in Ω. Since by (A_2), $f_{uu}(x, u) \geq 0$ a.e. in Ω, we obtain

$$f(x, u_1) \geq f(x, u_2) + f_u(x, u_2)(u_1 - u_2)$$

(5.7.6)

whenever $u_1 \geq u_2$, a.e. in Ω. Also, the function $h(x) = f(x, \eta) - f_u(x, \eta)\eta$ for any $\eta \in W^{1,p}(\Omega)$ such that $\alpha_0 \leq \eta \leq \beta_0$ satisfies, by (A_3), $h \in L^\delta(\Omega)$. Consequently, by Theorem 5.4.2, there exist unique weak solutions α_k, β_k $\in W_0^{1,p}(\Omega) \cap L^\infty(\Omega)$, $k = 1, 2, \ldots$, of the BVPs (5.7.2) and (5.7.3) respectively provided $\alpha_0 \leq \alpha_k \leq \beta_k \leq \beta_0$ for each $k \geq 1$. Now, we shall therefore show that

$$\alpha_0 \leq \alpha_1 \leq \ldots \leq \alpha_k \leq \beta_k \leq \ldots \leq \beta_1 \leq \beta_0 \text{ a.e. in } \Omega. \tag{5.7.7}$$

We will first prove

$$\alpha_0 \leq \alpha_1, \quad \beta_1 \leq \beta_0 \text{ and } \alpha_1 \leq \beta_1, \text{ a.e. in } \Omega. \tag{5.7.8}$$

To prove $\alpha_0 \leq \alpha_1$, we consider the variational problem

$$\int_\Omega | Du |^{p-2} Du Dv + c(x) u v = \int_\Omega F(x, u; \alpha_0) v \tag{5.7.9}$$

for each $v \in W^{1,p}(\Omega)$, $v \geq 0$, a.e. where $F(x, u; \alpha_0) = f(x, \alpha_0) + f_u(x, \alpha_0)(u - \alpha_0)$. Then α_0 is a lower solution of (5.7.9) because we have

$$\int_\Omega [| D\alpha_0 |^{p-2} D\alpha_0 Dv + c(x)\alpha_0 v] \leq \int_\Omega f(x, \alpha_0)v = \int_\Omega F(x, \alpha_0; \alpha_0)v.$$

Moreover, since α_1 is the unique solution of (5.7.2) for $k = 0$, we find

$$\int_\Omega [| D\alpha_1 |^{p-2} D\alpha_1 Dv + c(x)\alpha_1 v] = \int_\Omega F(x, \alpha_1; \alpha_0)v, \tag{5.7.10}$$

which may be considered as an upper solution of (5.7.9). Also $F(x, u; \alpha_0)$ satisfies condition (5.4.4) by (A_3) and hence by the Comparison Theorem 5.4.1 we get $\alpha_0(x) \leq \alpha_1(x)$, a.e. in Ω. A similar argument proves $\beta_1(x) \leq \beta_0(x)$ a.e. in Ω. To prove $\alpha_1 \leq \beta_1$, we see from (5.4.5) and (5.7.6) that β_1 satisfies

$$\int_\Omega [| D\beta_1 |^{p-2} D\beta_1 Dv + c(x)\beta_1 v] \tag{5.7.11}$$

$$= \int_\Omega [f(x, \beta_0) + f_u(x, \alpha_0)(\beta_1 - \beta_0)]v$$

$$\geq \int_\Omega [f(x, \alpha_0) + f_u(x, \alpha_0)(\beta_0 - \alpha_0 + \beta_1 - \beta_0)]v$$

$$= \int_\Omega F(x, \beta_1; \alpha_0)v,$$

which shows that β_1 is an upper solution of (5.7.9). Thus by Theorem 5.4.1, (5.7.10) and (5.7.11) yield $\alpha_1(x) \leq \beta_1(x)$, a.e. in Ω, proving (5.7.8).

Next we will show that if $\alpha_{k-1} \leq \alpha_k \leq \beta_k \leq \beta_{k-1}$, a.e. in Ω for some $k > 1$, then

$$\alpha_k \leq \alpha_{k+1} \leq \beta_{k+1} \leq \beta_k \text{ a.e. in } \Omega. \tag{5.7.12}$$

To show that $\alpha_k \leq \alpha_{k+1}$, let us consider the variational problem

$$\int_\Omega [|\, Du \,|^{p-2} \, DuDv + c(x)uv] = \int_\Omega F(x, u; \alpha_k)v \tag{5.7.13}$$

for each $v \in W_0^{1,p}(\Omega)$, $v \geq 0$, a.e. where

$$F(x, u; \alpha_k) = f(x, \alpha_k) + f_u(x, \alpha_k)(u - \alpha_k).$$

Since α_k is the weak solution of

$$\int_\Omega [|\, D\alpha_k \,|^{p-2} \, D\alpha_k Dv + c(x)\alpha_k v] = \int_\Omega [f(x, \alpha_{k-1}) + f_u(x, \alpha_{k-1})(\alpha_k - \alpha_{k-1})]v$$

for each $v \in W_0^{1,p}(\Omega)$, $v \geq 0$ a.e. in Ω, we get, because of (5.7.6) and the fact that $\alpha_{k-1} \leq \alpha_k$

$$\int_\Omega [|\, D\alpha_k \,|^{p-2} \, D\alpha_k Dv + c(x)\alpha_k v] \leq \int_\Omega f(x, \alpha_k)v \equiv \int_\Omega F(x, \alpha_k; \alpha_k)v,$$

showing α_k is a lower solution of (5.7.13). Furthermore, since α_{k+1} satisfies (5.7.2), it is clear, by the definition of $F(x, u; \alpha_k)$, that α_{k+1} may be taken as an upper solution of (5.7.13). Thus by Theorem 5.4.1, we have $\alpha_k(x) \leq \alpha_{k+1}(x)$ a.e. in Ω. Similarly, we can obtain $\beta_{k+1} \leq \beta_k$ a.e. in Ω. To prove that $\alpha_{k+1} \leq \beta_{k+1}$, we find that β_{k+1} satisfies, because of (5.7.5) and (5.7.6)

$$\int_\Omega [|\, D\beta_{k+1} \,|^{p-2} \, D\beta_{k+1} Dv + c(x)\beta_{k+1}v]$$

$$\geq \int_\Omega [f(x, \alpha_k) + f_u(x, \alpha_k)(\beta_k - \alpha_k + \beta_{k+1} - \beta_k)]v$$

$$= \int_\Omega F(x, \beta_{k+1}; \alpha_k)v,$$

for each $v \in W_0^{1,p}(\Omega)$, $v \geq 0$ a.e. Thus, by Theorem 5.4.1 since α_{k+1} satisfies (5.7.4), we get $\alpha_{k+1}(x) \leq \beta_{k+1}(x)$ a.e. in Ω, proving (5.7.12). Hence by induction (5.7.7) is valid.

As a result of (5.7.7), we have

$$\rho(x) = \lim_{k \to \infty} \alpha_k(x), \quad r(x) = \lim_{k \to \infty} \beta_k(x), \tag{5.7.14}$$

a.e in Ω. Moreover, $\alpha_k(x) \to \rho(x)$, $\beta_k(x) \to r(x)$ in $L^p(\Omega)$ which is guaranteed by the dominated convergence theorem and (5.7.7). We note also that α_k satisfies

$$\int_\Omega [| D\alpha_k |^{p-2} D\alpha_k Dv + c_0(x)\alpha_k v]$$

$$= \int_\Omega [f(x, \alpha_{k-1}) - f_u(x, \alpha_{k-1})\alpha_{k-1}]v$$

for each $v \in W_0^{1,p}(\Omega)$, $v \geq 0$ a.e. in Ω, where $c_0(x) = c(x) - f_u(x, \alpha_{k-1}) \geq c(x) - f_u(x, \beta_0)$. Using (5.4.2) with $v = \alpha_k$ and (A_3), it follows that

$$\| \alpha_k \|^p + N \int_\Omega | \alpha_k |^2 \leq \int_\Omega [f(x, \alpha_{k-1}) - f_u(x, \alpha_{k-1})\alpha_{k-1}]\alpha_k.$$

Since by (A_3), the right-hand side belongs to $L^\delta(\Omega)$, this implies that $\sup_k \| \alpha_k \|_{W_0^{1,p}(\Omega)} < \infty$. Similarly, $\sup_k \| \beta_k \|_{W_0^{1,p}(\Omega)} < \infty$. Hence there exist subsequences $\{\alpha_{k_j}\}$, $\{\beta_{k_j}\}$ that converge weakly to $\rho, r \in W_0^{1,p}(\Omega)$, respectively.

To show that (ρ, r) are weak solutions of (5.7.1), we fix $v \in W_0^{1,p}(\Omega)$ and find that

$$\int_\Omega [| D\alpha_k |^{p-2} D\alpha_k Dv + c(x)\alpha_k v] \tag{5.7.15}$$

$$= \int_\Omega [f(x, \alpha_{k-1} + f_u(x, \alpha_{k-1}(\alpha_k - \alpha_{k-1})]v,$$

and

$$\int_\Omega [| D\beta_k |^{p-2} D\beta_k Dv + c(x)\beta_k v] \tag{5.7.16}$$

$$= \int_\Omega [f(x, \beta_{k-1}) + f_u(x, \alpha_{k-1})(\beta_k - \beta_{k-1})]v.$$

Set $Q(Du) = | Du |^{p-2} Du$ and note that Q satisfies

$$\int_\Omega [Q(D\alpha_k) - Q(Dw)][D\alpha_k - Dw] \geq 0 \tag{5.7.17}$$

for each $w \in W^{1,p}(\Omega)$, $\alpha_k \neq w$, due to (5.4.2). Since $\{Q(D\alpha_k)\}$ is bounded in $L^p(\Omega)$, $Q(D\alpha_k) \to \xi$ weakly in $L^p(\Omega)$ for some $\xi \in L^p(\Omega)$. This implies taking the limit as $k \to \infty$, in (5.7.15)

$$\int_\Omega \xi Dv + c(x)\rho v = \int_\Omega f(x, \rho)v. \tag{5.7.18}$$

Setting $v = \alpha_k$ in (5.7.15) and substituting in (5.7.17), we arrive at

$$\int_\Omega [f(x, \alpha_{k-1}) + f_u(x, \alpha_{k-1})(\alpha_k - \alpha_{k-1})]\alpha_k - c(x)\alpha_k^2 - Q(D\alpha_k)Dw$$

$$-Q(Dw)[D\alpha_k - Dw] \geq 0.$$

As $k \to \infty$, we get

$$\int_\Omega f(x, \rho)\rho - c(x)\rho^2 - \xi \cdot Dw - Q(Dw)[D\rho - Dw] \geq 0. \qquad (5.7.19)$$

Set $v = \rho$ in (5.7.18). Then using (5.7.18) in (5.7.19), it follows that

$$\int_\Omega \xi \cdot D\rho - \xi \cdot D(w) - Q(Dw)[D(\rho - w)] \geq 0,$$

which leads to

$$\int_\Omega [\xi - Q(Dw)]D(\rho - w) \geq 0.$$

Choose $w = \rho - \lambda v$, $\lambda > 0$, so that one has

$$\int_\Omega [\xi - Q(D(\rho - \lambda v))Dv] \geq 0.$$

As $\lambda \to 0$, this results in

$$\int_\Omega [\xi - Q(D\rho)]Dv \geq 0. \qquad (5.7.20)$$

Setting $v = -v$, we find $\int_\Omega [\xi - Q(D\rho)]Dv \leq 0$ and thus we have

$$\int_\Omega [\xi - Q(D\rho)]Dv = 0.$$

Hence

$$\int \xi Dv = \int_\Omega Q(D\rho)Dv = \int_\Omega [f(x, \rho) - c(x)\rho]v$$

or equivalently,

$$\int_\Omega [Q(D\rho)Dv + c(x)\rho v] = \int_\Omega f(x, \rho)v$$

for $v \in W_0^{1,p}(\Omega)$, $v \geq 0$, showing that ρ is a weak solution of (5.7.1) in view of definition $Q(D\rho)$. Similar arguments show that r is a weak solution of (5.7.1).

In order to show that $\rho = r = u$ is the unique weak solution of (5.7.1), it is enough to prove that $r \leq \rho$ in Ω, since we already know that $\rho \leq r$ in Ω. Taking $\alpha_0 = r$, $\beta_0 = \rho$ and using Theorem 5.4.1, we get $r \leq \rho$ in Ω proving the claim.

Finally we need to estimate the rate of convergence of $\{\alpha_k\}$, $\{\beta_k\}$ to u respectively. We set $P_{k+1} = u - \alpha_{k+1}$ and note that $P_{k+1}(0) = 0$ on $\partial\Omega$. For each $v \in W_0^{1,p}(\Omega)$, $v \geq 0$ a.e., we get successively, using the mean value theorem and the increasing nature of $f_u(x, u)$,

$$\int_\Omega \{[\mid Du \mid^{p-2} Du - \mid D\alpha_{k+1} \mid^{p-2} D\alpha_{k+1}]Dv + c(x)P_{k+1}v\}$$

$$= \int_\Omega [f(x, u) - f(x, \alpha_k) - f_u(x, \alpha_k)(\alpha_{k+1} - \alpha_k)]v$$

$$\leq \int_\Omega \{[f_u(x, u) - f_u(x, \alpha_k)]P_k + f_u(x, \alpha_k)P_{k+1}\}v$$

$$= \int_\Omega [f_{uu}(x, \sigma)P_k^2 + f_u(x, \alpha_k)P_{k+1}]v.$$

Taking $v = P_{k+1}$ and using (A_3), we get, since by assumption $0 < N \leq c(x) - f_u(x, \beta_0) \leq c(x) - f_u(x, \alpha_k)$,

$$\int_\Omega [[\mid Du \mid^{p-2} Du - \mid D\alpha_{k+1} \mid^{p-2} D\alpha_k]DP_{k+1} + NP_{k+1}^2] \leq N_0 \int_\Omega P_k^2 P_{k+1}$$

where $\mid f_{uu}(x, u) \mid_{L^\infty(\Omega)} \leq N_0$ for $\alpha_0 \leq u \leq \beta_0$. If $p \geq 2$, we can use the estimate (5.4.3) to arrive at

$$c(p) \parallel P_{k+1} \parallel^p \;\leq\; N_0 \parallel (P_k)^2 \parallel_{L^{\frac{np}{n(p-1)+p}(\Omega)}} \parallel P_{k+1} \parallel_{L^{p^*}(\Omega)}$$

$$\leq\; \tilde{N}_0 \parallel (P_k^2) \parallel_{L^\delta(\Omega)} \parallel P_{k+1} \parallel$$

for a suitable constant $\tilde{N}_0$. It then follows that

$$\parallel P_{k+1} \parallel^{p-1} \leq N^0 \parallel (P_k^2) \parallel_{W_0^{1,\delta}(\Omega)},$$

where $\tilde{N}_0$ is a suitable constant. One can prove a similar estimate for $Q_{k+1} = \beta_{k+1} - u$. The proof of the theorem is complete.

We remark that for $1 < p < 2$, the problem of proving rapid convergence is open.

In the case of problems in ordinary differential equations, the quadratic convergence is defined as $\max_{0 \leq t \leq T} \mid P_{k+1}(t) \mid \leq c \max_{0 \leq t \leq T} \mid P_k(t) \mid^2$, for some constant c.

5.8 Generalized Quasilinearization (GQ): Degenerate Problems

We shall continue to discuss the method of generalized quasilinearization and extend it, in this section, to the degenerate semilinear BVP (5.5.11). The necessary preliminary notation and the results described in Section 5.5 such as the concept of weighted Sobolev spaces, existence of solutions of the linear BVP (5.5.4) and the comparison theorems will be utilized without further comment.

For convenient ready reference, let us restate (5.5.11) in the following form.

$$\left[\begin{array}{ll} Lu = f(t,u) + g(t,u) & \text{in } \Omega, \\ \quad\quad u = 0 & \text{on } \partial\Omega, \end{array}\right. \tag{5.8.1}$$

where $f, g : \Omega \times R \to R$ are Caratheodory functions. For any $u \in H_0^1(\Omega, w)$, assume that $f, g, f_u, g_u \in L^1(\Omega)$. We shall prove the following result relative to (5.8.1).

Theorem 5.8.1 *Assume that*

(A_1) $\alpha_0, \beta_0 \in H^1(\Omega, w)$ *are lower and upper solutions of* (5.8.1) *such that* $\alpha_0(x) \le \beta_0(x)$ *a.e. in* Ω;

(A_2) f, g *satisfy, in addition, that* f_u, g_u, f_{uu}, g_{uu} *exist and are Caratheodory functions,* $f_{uu}(x, u) \ge 0$, $g_{uu}(x, u) \le 0$ *a.e. in* Ω, *and* $f_{uu}, g_{uu} \in L^\infty$ *for* $\alpha_0 \le u \le \beta_0$;

(A_3) $0 < N \le c(x) - [f_u(x, \beta_0) + g_u(x, \alpha_0)]$ *a.e. in* Ω *and for any* μ, $\eta \in H^1(\Omega, w)$ *with* $\alpha_0 \le \mu \le \eta \le \beta_0$, *the function* $h(x) = f(x, \eta)$ $+g(x, \eta) - f_u(x, \mu)\eta - g_u(x, \eta)\eta$ *belongs to* $L^2(\Omega)$.

Then there exist monotone sequences $\{\alpha_n(x)\}$, $\{\beta_n(x)\}$, $\in H_0^1(\Omega, w)$ *such that* $\alpha_n \to \rho$, $\beta_n \to r$ *weakly in* $H_0^1(\Omega, w)$ *with* $\rho, r \in H_0^1(\Omega, w)$, *where* $\rho = r = u$ *is the unique solution of* (5.8.1) *satisfying* $\alpha_0 \le u \le \beta_0$ *a.e. in* Ω *and the convergence is quadratic.*

Proof Consider the following linear BVPs for $k = 1, 2, \ldots$,

$$\left[\begin{array}{ll} L\alpha_{k+1} &= F(x, \alpha_{k+1}; \alpha_k, \beta_k) \text{ in } \Omega \\ \quad \alpha_{k+1} &= 0 \text{ on } \partial\Omega, \end{array}\right. \tag{5.8.2}$$

$$\left[\begin{array}{ll} L\beta_{k+1} &= G(x, \beta_{k+1}; \alpha_k, \beta_k) \text{ in } \Omega \\ \quad \beta_{k+1} &= 0 \text{ on } \partial\Omega, \end{array}\right. \tag{5.8.3}$$

whose bilinear forms associated with the operator L are given by

$$B[\alpha_{k+1}, v] = \int_\Omega F(x, \alpha_{k+1}; \alpha_k, \beta_k)v \qquad (5.8.4)$$

and

$$B[\beta_{k+1}, v] = \int_\Omega G(x, \beta_{k+1}; \alpha_k, \beta_k)v, \qquad (5.8.5)$$

for each $v \in H_0^1(\Omega, w)$, $v \geq 0$ a.e. in Ω, where for each $k = 1, 2, \ldots$

$$F(x, u; \alpha_k, \beta_k) = f(x, \alpha_k) + g(x, \alpha_k) + f_u(x, \alpha_k)(u - \alpha_k) + g_u(x, \beta_k)(u - \alpha_k),$$
$$(5.8.6)$$

and

$$G(x, u; \alpha_k, \beta_k) = f(x, \beta_k) + g(x, \beta_k) + f_u(x, \alpha_k)(u - \beta_k) + g_u(x, \beta_k)(u - \beta_k).$$
$$(5.8.7)$$

We note that the assumptions $f_{uu}(x, u) \geq 0$ and $g_{uu}(x, u) \leq 0$ a.e. in Ω imply the inequalities

$$f(x, u) \geq f(x, v) + f_u(x, v)(u - v), \qquad (5.8.8)$$

and

$$g(x, u) \geq g(x, v) + g_u(x, u)(u - v), \text{ for } u \geq v \text{ a.e. in } \Omega. \qquad (5.8.9)$$

Since $f_u(x, u)$ is nondecreasing in u and $g_u(x, u)$ is nonincreasing in u, a.e. in Ω, by (A_2) we find that $c(x) - f_u(x, \eta) - g_u(x, \eta) \geq c(x) - f_u(x, \beta_0) - g_u(x, \alpha_0) \geq N > 0$, a.e. in Ω. Hence for each $\mu, \eta \in H^1(\Omega, w)$ with $\alpha_0 \leq \mu \leq \eta \leq \beta_0$, $h(x) \in L^2(\Omega)$ where $h(x) = f(x, \eta) + g(x, \eta) - f_u(x, \mu)\eta - g_u(x, \eta)\eta$ by (A_3). Consequently, there exist a unique weak solution of (5.8.2) and (5.8.3) for each $k = 1, 2, \ldots$ provided $\alpha_0 \leq \alpha_k \leq \beta_k \leq \beta_0$ a.e. in Ω. Our aim is therefore to show that the weak solutions $\alpha_k, \beta_k \in H_0^1(\Omega, w)$ satisfy

$$\alpha_0 \leq \alpha_1 \leq \ldots \leq \alpha_k \leq \beta_k \leq \ldots \leq \beta_1 \leq \beta_0 \text{ a.e. in } \Omega. \qquad (5.8.10)$$

We will first show that

$$\alpha_0 \leq \alpha_1 \leq \beta_1 \leq \beta_0 \text{ a.e. in } \Omega. \qquad (5.8.11)$$

Since α_0 is a lower solution of (5.8.1), we see that

$$B[\alpha_0, v] \leq \int_\Omega F(x, \alpha_0; \alpha_0, \beta_0)v$$

and

$$B[\alpha_1, v] = \int_\Omega F(x, \alpha_1; \alpha_0, \beta_0) v$$

for each $v \in H_0^1(\Omega, w)$, $v \geq 0$ a.e. in Ω. Also F satisfies condition (5.5.9) with $K(x) = f_u(x, \alpha_0) + g_u(x, \beta_0)$ and by (A_3), $0 < N \leq c(x) - f_u(x, \alpha_0)$ $-g_u(x, \beta_0)$. Hence Theorem 5.5.2 yields that $\alpha_0(x) \leq \alpha_1(x)$ a.e. in Ω. Similarly, we can show $\beta_1(x) \leq \beta_0(x)$ a.e. in Ω. Next we show that $\alpha_1 \leq \beta_0$ a.e. in Ω and $\alpha_0 \leq \beta_1$ a.e. in Ω. To show $\alpha_1 \leq \beta_0$, we see using (5.8.8), (5.8.9) that

$$\begin{aligned}
B[\alpha_1, v] &= \int_\Omega F(x, \alpha_1; \alpha_0, \beta_0) v \\
&\leq \int_\Omega [f(x, \beta_0) - f_u(x, \alpha_0)(\beta_0 - \alpha_0) + g(x, \beta_0) \\
&\quad -g_u(x, \beta_0)(\beta_0 - \alpha_0) + f_u(x, \alpha_0)(\alpha_1 - \alpha_0) + g_u(x, \beta_0)(\alpha_1 - \alpha_0)] v, \\
&\leq \int_\Omega [f(x, \beta_0) + g(x, \beta_0) + f_u(x, \alpha_0)[\alpha_1 + \alpha_0 - \beta_0] \\
&\quad + g_u(x, \beta_0)[\alpha_0 - \beta_0 + \alpha_1 - \alpha_0] v \\
&\leq \int_\Omega G(x, \alpha_1; \alpha_0, \beta_0) v.
\end{aligned}$$

Also, $B[\beta_0, v] \geq \int_\Omega [f(x, \beta_0)\ g(x, \beta_0)] v = \int_\Omega G(x, \beta_0; \alpha_0, \beta_0) v$. Hence Theorem 5.5.2 shows that $\alpha_1 \leq \beta_0$ a.e. in Ω. A similar argument proves that $\alpha_0 \leq \beta_1$ a.e. in Ω. To prove $\alpha_1(x) \leq \beta_1(x)$ a.e. in Ω, we use (5.8.8), (5.8.9) and the fact $g_u(x, u)$ is nonincreasing in u, to get

$$\begin{aligned}
B[\alpha_1, v] &= \int_\Omega F(x, \alpha_1; \alpha_0, \beta_0) v \qquad\qquad (5.8.12) \\
&\leq \int_\Omega [f(x, \alpha_1) + g(x, \alpha_1) + g_u(x, \alpha_1)(\alpha_0 - \alpha_1) + g_u(x, \beta_0)(\alpha_1 - \alpha_0)] v \\
&\leq \int_\Omega [f(x, \alpha_1) + g(x, \alpha_1) + [g_u(x, \beta_0) - g_u(x, \alpha_1)](\alpha_1 - \alpha_0)] v \\
&\leq \int_\Omega [f(x, \alpha_1) + g(x, \alpha_1)] v \equiv \int_\Omega F(x, \alpha_1; \alpha_1, \beta_1) v
\end{aligned}$$

for each $v \in H_0^1(\Omega, w)$, $v \geq 0$ a.e. in Ω. Similarly,

$$
\begin{aligned}
B[\beta_1, v] &= \int_\Omega G(x, \beta_1; \alpha_0, \beta_0) v && \text{(5.8.13)} \\
&\geq \int_\Omega [f(x, \beta_1) + f_u(x, \beta_1)(\beta_0 - \beta_1) + f_u(x, \alpha_0)(\beta_1 - \beta_0) + g(x, \beta_1)] v \\
&= \int_\Omega [f(x, \beta_1) + [-f_u(x, \beta_1) + f_u(x, \alpha_0)(\beta_1 - \beta_0) + g(x, \beta_1)] v \\
&\geq \int_\Omega [f(x, \beta_1) + g(x, \beta_1)] v \equiv \int_\Omega F(x, \beta_1; \alpha_1, \beta_1) v,
\end{aligned}
$$

because of the fact that $f_u(x, u)$ is nondecreasing in u and $\alpha_0 \leq \beta_1$. It then follows by Theorem 5.5.2 that $\alpha_1(x) \leq \beta_1(x)$ a.e. in Ω proving (5.8.11).

Next we shall show that if $\alpha_{k-1} \leq \alpha_k \leq \beta_k \leq \beta_{k-1}$ for some $k > 1$, then we have

$$
\alpha_k \leq \alpha_{k+1} \leq \beta_{k+1} \leq \beta_k \text{ a.e. in } \Omega. \tag{5.8.14}
$$

By following the arguments employed to prove (5.8.12) and (5.8.13), we can show, because of the assumption $\alpha_{k-1} \leq \alpha_k \leq \beta_k \leq \beta_{k-1}$, that

$$
B[\alpha_k, v] \leq \int_\Omega F(x, \alpha_k; \alpha_k, \beta_k) v, \tag{5.8.15}
$$

and

$$
B[\beta_k, v] \geq \int_\Omega G(x, \beta_k; \alpha_k, \beta_k) v, \tag{5.8.16}
$$

for each $v \in H_0^1(\Omega, w)$ $v \geq 0$ a.e. in Ω. Moreover, $\alpha_{k+1}, \beta_{k+1}$ satisfy

$$
B[\alpha_{k+1}, v] = \int_\Omega F(x, \alpha_{k+1}; \alpha_k, \beta_k) v
$$

and

$$
B[\beta_{k+1}, v] = \int_\Omega G(x, \beta_{k+1}; \alpha_k, \beta_k) v,
$$

for each $v \in H_0^1(\Omega, w)$, $v \geq 0$ a.e. in Ω. We therefore obtain by Theorem 5.5.2, $\alpha_k \leq \alpha_{k+1}$ and $\beta_{k+1} \leq \beta_k$ a.e. in Ω. Next we show that $\alpha_{k+1} \leq \beta_k$

a.e. in Ω. We find using (5.8.8), (5.8.9) that

$$
\begin{aligned}
B[\alpha_{k+1}, v] &= \int_\Omega F(x, \alpha_{k+1}; \alpha_k, \beta_k) v \\
&\leq \int_\Omega [f(x, \beta_k) - f_u(x, \alpha_k)(\beta_k - \alpha_k) + g_u(x, \beta_k)(\beta_k - \alpha_k) \\
&\quad + g(x, \beta_k) + f_u(x, \alpha_k)(\alpha_{k+1} - \alpha_k) + g_u(x, \beta_k)(\alpha_{k+1} - \alpha_k)] v \\
&\leq \int_\Omega [f(x, \beta_k) + g(x, \beta_k) + f_u(x, \alpha_k)(-\beta_k + \alpha_k + \alpha_{k-1} - \alpha_k) \\
&\quad + g_u(x, \beta_k)(\alpha_k - \beta_k + \alpha_{k+1} - \alpha_k)] v \\
&= \int_\Omega G(x, \alpha_{k+1}; \alpha_k, \beta_k) v.
\end{aligned}
$$

This together with (5.8.16) yields by Theorem 5.5.2 that $\alpha_{k+1} \leq \beta_k$ a.e. in Ω. Similarly, one can obtain $\alpha_k \leq \beta_{k+1}$ a.e. in Ω by using (5.8.15) and

$$
\begin{aligned}
B[\beta_{k+1}, v] &= \int_\Omega G(x, \beta_{k+1}; \alpha_k, \beta_k) v \\
&\geq \int_\Omega [f(x, \alpha_k) + f_u(x, \alpha_k)(\beta_k - \alpha_k) + g(x, \alpha_k) + f_u(x, \beta_k)(\beta_k - \alpha_k) \\
&\quad + f_u(x, \alpha_k)(\beta_{k+1} - \beta_1) + g_u(x, \beta_k)(\beta_{k+1} - \beta_k)] v \\
&= \int_\Omega [f(x, \alpha_k) + g(x, \alpha_k) + f_u(x, \alpha_k)(\beta_k - \alpha_k + \beta_{k+1} - \beta_k) \\
&\quad + g_u(x, \beta_k)(\beta_k - \alpha_k + \beta_{k+1} - \beta_k)] v \\
&= \int_\Omega F(x, \beta_{k+1}; \alpha_k, \beta_k) v.
\end{aligned}
$$

To prove $\alpha_{k+1} \leq \beta_{k+1}$ a.e. in Ω, we need to show that α_{k+1} and β_{k+1} are lower and upper solutions satisfying

$$
B[\alpha_{k+1}, v] \leq \int_\Omega F(x, \alpha_{k+1}; \alpha_k, \beta_k) v,
$$

and

$$
B[\beta_{k+1}, v] \geq \int_\Omega F(x, \beta_{k+1}; \alpha_k, \beta_1) v,
$$

for each $v \in H_0^1(\Omega, w)$, $v \geq 0$ a.e. This precisely uses the same arguments as we did in proving (5.8.12) and (5.8.13) replacing α_0, β_0, α_1, β_1 by α_k, β_k, α_{k+1}, β_{k+1} and using (5.8.8), (5.8.9) and the monotone character f_u and g_u. Consequently, we get by Theorem 5.5.2 that $\alpha_{k+1} \leq \beta_{k+1}$ a.e. in Ω, proving

(5.8.14). Thus by induction, it follows that (5.8.10) is valid. Therefore, the pointwise limits

$$\rho(x) = \lim_{k \to \infty} \alpha_k(x), \ r(x) = \lim_{k \to \infty} \beta_k(x)$$

exist a.e. in Ω. Moreover, $\alpha_k(x) \to \rho(x)$, $\beta(x) \to r(x)$ in $L^2(\Omega)$ by the dominated convergence theorem and (5.8.10). We note that α_k satisfies

$$\int_\Omega [\sum_{i,j=1}^n a_{ij}(x)\alpha_{k,x_i}v_{x_j} + c_0(x)\alpha_k v]dx$$

$$= \int_\Omega [f(x,\alpha_{k-1}) - f_u(x,\alpha_{k-1})\alpha_{k-1}$$

$$+g(x,\alpha_{k-1}) - g_u(x,\beta_{k-1})(\alpha_k - \alpha_{k-1})]v,$$

for each $v \in H_0^1(\Omega, w)$, $v \geq 0$ a.e. in Ω, where $c_0(x) = c(x) - f_u(x,\alpha_{k-1})$ $-g_u(x,\beta_{k-1})$. We now use the ellipticity condition and (A_3) with $v = \alpha_k$, to get

$$\int_\Omega [w(x) \mid \alpha_{k,x} \mid^2 +N \mid \alpha_k \mid^2]dx$$

$$\leq \int_\Omega [f(x,\alpha_{k-1}) - f_u(x,\alpha_{k-1})\alpha_{k-1}$$

$$+g(x,\alpha_{k-1}) - g_u(x,\beta_{k-1})\alpha_{k-1}]\alpha_k dx.$$

We then get, since by (A_3) the integrand on the right-hand side belongs to $L^2(\Omega)$, in view of the fact that $\alpha_{k-1}, \beta_{k-1} \in H_0^1(\Omega, w)$, the estimate $\sup_k \| \alpha_k \|_{H_0^1(\Omega,w)} \leq \infty$. Similarly $\sup_k \| \beta \|_{H_0^1(\Omega,w)} < \infty$. Hence there exist subsequences $\{\alpha_{k_j}(x)\}$, $\{\beta_{k_j}(x)\}$ which converge weakly to $\rho, r \in H_0^1(\Omega, w)$, respectively. To verify that ρ, r are weak solutions of (5.8.1), we fix $v \in H_0^1(\Omega, w)$ and find that $\alpha_{k+1}, \beta_{k+1}$ satisfy (5.8.4) and (5.8.5), with F and G defined by (5.8.6) and (5.8.7). Taking the limit as $k \to \infty$, we obtain

$$B[\rho, v] = \int_\Omega [f(x,\rho) + g(x,\rho)]vdx$$

and

$$B[r, v] = \int_\Omega [f(x,r) + g(x,r)]vdx,$$

showing (ρ, r) are weak solutions (5.8.1).

To show that $\rho = r = u$ is the unique solution of (5.8.1), it is enough to prove that $r \leq \rho$ a.e. in Ω, since we know that $\rho \leq r$ a.e. in Ω. Taking

$\alpha_0 = r$, $\beta_0 = \rho$ and using Theorem 5.5.2 we get $r \le \rho$ a.e. in Ω, proving the claim.

To prove quadratic convergence of the sequences $\{\alpha_k\}$, $\{\beta_k\}$ to u respectively, we consider $p_{k+1} = u - \alpha_{k+1}$ so that $p_k(0) = 0$ on $\partial\Omega$. Then we have for $v \in H_0^1(\Omega, w)$ $v \ge 0$ a.e. in Ω,

$$
\begin{aligned}
B[p_{k+1}, v] &= \int_\Omega [\tilde{f}(x, u) + \tilde{g}(x, u) - \tilde{f}(x, \alpha_k) - \tilde{f}_u(x, \alpha_k)(\alpha_{k+1} - \alpha_k) \\
&\quad - \tilde{g}(x, \alpha_k) - \tilde{g}_u(x, \beta_k)(\alpha_{k+1} - \alpha_k)]v \\
&\le \int_\Omega [(\tilde{f}_u(x, u) - \tilde{f}_u(x, \alpha_k))p_k - (\tilde{g}_u(x, \beta_k) - \tilde{g}_u(x, \alpha_k))p_k \\
&\quad + (\tilde{f}_u(x, \alpha_k) + g_u(x, \beta_k))p_{k+1}]v \\
&= \int_\Omega [\tilde{f}_{uu}(x, \xi_1)p_k^2 - \tilde{g}_{uu}(x, \sigma_1)(\beta_k - \alpha_k)p_k \\
&\quad + (\tilde{f}_u(x, \alpha_k) + \tilde{g}_u(x, \beta_k))p_{k+1}]v,
\end{aligned}
$$

where $\alpha_k \le \xi_1 \le u$, $\alpha_k \le \sigma_1 \le \beta_k$. But

$$
\begin{aligned}
-\tilde{g}_{uu}(x, \sigma_1)[\beta_k - \alpha_k]p_k &\le N_2[q_k + p_k]p_k \\
&\le N_2[p_k^2 + p_k q_k] \\
&\le \frac{3}{2} N_2 p_k^2 + \frac{1}{2} N_2 q_k^2,
\end{aligned}
$$

where $|\tilde{g}_{uu}(x, u)|_{L^\infty(\Omega)} \le N_2$, $q_k = \beta_k - u$. Thus, we get

$$
\int_\Omega \sum_{i,j=1}^n a_{ij}(x)(p_{k+1})_{x_i} v_{x_j} + c_0(x)p_{k+1}v \le \int_\Omega [(N_1 + \frac{3}{2}N_2)p_k^2 + \frac{1}{2}N_2 q_k^2]v
$$

where $|\tilde{f}_{uu}(x, u)|_{L^\infty(\Omega)} \le N$. Taking $v = p_{k+1}$ and using the ellipticity condition, we arrive at

$$
\int_\Omega [w(x) |(p_{k+1})_x|^2 + N|p_{k+1}|^2] \le N_0 \int_\Omega (|p_k^2| + |q_k^2|)|p_{k+1}|;
$$

where $N_0 = N_1 + \frac{3}{2}N_2$. Let $\theta = \min(1, N)$. Then

$$
\begin{aligned}
\theta \|p_{k+1}\|_{H_0^1(\Omega, w)}^2 &\le N_0(\frac{2^{\frac{1}{2}}}{\theta}[\|p_k^2\|_{L^2(\Omega)} + \|q_k^2\|_{L^2(\Omega)}](\frac{\theta}{2})^{\frac{1}{2}}\|p_{k+1}\|_{L^2(\Omega)} \\
&\le \frac{1}{2}[\frac{2N_0^2}{\theta}\|p_k^2\|_{L^2(\Omega)}^2 + \frac{\theta}{2}\|p_{k+1}\|_{L^2(\Omega)}^2] \\
&\quad + \frac{1}{2}[\frac{2N_0^2}{\theta}\|q_k^2\|_{L^2(\Omega)}^2 \frac{\theta}{2}\|p_{k+1}\|_{L^2(\Omega)}^2], \\
&= \frac{N_0^2}{\theta}[\|p_k^2\|_{L^2(\Omega)}^2 + \|q_k^2\|_{L^2(\Omega)}^2] + \frac{\theta}{2}\|p_{k+1}\|_{L^2(\Omega)}^2.
\end{aligned}
$$

Thus, we get

$$\frac{\theta}{2}\|p_{k+1}\|^2_{H^1_0(\Omega,w)} \le \frac{N_0^2}{\theta}[\|p_k^2\|^2_{H^1_0(\Omega,w)} + \|q_k^2\|^2_{H^1_0(\Omega,w)}],$$

or equivalently,

$$\|p_{k+1}\|^2_{H^1_0(\Omega)} \le \frac{2N_0^2}{\theta^2}[\|p_k^2\|^2_{H^1_0(\Omega,w)} + \|q_k^2\|^2_{H^1_0(\Omega,w)}]. \qquad (5.8.17)$$

One can get an estimate similar to (5.8.17) for $q_{k+1} = \beta_{k+1} - u$. We omit the details.

Some remarks are now in order.

Remarks 5.8.1

(i) If $g(x, u) \equiv 0$ and the problem (5.8.1) is nondegenerate, we get the result for $f(x, u)$ convex.

(ii) If $f(x, u) \equiv 0$ and the problem (5.8.1) is nondegenerate, we get the dual result for $g(x, u)$ concave.

(iii) Consider the case when $g(x, u) \equiv 0$. Suppose that $f(x, u)$ is not convex, but $f(x, u) + g(x, u)$ is convex where $g(x, u)$ is convex. Then the problem can be recast into (5.8.1) with $\tilde{f}(x, u) = f(x, u) + g(x, u)$ and $\tilde{g}(x, u) = -g(x, u)$ so that $\tilde{f}$ and $\tilde{g}$ satisfy all the assumptions of Theorem 5.8.1. Consequently, the problem when $\tilde{f}(x, u)$ is not convex is included in Theorem 5.8.1 as a special case which is a new result.

(iv) The observation in (iii) is also valid when BVPs (5.5.6) and (5.8.1) are nondegenerate.

(v) A similar conclusion follows giving a dual result when f in (5.5.6) is not concave.

5.9   Notes and Comments

Only recently, the approach of using weak lower and upper solutions combined with the monotone iterative technique for the solvability of nonlinear elliptic problems is attempted in the framework of weak or generalized solutions. See for example, Diaz [24]. For the exposition of this approach provided in Section 5.2 and Theorems 5.2.2 and 5.2.5, we refer to Lakshmikantham and Leela [57]. The existence result for the linear elliptic BVP

(5.2.6), namely, Theorem 5.2.3, is a special case of a general result given in Gilbarg and Trudinger [34]; see also Evans [32]. All the results in Sections 5.3, 5.4, and 5.5 dealing with the monotone iterative technique in the unified setting for semilinear, quasilinear and degenerate elliptic BVPs, are taken from Köksal and Lakshmikantham [44], [43], [41]. For the existence theorem for linear degenerate elliptic BVPs, Theorem 5.5.2, see Guglielmino and Nicolosi [36] and Murthy and Stampachia [71]. The contents of Section 5.6, relative to the method of generalized quasilinearization of semilinear elliptic BVPs, are adapted from the work of Lakshmikantham and Vatsala [62], whereas the extensions of the same methodology to quasilinear and degenerate elliptic BVPs discussed in Sections 5.7 and 5.8 are due to Lakshmikantham and Leela [54], [58].

In Drábek and Hernandez [26], a general version of the monotone iterative technique via lower and upper solutions is considered for quasilinear (*p*-Laplacian) elliptic BVPs, which extends earlier results. See also Kura [46], Duel and Hess [30]. Refer to Cal and Heikkilä [15] for similar results for discontinuous nonlinear elliptic problems where the usual monotone method does not work. For details on weighted Sobolev spaces, see Kufner [45]. For allied results, see Dancer and Sweers [22], Le [63], Le and Schmitt [64] and Kura [46].

Chapter 6

Parabolic Equations

6.1 Introduction

This chapter is dedicated to the development of the general methodology of the monotone iterative technique and generalized quasilinearization to second-order parabolic differential equations of several types.

We start in Section 6.2 with necessary notation, the definition of weak solutions and lower, upper solutions and prove needed comparison theorems. We also state an existence and uniqueness result for weak solutions to linear parabolic IBVPs and then extend the monotone iterative technique for semilinear parabolic IBVPs, where the nonlinearity admits a splitting of the difference of two monotone functions so that several results are included as special cases. In the same set-up, we then investigate the method of generalized quasilinearization for semilinear parabolic IBVPs in Section 6.3. Sections 6.4 and 6.5 are devoted to the study of nonlocal semilinear parabolic IBVPs. In Section 6.4, we introduce the problem, describe the required notation and prove existence and comparison results for nonlocal problems. Section 6.5 establishes the method of generalized quasilinearization for nonlocal problems so as to include several particular cases of interest. Sections 6.6 and 6.7 deal with quasilinear parabolic IBVPs, where the quasilinear differential operator employed is of Leray–Lions type. As a result, one needs to develop the required existence and comparison results and that is precisely what is accomplished in Section 6.6. Using the necessary ingredients that are now available in Section 6.6, Section 6.7 describes the methodology of generalized quasilinearization.

6.2   Monotone Iterative Technique

In Chapter 5, we have developed a unified approach to the monotone iterative technique for a variety of elliptic problems. The main idea is to consider
the situation when the nonlinear term involved admits a splitting of a difference of two monotone functions or equivalently, of a sum of two functions,
one of which is increasing and the other decreasing. This setting includes
several known results as well as providing some interesting new ones. In this
section, we shall extend this approach to semilinear parabolic problems.

   Let $\Omega \subset R^N$ be a bounded domain with Lipschitz boundary $\partial\Omega$, $Q =
\Omega \times (0,T)$ and $\Gamma = \partial\Omega \times (0,T)$, $T > 0$.   Consider the following initial
boundary value problem (IBVP)

$$\left[\begin{array}{ll} u_t + Lu = f(x,t,u) + g(x,t,u) & \text{in } Q, \\ u = 0 \ \text{ on } \Gamma \text{ and } u(x,0) = 0 & \text{in } \Omega, \end{array} \right. \tag{6.2.1}$$

where L denotes the second-order partial differential operator in the divergence form

$$Lu = -\sum_{i,j=1}^{N} \frac{\partial}{\partial x_i}\left(a_{ij}(x,t)\frac{\partial u}{\partial x_j} \right) + c(x,t)u,$$

with coefficients $a_{ij}, c \in L^\infty(Q)$ and

$$\sum_{i,j}^{N} a_{ij}(x,t)\xi_i\xi_j \geq \mu \mid \xi \mid^2, \text{ a.e. in } Q, \mu > 0.$$

The functions $f,g\colon Q \times R \to R$ are Caratheodory type, that is, f,g are
measurable in $(x,t) \in Q$ for each $u \in R$ and continuous in u for a.e. $(x,t) \in
Q$.

   Let $H^1(\Omega)$ be the usual Sobolev space of square integrable functions and
let $(H^1(\Omega))^*$ denote its dual space. Then by identifying $L^2(\Omega)$ with its dual
space, $H^1(\Omega) \subset L^2(\Omega) \subset (H^1(\Omega))^*$ forms an evolution triple with all the embeddings being continuous, dense and compact. Let $V = L^2(0,T;H^1(\Omega))$,
its dual space $V^* = L^2(0,T;(H^1(\Omega))^*)$, and define W by

$$W = [w \in V\colon \frac{\partial w}{\partial t} \in V^*],$$

where the derivative $\frac{\partial}{\partial t}$ is understood in the sense of vector-valued distributions and characterized by

$$\int_0^T u'(t)\phi(t)dt = -\int_0^T u(t)\phi'(t)dt, \text{ for all } \phi \in C_0^\infty(0,T).$$

The space W endowed with the norm

$$\| w \|_W = \| w \|_V + \| \frac{\partial w}{\partial t} \|_{V^*},$$

is a Banach space which is separable and reflexive due to the separability and reflexivity of V and V^* respectively. It is well known that the embedding $W \subset C([0,T]; L^2(\Omega))$ is continuous. In view of the fact that $H^1(\Omega) \subset L^2(\Omega)$ is compactly embedded, we have a compact embedding of $W \subset L^2(Q) \equiv L^2(0,T; L^2(\Omega))$. We let $H_0^1(\Omega)$ be the subspace of $H^1(\Omega)$ whose elements have generalized homogeneous boundary values, and denote by $H^{-1}(\Omega)$ its dual space. Then $H_0^1(\Omega) \subset L^2(\Omega) \subset H^{-1}(\Omega)$ forms an evolution triple also, and all statements made above remain valid in this situation when $V_0 = L^2(0,T; H_0^1(\Omega))$, $V_0^* = L^2(0,T; H^{-1}(\Omega))$ and $W_0 = [w \in V_0 \colon \frac{\partial w}{\partial t} \in V_0^*]$.

We denote the duality pairing between the elements of V^* and V (respectively V_0^* and V_0) by $\langle \cdot, \cdot \rangle$. The bilinear form associated with the operator L is given by

$$B[u,v] = \sum_{i,j}^{N} \int_Q [a_{ij}(x,t)u_{x_i}v_{x_j} + c(x,t)uv]dxdt,$$

for $u, v \in V$. Then the IBVP (6.2.1) takes the variational form

$$\langle u_t, v \rangle + B[u,v] = \int_Q [Fu + Gu]vdxdt, \qquad (6.2.2)$$

for all $v \in V_0$, where

$$Fu = f(x,t,u), \quad Gu = g(x,t,u)$$

are the Nemytski operators related to f and g, respectively.

A partial ordering in $L^2(Q)$ is defined by $u \le w$ iff $w - u \in L_+^2(Q)$, the set of all nonnegative elements of $L^2(Q)$. This induces a corresponding partial ordering also in W which is a subset of $L^2(Q)$. Hence if $\underline{u}, \bar{u} \in W$ with $\underline{u} \le \bar{u}$, then

$$[\underline{u}, \bar{u}] = [u \in W \colon \underline{u} \le u \le \bar{u}]$$

denotes the order interval generated by $\underline{u}$, $\bar{u}$. Let $(\cdot, \cdot)$ denote the inner product in $L^2(Q)$. Then (6.2.2) takes the form

$$\langle u_t, u \rangle + B[u,v] = (Fu + Gu, v) \quad \text{for all } v \in V_0. \qquad (6.2.3)$$

Definition 6.2.1 *A function $u \in W_0$ is said to be a weak solution of* (6.2.1) *if $Fu + Gu \in L^2(Q)$, $u(x,0) = 0$ in Ω and* (6.2.3) *holds.*

Definition 6.2.2 *The functions* $u, w \in W_0$ *are said to be coupled weak solutions of (6.2.1) if* $Fu + Gw \in L^2(Q)$, $Fw + Gu \in L^2(Q)$, *and*

(i) $u(x, 0) = 0 = w(x, 0)$ *in* Ω,

(ii) $\langle u_t, v \rangle + B[u, v] = (Fu + Gw, v)$ *and* $\langle w_t, v \rangle + B[w, v] = (Fw + Gu, v)$
hold

for each $v \in V_0$.

Definition 6.2.3 *The functions* $\alpha, \beta \in W$ *are said to be coupled weak lower and upper solutions of (6.2.1) if* $F\alpha + G\beta \in L^2(Q)$, $F\beta + G\alpha \in L^2(Q)$ *and*

(i) $\alpha(x, 0) \leq 0 \leq \beta(x, 0)$ *in* Ω *and* $\alpha(x, t) \leq 0 \leq \beta(x, t)$ *on* Γ,

(ii) $\langle \alpha_t, v \rangle + B[\alpha, v] \leq (F\alpha + G\beta, v)$ *and* $\langle \beta_t, v \rangle + B[\beta, v] \geq (F\beta + G\alpha, v)$

hold for each $v \in V_0 \cap L_+^2(Q)$.

We can now prove the following comparison result.

Theorem 6.2.1 *Assume that* $\alpha, \beta \in W$ *are coupled lower and upper solutions of (6.2.1) and* f, g *satisfy the relations*

$$\left[\begin{array}{ll} f(x, t, u_1) - f(x, t, u_2) & \leq d_1(x, t)(u_1 - u_2), \\ g(x, t, u_1) - g(x, t, u_2) & \geq -d_2(x, t)(u_1 - u_2), \end{array} \right. \tag{6.2.4}$$

whenever $u_1 \geq u_2$ *for some* $d_1, d_2 \in L_+^\infty(Q)$. *Then we have* $\alpha \leq \beta$.

Proof By the definition of coupled lower and upper solutions of (6.2.1), we get

$$\alpha - \beta \leq 0 \text{ on } \Gamma, \quad \alpha(x, 0) - \beta(x, 0) \leq 0 \text{ in } \Omega,$$

and

$$\langle (\alpha - \beta)_t, v \rangle + B[\alpha - \beta, v] \leq (F\alpha - F\beta + G\beta - G\alpha, v)$$

for all $v \in V_0 \cap L_+^2(Q)$. Choose $v = (\alpha - \beta)^+ = \max(\alpha - \beta, 0) \in V_0 \cap L_+^2(Q)$ so that $(\alpha - \beta)^+(x, 0) = 0$ and using (6.2.4) and the ellipticity condition, we get, for any $s \in [0, T]$, the following inequality

$$\frac{1}{2} \parallel (\alpha - \beta)^+ (\cdot, s) \parallel^2_{L^2(\Omega)} + \mu \parallel \nabla(\alpha - \beta)^+ \parallel^2_{L^2(Q_s)}$$

$$\leq \parallel d_1 + d_2 - c \parallel_{L^\infty(Q)} \int_{Q_s} ((\alpha - \beta)^+)^2 dx dt$$

where $Q_s = \Omega \times (0, s) \subset Q$. Now set $y(s) = \| (\alpha - \beta)^+(\cdot, s) \|^2_{L^2(\Omega)}$ to obtain

$$y(s) \leq 2 \| d_1 + d_2 - c \|_{L^\infty(\Omega)} \int_0^s y(t)dt, \ s \in [0, T].$$

Since $\alpha - \beta \in C[[0, T]; L^2(\Omega)]$, $y(s) \geq 0$ and $y(0) = 0$, Gronwall's inequality implies that $y(s) = 0$ for any $s \in [0, T]$ which means $(\alpha - \beta)^+(x, t) = 0$ a.e. in Q, that is, $\alpha \leq \beta$, proving the claim.

The following corollary is also needed in our discussion.

Corollary 6.2.1 *For any $p \in W$ satisfying $p(x, t) \leq 0$ on Γ, $p(x, 0) \leq 0$ in Ω and*

$$\langle p_t, v \rangle + B[p, v] \leq 0 \text{ for all } v \in V_0 \cap L^2_+(Q),$$

we have $p \leq 0$.

The following standard existence and uniqueness result relative to linear IBVPs

$$u_t + Lu = h \text{ in } Q, u(x, t) = 0 \text{ on } \Gamma, \text{ and } u(x, 0) = \psi(x) \text{ in } \Omega, \quad (6.2.5)$$

where $\psi \in L^2(\Omega)$ and $h \in L^2(Q) \subset V^*$, is required for our discussion.

Theorem 6.2.2 *There exists a unique weak solution $u \in W_0$ to the IBVP (6.2.5), and for some constant $\gamma > 0$*

$$\| u \|_W \leq \gamma(\| \psi \|_{L^2(\Omega)} + \| h \|_{L^2(Q)}).$$

We shall prove the following result which includes several interesting special cases.

Theorem 6.2.3 *Assume that*

(A_1) *α_0, β_0 are coupled weak lower and upper solutions of (6.2.1) such that $\alpha_0 \leq \beta_0$ a.e. in Q;*

(A_2) *$f, g : Q \times R \to R$ are Caratheodory functions such that $f(x, t, u)$ is nondecreasing in u and $g(x, t, u)$ is nonincreasing in u for $(t, x) \in Q$, a.e.;*

(A_3) *for any $\eta, \mu \in [\alpha_0, \beta_0]$, $\eta, \mu \in W$, $h \in L^2(Q)$ where $h(x, t) = f(x, t, \eta) + g(x, t, \mu)$.*

Then there exist monotone sequences $\{\alpha_k(x,t)\}$, $\{\beta_k(x,t)\} \in W_0$ such that $\alpha_k \to \rho$, $\beta_k \to r$ in W_0 and ρ, r are the coupled weak minimal and maximal solutions of (6.2.1).

Proof   Consider the following IBVPs for each $k = 1, 2, \ldots$

$$\left[\begin{array}{l} \frac{\partial \alpha_k}{\partial t} + L\alpha_k = f(x,t,\alpha_{k-1}) + g(x,t,\beta_{k-1}) \quad \text{in} \quad Q, \\ \alpha_k(x,t) = 0 \quad \text{on} \quad \Gamma \text{ and } \quad \alpha_k(x,0) = 0 \quad \text{in} \quad \Omega, \end{array} \right. \tag{6.2.6}$$

and

$$\left[\begin{array}{l} \frac{\partial \beta_k}{\partial t} + L\beta_k = f(x,t,\beta_{k-1}) + g(x,t,\alpha_{k-1}) \quad \text{in} \quad Q, \\ \beta_k(x,t) = 0 \quad \text{on} \quad \Gamma \quad \text{and } \quad \beta_k(x,0) = 0 \quad \text{in} \quad \Omega, \end{array} \right. \tag{6.2.7}$$

whose variational forms associated with (6.2.6) and (6.2.7) are given by

$$\left\langle \frac{\partial \alpha_k}{\partial t}, v \right\rangle + B[\alpha_k, v] = (F\alpha_{k-1} + G\beta_{k-1}, v), \tag{6.2.8}$$

and

$$\left\langle \frac{\partial \beta_k}{\partial t}, v \right\rangle + B[\beta_k, v] = (F\beta_{k-1} + G\alpha_{k-1}, v) \tag{6.2.9}$$

for all $v \in V_0$.

By (A_3), $h \in L^2(Q)$ where $h(x,t) = f(x,t,\alpha_0) + g(x,t,\beta_0)$. Hence by Theorem 6.2.2 there is a unique weak solution $\alpha_1 \in W_0$ of (6.2.6) for $k = 1$. In the same way, we get the existence of a unique weak solution $\beta_1 \in W_0$ of (6.2.7). We shall first show that

$$\alpha_0 \le \alpha_1 \le \beta_1 \le \beta_0. \tag{6.2.10}$$

To prove $\alpha_0 \le \alpha_1$, we find that α_1 satisfies the relation

$$\left\langle \frac{\partial \alpha_1}{\partial t}, v \right\rangle + B[\alpha_1, v] = (F\alpha_0 + G\beta_0, v), \tag{6.2.11}$$

and note that α_0 satisfies

$$\left\langle \frac{\partial \alpha_0}{\partial t}, v \right\rangle + B[\alpha_0, v] \le (F\alpha_0 + G\beta_0, v), \tag{6.2.12}$$

for each $v \in V_0 \cap L_+^2(Q)$ and $\alpha_0(x,t) \le 0$ on Γ, $\alpha_0(x,0) \le 0$ in Ω. Let $p = \alpha_0 - \alpha_1$. Then p satisfies $p(x,t) \le 0$ on Γ, $p(x,0) \le 0$ in Ω and

$$\left\langle \frac{\partial p}{\partial t}, v \right\rangle + B[p, v] \le 0 \quad \text{for all } v \in V_0 \cap L_+^2(Q).$$

Hence by Corollary 6.2.1, we get $p \leq 0$, that is $\alpha_0 \leq \alpha_1$. A similar reasoning proves that $\beta_1 \leq \beta_0$. To show that $\alpha_1 \leq \beta_1$, we see from (6.2.9) that β_1 satisfies

$$\left\langle \frac{\partial \beta_1}{\partial t}, v \right\rangle + B[\beta_1, v] = (F\beta_0 + G\alpha_0, v) \geq (F\alpha_0 + G\beta_0, v) \qquad (6.2.13)$$

for each $v \in V_0 \cap L^2_+(Q)$, using the monotone character of F and G. By Corollary 6.2.1 we readily obtain, in view of (6.2.11) and (6.2.12), $\alpha_1 \leq \beta_1$, proving (6.2.10).

Next we shall prove that if for some $j > 1$,

$$\alpha_0 \leq \alpha_{j-1} \leq \alpha_j \leq \beta_j \leq \beta_{j-1} \leq \beta_0, \qquad (6.2.14)$$

then

$$\alpha_j \leq \alpha_{j+1} \leq \beta_{j+1} \leq \beta_j. \qquad (6.2.15)$$

Because of (A_3) and the relation (6.2.14), Theorem 6.2.2 guarantees the existence of unique weak solutions $\alpha_j, \beta_j, \alpha_{j+1}, \beta_{j+1} \in W_0$ of (6.2.6) and (6.2.7) with $k = j, k = j + 1$. To show that $\alpha_j \leq \alpha_{j+1}$, consider

$$\left\langle \frac{\partial \alpha_{j+1}}{\partial t}, v \right\rangle + B[\alpha_{j+1}, v] = (F\alpha_j + G\beta_j, v) \qquad (6.2.16)$$

and

$$\left\langle \frac{\partial \alpha_j}{\partial t}, v \right\rangle + B[\alpha_j, v] = (F\alpha_{j-1} + G\beta_{j-1}, v) \leq (F\alpha_j + G\beta_j, v) \qquad (6.2.17)$$

for each $v \in V_0 \cap L^2_+(Q)$. In (6.2.17), we have used the monotone nature of F, G and (6.2.14). Corollary 6.2.1 then yields that $\alpha_j \leq \alpha_{j+1}$. Similarly, we can show that $\beta_{j+1} \leq \beta_j$. To prove $\alpha_{j+1} \leq \beta_{j+1}$, we see, using the monotone character of F, G and (6.2.14), that

$$\left\langle \frac{\partial \beta_{j+1}}{\partial t}, v \right\rangle + B[\beta_{j+1}, v] = (F\beta_j + G\alpha_j, v) \geq (F\alpha_j + G\beta_j, v) \qquad (6.2.18)$$

for each $v \in V_0 \cap L^2_+(Q)$. The relation (6.2.18), together with (6.2.16), prove by Corollary 6.2.1 that $\alpha_{j+1} \leq \beta_{j+1}$, showing that (6.2.15) is valid. Hence by induction, we arrive at

$$\alpha_0 \leq \alpha_1 \leq \alpha_2 \leq \ldots \leq \alpha_k \leq \beta_k \leq \ldots \leq \beta_2 \leq \beta_1 \leq \beta_0, \qquad (6.2.19)$$

for all k. By the monotonicity of iterates $\{\alpha_k\}, \{\beta_k\}$ and (6.2.19), there exist pointwise limits

$$\rho(x,t) = \lim_{k\to\infty} \alpha_k(x,t), \ r(x,t) = \lim_{k\to\infty} \beta_k(x,t), \ \text{for a.e. in } Q. \quad (6.2.20)$$

Furthermore, since $\alpha_k, \beta_k \in [\alpha_0, \beta_0]$, it follows by Lebesgue's dominated convergence theorem that

$$\alpha_k \to \rho \ \text{and} \ \beta_k \to r \ \text{in} \ L^2(Q). \quad (6.2.21)$$

By (A_3), we have for any $\eta, \mu \in [\alpha_0, \beta_0]$, $F\eta + G\mu$ is continuous and bounded as a mapping from $[\alpha_0, \beta_0] \subset L^2(Q) \to L^2(Q)$ and therefore in view of (6.2.21), it follows that

$$F\alpha_k \to F\rho, \ G\beta_k \to Gr, \ F\beta_k \to F\rho, \ G\alpha_k \to G\rho \ \text{in} \ L^2(Q) \ \text{as} \ k \to \infty. \quad (6.2.22)$$

Since α_k, β_k are solutions of the linear IBVPs (6.2.6) and (6.2.7) with homogeneous initial boundary values, the estimate of Theorem 6.2.2 gives

$$\| \alpha_k \|_W \leq \gamma \| F\alpha_{k-1} + G\beta_{k-1} \|_{L^2(Q)},$$

and

$$\| \beta_k \|_W \leq \gamma \| F\beta_{k-1} + G\alpha_{k-1} \|_{L^2(Q)}.$$

Due to (6.2.22), the foregoing estimates imply the strong convergence of the sequences $\{\alpha_k\}, \{\beta_k\}$ in W whose limits must be ρ and r respectively, because of the compact embedding $W \subset L^2(Q)$. Since α_k, β_k for $k \geq 1$ belong to the subset $M = [u \in W: u(x,0) = 0]$ which is closed in W, it follows that their limits $\rho, r \in M$, which are the limits satisfying also homogeneous initial and boundary conditions. The convergence result (6.2.22) together with

$$\alpha_k \to \rho, \beta_k \to r \ \text{in} \ W \ \text{as} \ k \to \infty,$$

permit us to pass to the limit in the corresponding variational forms (6.2.8) and (6.2.9) as $k \to \infty$. This yields

$$\left\langle \frac{\partial \rho}{\partial t}, v \right\rangle + B[\rho, v] = (F\rho + Gr, v),$$

and

$$\left\langle \frac{\partial r}{\partial t}, v \right\rangle + B[r, v] = (Fr + G\rho, v),$$

for all $v \in V_0$, showing (ρ, r) are coupled weak solutions of (6.2.1).

Let $u \in [\alpha_0, \beta_0]$ be any weak solution of (6.2.1). Then we show that $\alpha_1 \leq u \leq \beta_1$. Since α_1 satisfies (6.2.11), we get, using monotone nature of F and G,

$$\left\langle \frac{\partial \alpha_1}{\partial t}, v \right\rangle + B[\alpha_1, v] = (F\alpha_0 + G\beta_0, v) \leq (Fu + Gu, v),$$

for each $v \in V_0 \cap L^2_+(Q)$. Since u is a weak solution of (6.2.1) we have

$$\left\langle \frac{\partial u}{\partial t}, v \right\rangle + B[u, v] = (Fu + Gu, v)$$

for each $v \in V_0 \cap L^2_+(Q)$. Corollary 6.2.1 therefore yields $\alpha_1 \leq u$. A similar argument shows that $u \leq \beta_1$. Next we suppose that for some $j > 1$,

$$\alpha_0 \leq \alpha_j \leq u \leq \beta_j \leq \beta_0.$$

Then utilizing the monotone character of F and G, we obtain

$$\left\langle \frac{\partial \alpha_{j+1}}{\partial t}, v \right\rangle + B[\alpha_{j+1}, v] = (F\alpha_j + G\beta_j, v) \leq (Fu + Gu, v)$$

for each $v \in V_0 \cap L^2_+(Q)$. It then follows from Corollary 6.2.1 that $\alpha_{j+1} \leq u$. In the same way, we can prove that $u \leq \beta_{j+1}$. Thus we arrive at

$$\alpha_0 \leq \alpha_j \leq \alpha_{j+1} \leq u \leq \beta_{j+1} \leq \beta_j \leq \beta_0.$$

Hence by induction $\alpha_k \leq u \leq \beta_k$ for all k and consequently, taking the limit as $k \to \infty$, we get $\rho \leq u \leq r$, proving that ρ, r are coupled weak minimal and maximal solutions of (6.2.1). The proof is now complete.

Corollary 6.2.2 *In addition to the assumptions of Theorem 6.2.3, if we suppose that f, g satisfy the conditions (6.2.4) of Theorem 6.2.1, then $u = \rho = r$ is the unique weak solution of (6.2.1).*

Proof   Since $\rho \leq r$, it is sufficient to show that $r \leq \rho$. But this follows right away from Theorem 6.2.1 if we set $\alpha_0 = r$, $\beta_0 = \rho$. Thus the claim of the corollary is true.

Theorem 6.2.3 includes several interesting special cases and therefore presents a unified framework for the monotone iterative technique.

Remarks 6.2.1

(i) If in Theorem 6.2.3, $g(x, t, u) \equiv 0$, then we get the result where ρ, r are weak minimal and maximal solutions of (6.2.1).

(ii) If $f(x, t, u) \equiv 0$ in Theorem 6.2.3, then we have a result corresponding to the nonincreasing case.

(iii) If $g(x, t, u) \equiv 0$ and $f(x, t, u)$ is not nondecreasing in u but $\tilde{f}(x, t, u) = f(x, t, u) + d(x, t)u$ is nondecreasing in u where $d \in L_+^\infty(Q)$, then we consider the IBVP

$$\frac{\partial u}{\partial t} + Lu + du = \tilde{f}(x, t, u) \quad \text{in } Q, \tag{6.2.23}$$

$$u = 0 \text{ on } \Gamma \text{ and } u(x, 0) = 0 \text{ on } \Omega.$$

It is easy to verify that α_0, β_0 are weak lower and upper solutions of (6.2.23). Theorem 6.2.3 applied to (6.2.23) yields the same conclusion as in (i) for the IBVP (6.2.1).

(iv) If $f(x, t, u) \equiv 0$ and $g(x, t, u)$ is not nonincreasing in u but $\tilde{g}(x, t, u) = g(x, t, u) - d(x, t)u$ is nonincreasing in u, where $d \in L_+^\infty(Q)$, then we consider the IBVP

$$\frac{\partial u}{\partial t} + Lu - du = \tilde{g}(x, t, u) \quad \text{in } Q, \tag{6.2.24}$$

$$u = 0 \text{ on } \Gamma \text{ and } u(x, 0) = 0 \text{ on } \Omega$$

If we assume that α_0, β_0 are coupled weak lower and upper solutions of (6.2.24), Theorem 6.2.3 guarantees the same conclusion as in (ii) for the IBVP (6.2.24). In addition, if the uniqueness conditions (6.2.4) are satisfied, then application of Theorem 6.2.3 yields the same conclusion for the IBVP (6.2.1) when the right-hand side function does not satisfy the monotonicity requirement.

(v) If in Theorem 6.2.3, $g(x, t, u)$ is nonincreasing in u and $f(x, t, u)$ is not nondecreasing in u but $\tilde{f}(x, t, u) = f(x, t, u) + d(x, t)u$ is nondecreasing in u where $d \in L_+^\infty(Q)$, then we consider the IBVP

$$\frac{\partial u}{\partial t} + Lu + du = \tilde{f}(x, t, u) + g(x, t, u) \quad \text{in } Q, \tag{6.2.25}$$

$$u = 0 \text{ on } \Gamma \text{ and } u(x, 0) \text{ on } \Omega.$$

Clearly, α_0, β_0 are coupled weak lower and upper solutions of (6.2.25) and hence the conclusion of Theorem 6.2.3 applied to (6.2.25) holds for the IBVP (6.2.1).

(vi) Suppose that $f(x,t,u)$ is nondecreasing in u and $g(x,t,u)$ is not non-increasing in u but $\tilde{g}\,(x,t,u) = g(x,t,u) - d(x,t)u$ is nonincreasing in u, with $d \in L^\infty_+(Q)$. Then consider the IBVP

$$\frac{\partial u}{\partial t} + Lu - du = f(x,t,u) + \tilde{g}(x,t,u) \text{ in } Q, \ u = 0 \text{ on } \Gamma \quad (6.2.26)$$

$$\text{and } u(x,0) = 0 \text{ on } \Omega$$

Again if we assume that α_0, β_0 are coupled weak lower and upper solutions of (6.2.26), Theorem 6.2.3 applied to IBVP (6.2.26) assures the conclusion of Theorem 6.2.3 for the IBVP (6.2.26). Moreover, if the uniqueness conditions (6.2.4) hold then the conclusion of Theorem 6.2.3 is obtained for the IBVP (6.2.1).

(vii) Suppose that in Theorem 6.2.3 both f and g are not monotone but $\tilde{f}(x,t,u) = f(x,t,u) + d_1(x,t)u$ is nondecreasing in u and $\tilde{g}(x,t,u) = g(x,t,u) - d_2(x,t)u$ is nonincreasing in u, where $d_1, d_2 \in L^\infty_+(Q)$. Then we consider the IBVP

$$\frac{\partial u}{\partial t} + Lu + d_1 u - d_2 u = \tilde{f}(x,t,u) + \tilde{g}(x,t,u) \text{ in } Q, \ u = 0 \text{ on } \Gamma$$

$$\text{and } u(x,0) = 0 \text{ on } \Omega. \quad (6.2.27)$$

If α_0, β_0 are coupled weak lower and upper solutions of (6.2.27), Theorem 6.2.3 guarantees the same conclusion as in Theorem 6.2.3 for IBVP (6.2.27). And the conclusion remains the same for the IBVP (6.2.1) if the uniqueness condition (6.2.4) is satisfied by $\tilde{f}$ and $\tilde{g}$.

6.3 Generalized Quasilinearization

We shall concentrate, in this section, on the method of generalized quasilinearization for semilinear parabolic problems and prove a general result which includes several cases of interest. We shall utilize the notation regarding the spaces H^1, V, V_0, W, H^1_0, V_0, V_0^* and W_0 as described in Section 6.2 without repeating. Also, other necessary results will be given for convenience in the present set-up.

Let $\Omega \subset \mathbb{R}^N$ be a bounded domain with Lipschitz boundary $\partial\Omega$, $Q = \Omega \times (0,T)$ and $\Gamma = \partial\Omega \times (0,T)$, $T > 0$. Consider the following initial boundary value problem (IBVP)

$$\left[\begin{array}{ll} \frac{\partial u(x,t)}{\partial t} + Au(x,t) = f(x,t,u(x,t)) & \text{in } Q \\ u(x,t) = 0 \text{ on } \Gamma, \text{ and } u(x,0) = 0 & \text{in } \Omega \end{array} \right. \quad (6.3.1)$$

where A is a second-order strongly elliptic differential operator in the form

$$Au(x,t) = - \sum_{i,j=1}^{N} \frac{\partial}{\partial x_i} \left(a_{ij}(x,t) \frac{\partial u}{\partial x_j} \right),$$

with coefficients $a_{ij} \in L^{\infty}(Q)$ satisfying, for all $\xi = (\xi_1, \ldots, \xi_N) \in \mathbb{R}^N$,

$$\sum_{i,j=1}^{N} a_{ij}(x,t)\xi_i\xi_j \geq \mu|\xi|^2, \text{ for a.e. } (x,t) \in Q,$$

where μ is some positive constant. The nonlinear right-hand side $f : Q \times \mathbb{R} \to \mathbb{R}$ is a Caratheodory function, that is f is measurable in $(x,t) \in Q$ for each $u \in \mathbb{R}$ and continuous in u for a.e. $(x,t) \in Q$.

We denote the duality pairing between the elements of V_0^* and V_0 by $\langle \cdot, \cdot \rangle$, and define the bilinear form B associated with the operator A by

$$B[u,\varphi] = \sum_{i,j=1}^{N} \int_Q a_{ij}(x,t) \frac{\partial u}{\partial x_i} \frac{\partial \varphi}{\partial x_j} \, dxdt,$$

for all $u, \varphi \in V$. This is related to the operator $A : V \to V^*$ by $\langle Au, \varphi \rangle = B[u,\varphi]$, $u, \varphi \in V$. A partial ordering in $L^2(Q)$ is defined by $u \leq w$ if and only if $w - u$ belongs to the set $L_+^2(Q)$ of all nonnegative elements of $L^2(Q)$. This induces a corresponding partial ordering also in the subspace W of $L^2(Q)$, and if $\underline{u}, \bar{u} \in W$ with $\underline{u} \leq \bar{u}$ then $[\underline{u}, \bar{u}] := \{u \in W \mid \underline{u} \leq u \leq \bar{u}\}$ denotes the order interval formed by $\underline{u}$ and $\bar{u}$. Let $(\cdot, \cdot)$ denote the inner product in $L^2(Q)$, and denote by F the Nemytski operator related with the function f by $Fu(x,t) = f(x,t,u(x,t))$.

Definition 6.3.1 *A function $u \in W_0$ is called a weak solution of (6.3.1) if $Fu \in L^2(Q)$ and*

    (i) $u(x,0) = 0$ in Ω,

    (ii) $\langle \frac{\partial u}{\partial t}, \varphi \rangle + B[u,\varphi] = (Fu, \varphi)$ for all $\varphi \in V_0$.

Definition 6.3.2 *A function $\alpha \in W$ is said to be a weak lower solution of (6.3.1) if $F\alpha \in L^2(Q)$ and*

    (i) $\alpha(x,t) \leq 0$ on Γ, $\alpha(x,0) \leq 0$ in Ω,

    (ii) $\langle \frac{\partial \alpha}{\partial t}, \varphi \rangle + B[\alpha,\varphi] \leq (F\alpha, \varphi)$ for all $\varphi \in V_0 \cap L_+^2(Q)$.

A *weak upper solution* of (6.3.1) is defined similarly by reversing the inequalities. We can now prove the following comparison result.

Theorem 6.3.1 *Let α and β be lower and upper solutions of (6.3.2), respectively. Suppose further that there is a nonnegative function $K \in L_+^\infty(Q)$ such that*

$$f(x, t, u_1) - f(x, t, u_2) \leq K(x, t)(u_1 - u_2) \tag{6.3.2}$$

whenever $u_1 \geq u_2$. Then we have $\alpha \leq \beta$ in Q.

Proof   The definition of lower and upper weak solutions of (6.3.1) yields

$$\alpha(x, t) - \beta(x, t) \leq 0 \text{ on } \Gamma, \text{ and } \alpha(x, 0) - \beta(x, 0) \leq 0 \text{ in } \Omega,$$

and

$$\left\langle \frac{\partial(\alpha - \beta)}{\partial t}, \varphi \right\rangle + B[\alpha - \beta, \varphi] \leq (F\alpha - F\beta, \varphi) \tag{6.3.3}$$

for all $\varphi \in V_0 \cap L_+^2(Q)$. Taking the special test function $\varphi = (\alpha - \beta)^+ :=$ $\max(\alpha - \beta, 0) \in V_0 \cap L_+^2(Q)$, then $(\alpha - \beta)^+(x, 0) = 0$ and by using (6.3.2) and the strong ellipticity of A we get from the weak formulation (6.3.3) for any $\tau \in (0, T]$ the following inequality

$$\frac{1}{2}\|(\alpha - \beta)^+(\cdot, \tau)\|_{L^2(\Omega)}^2 \quad + \quad \mu \|\nabla(\alpha - \beta)^+\|_{L^2(Q_\tau)}^2 \tag{6.3.4}$$

$$\leq \quad \|K\|_{L^\infty(Q)} \int_{Q_\tau} ((\alpha - \beta)^+)^2 \, dx dt,$$

where $Q_\tau := \Omega \times (0, \tau) \subset Q$. Thus, by setting

$$y(\tau) = \|(\alpha - \beta)^+(\cdot, \tau)\|_{L^2(\Omega)}^2$$

from (6.3.4) we obtain the inequality

$$y(\tau) \leq 2\|K\|_{L^\infty(Q)} \int_0^\tau y(t) \, dt, \quad \text{for all } \tau \in [0, T]. \tag{6.3.5}$$

Since $\alpha - \beta \in C([0, T]; L^2(\Omega))$ and $y(\tau) \geq 0$, we get from (6.3.5) by applying Gronwall's lemma, $y(\tau) = 0$ for any $\tau \in [0, T]$, which implies $(\alpha - \beta)^+(x, t) = 0$ for a.e. $(x, t) \in Q$, i.e., $\alpha \leq \beta$, proving the claim.
    The following corollary is useful in our discussion.

Corollary 6.3.1 *Let $c \in L^\infty(Q)$. Then for any $p \in W$ satisfying $p(x,t) \le 0$ on Γ, $p(x,0) \le 0$ in Ω and*

$$\left\langle \frac{\partial p}{\partial t}, \varphi \right\rangle + B[p, \varphi] + (cp, \varphi) \le 0 \quad \text{for all } \varphi \in V_0 \cap L^2_+(Q),$$

we have $p \le 0$ in Q.

For any $c \in L^\infty(Q)$ the bilinear form $B[u, v] + c(u, v)$ of Corollary 6.3.1 is related to the linear operator $A + cI$, and the following standard existence and uniqueness result holds for the linear IBVP

$$\frac{\partial u}{\partial t} + Au + cu = h \text{ in } Q, \quad u(x,t) = 0 \text{ on } \Gamma \text{ and } u(x,0) = \psi(x) \text{ in } \Omega,$$

$$(6.3.6)$$

where $\psi \in L^2(\Omega)$ and $h \in L^2(Q) \subset V_0^*$.

Theorem 6.3.2 *There exists a unique weak solution $u \in W_0$ to the IBVP (6.3.6), and there is a constant $\nu > 0$ such that*

$$\|u\|_W \le \nu \left(\|\psi\|_{L^2(\Omega)} + \|h\|_{L^2(Q)} \right).$$

Remarks 6.3.1

(i) Note that unlike in the corresponding elliptic case no sign condition on the coefficient $c \in L^\infty(Q)$ in Corollary 6.3.1 and Theorem 6.3.2 is needed, because of the possibility of applying the transformation $u = e^{\lambda t} w$ (also called the exponential shift) with $\lambda \in \mathbb{R}$ appropriately chosen.

(ii) Homogeneous initial and boundary conditions have been taken only for the sake of simplicity. Any inhomogeneous initial and boundary conditions given by the initial values and traces, respectively, of some function $w \in W$ can be reduced to homogeneous ones by an appropriate translation.

(iii) A fully linear parabolic operator in the form

$$\frac{\partial u}{\partial t} - \sum_{i,j=1}^{N} \frac{\partial}{\partial x_i}\left(a_{ij}(x,t)\frac{\partial u}{\partial x_j}\right) + \sum_{i=1}^{N} b_i(x,t)\frac{\partial u}{\partial x_i} + c(x,t)u$$

can be treated without any additional difficulties.

We shall investigate, instead of (6.3.1), the following IBVP

$$\left[\begin{array}{ll} \frac{\partial u(x,t)}{\partial t} + Au(x,t) = f(x,t,u(x,t)) + g(x,t,u(x,t)) & \text{in } Q, \\ u(x,t) = 0 \text{ on } \Gamma \text{ and } u(x,0) = 0 & \text{in } \Omega \end{array}\right. \qquad (6.3.7)$$

where $f, g: Q \times \mathbb{R} \to \mathbb{R}$ are Caratheodory functions.

We shall prove the following result which includes several interesting special cases.

Theorem 6.3.3 *Assume that*

(A1) α_0 *and* β_0 *are lower and upper solutions of* (6.3.7), *respectively, such that* $\alpha_0 \le \beta_0$, *and, in addition,* $\alpha_0, \beta_0 \in L^4(Q)$.

(A2) $f, g: Q \times \mathbb{R} \to \mathbb{R}$ *are Caratheodory functions such that* f_u, g_u, f_{uu}, g_{uu} *exist and are Caratheodory functions, and* $f_{uu}(x,t,u) \ge 0$, $g_{uu}(x,t,u) \le 0$ *for all* $u \in [\alpha_0(x,t), \beta_0(x,t)]$ *and for a.e.* $(x,t) \in Q$.

(A3) *There is some positive constant* C *such that*

$$\| f_u(\cdot, \cdot, \eta), \ g_u(\cdot, \cdot, \eta), \ f_{uu}(\cdot, \cdot, \eta), \ g_{uu}(\cdot, \cdot, \eta) \|_{L^\infty(Q)} \le C \qquad (6.3.8)$$

for any $\eta \in [\alpha_0, \beta_0]$.

Then there exist monotone sequences $(\alpha_k)_{k=1}^\infty$, $(\beta_k)_{k=1}^\infty \subset W_0$ *such that* $\alpha_k \to \varrho$, $\beta_k \to r$ *in* W_0, *where* $\varrho = r = u$ *is the unique weak solution of* (6.3.7) *satisfying* $\alpha_0 \le u \le \beta_0$, *and the convergence is quadratic.*

Proof   We prove our result in several steps.

(a) **Generalized quasilinearization and iterative schemes** We introduce linearizations of $f + g$ in the form

$$\begin{aligned} G(x,t,u;\alpha,\beta) \ := \ & f(x,t,\beta) + g(x,t,\beta) + f_u(x,t,\alpha)(u - \beta) \\ & + g_u(x,t,\beta)(u - \beta) \end{aligned} \qquad (6.3.9)$$

and

$$\begin{aligned} F(x,t,u;\alpha,\beta) \ := \ & f(x,t,\alpha) + g(x,t,\alpha) + f_u(x,t,\alpha)(u - \alpha) \\ & + g_u(x,t,\beta)(u - \alpha), \end{aligned} \qquad (6.3.10)$$

and consider the following related linear iteration schemes for $k = 0, 1, 2, \ldots$

$$\left[\begin{array}{ll} \frac{\partial \beta_{k+1}}{\partial t} + A\beta_{k+1} = G(x,t,\beta_{k+1};\alpha_k,\beta_k) & \text{in } Q, \\ \beta_{k+1}(x,t) = 0 \text{ on } \Gamma \text{ and } \beta_{k+1}(x,0) = 0 & \text{in } \Omega, \end{array}\right. \qquad (6.3.11)$$

and

$$\left[\begin{array}{l} \frac{\partial \alpha_{k+1}}{\partial t} + A\alpha_{k+1} = F(x,t,\alpha_{k+1};\alpha_k,\beta_k) \quad \text{in } Q, \\ \alpha_{k+1}(x,t) = 0 \quad \text{on } \Gamma \text{ and } \alpha_{k+1}(x,0) = 0 \quad \text{in } \Omega, \end{array} \right. \qquad (6.3.12)$$

whose variational forms associated with (6.3.11) and (6.3.12) are given by

$$\left\langle \frac{\partial \beta_{k+1}}{\partial t}, \varphi \right\rangle + B[\beta_{k+1}, \varphi] = \int_Q G(x,t,\beta_{k+1};\alpha_k,\beta_k)\, \varphi\, dxdt, \quad (6.3.13)$$

and

$$\left\langle \frac{\partial \alpha_{k+1}}{\partial t}, \varphi \right\rangle + B[\alpha_{k+1}, \varphi] = \int_Q F(x,t,\alpha_{k+1};\alpha_k,\beta_k)\, \varphi\, dxdt, \quad (6.3.14)$$

for all $\varphi \in V_0$. We shall show that the weak solutions β_k and α_k of (6.3.11) and (6.3.12), respectively, are uniquely defined and satisfy

$$\alpha_0 \le \alpha_1 \le \alpha_2 \le \cdots \le \alpha_k \le \cdots \le \beta_k \le \cdots \le \beta_2 \le \beta_1 \le \beta_0. \quad (6.3.15)$$

Let us consider $k = 0$. Then according to (6.3.9), $G(x,t,u;\alpha_0,\beta_0)$ is of the form

$$G(x,t,u;\alpha_0,\beta_0) = c(x,t)u + h(x,t)$$

where $c(x,t) = f_u(x,t,\alpha_0) + g_u(x,t,\beta_0)$, and

$$h(x,t) = f(x,t,\beta_0) + g(x,t,\beta_0) - f_u(x,t,\alpha_0)\beta_0 - g_u(x,t,\beta_0)\beta_0,$$

which implies by (A3) that $c \in L^\infty(Q)$ and $h \in L^2(Q)$. Thus by Theorem 6.3.2 there is a unique solution $\beta_1 \in W_0$ of (6.3.11). In just the same way one gets the existence of a unique solution $\alpha_1 \in W_0$ of (6.3.12). Let us next prove that

$$\alpha_0 \le \alpha_1 \le \beta_1 \le \beta_0. \qquad (6.3.16)$$

Since α_0 is a lower solution of (6.3.7) we see that

$$\left\langle \frac{\partial \alpha_0}{\partial t}, \varphi \right\rangle + B[\alpha_0, \varphi] \le \int_Q F(x,t,\alpha_0;\alpha_0,\beta_0)\, \varphi\, dxdt,$$

for all $\varphi \in V_0 \cap L^2_+(Q)$, and $\alpha_0(x,t) \le 0$ on Γ, $\alpha_0(x,0) \le 0$ in Ω. By definition the iterate α_1 is the unique solution of (6.3.12) for $k = 0$, and thus α_1 satisfies homogeneous initial and boundary conditions and the variational form (6.3.14). Let $p = \alpha_0 - \alpha_1$. Then p satisfies $p(x, t) \leq 0$ on Γ, $p(x, 0) \leq 0$ in Ω and

$$\left\langle \frac{\partial p}{\partial t}, \varphi \right\rangle + B[p, \varphi] + (-cp, \varphi) \leq 0 \text{ for all } \varphi \in V_0 \cap L^2_+(Q),$$

where $c(x, t) = f_u(x, t, \alpha_0) + g_u(x, t, \beta_0)$, and thus $c \in L^\infty(Q)$ by (A3). Hence Corollary 6.3.1 yields $p \leq 0$, i.e., $\alpha_0 \leq \alpha_1$. Similarly, we can show that $\beta_1 \leq \beta_0$. Next we show that $\alpha_1 \leq \beta_0$ and $\alpha_0 \leq \beta_1$. To show that $\alpha_1 \leq \beta_0$, we use the following inequalities which are immediate consequences of $f_{uu}(x, t, u) \geq 0$ (f convex in u) and $g_{uu}(x, t, u) \leq 0$ (g concave in u) imposed by assumption (A2):

$$f(x, t, u) \geq f(x, t, v) + f_u(x, t, v)(u - v), \tag{6.3.17}$$

$$g(x, t, u) \geq g(x, t, v) + g_u(x, t, v)(u - v) \tag{6.3.18}$$

for all $u, v \in [\alpha_0, \beta_0]$. By (6.3.17) and (6.3.18) we obtain, for all $\varphi \in V_0 \cap L^2_+(Q)$

$$\left\langle \frac{\partial \alpha_1}{\partial t}, \varphi \right\rangle + B[\alpha_1, \varphi] = \int_Q F(x, t, \alpha_1; \alpha_0, \beta_0) \, \varphi \, dx dt \tag{6.3.19}$$

$$\leq \int_Q \Big[f(x, t, \beta_0) - f_u(x, t, \alpha_0)(\beta_0 - \alpha_0)$$
$$+ g(x, t, \beta_0) - g_u(x, t, \beta_0)(\beta_0 - \alpha_0)$$
$$+ f_u(x, t, \alpha_0)(\alpha_1 - \alpha_0)$$
$$+ g_u(x, t, \beta_0)(\alpha_1 - \alpha_0) \Big] \, \varphi \, dx dt$$

$$\leq \int_Q [f(x, t, \beta_0) + g(x, t, \beta_0)$$
$$+ f_u(x, t, \alpha_0)(\alpha_1 - \beta_0)$$
$$+ g_u(x, t, \beta_0)(\alpha_1 - \beta_0)] \, \varphi \, dx dt$$

$$= \int_Q G(x, t, \alpha_1; \alpha_0, \beta_0) \, \varphi \, dx dt.$$

Since β_0 is an upper solution of (6.3.7) it satisfies $\beta_0(x, t) \geq 0$ on Γ, $\beta_0(x, 0) \geq 0$ in Ω and

$$\left\langle \frac{\partial \beta_0}{\partial t}, \varphi \right\rangle + B[\beta_0, \varphi] \geq \int_Q G(x, t, \beta_0; \alpha_0, \beta_0) \, \varphi \, dx dt \tag{6.3.20}$$

for all $\varphi \in V_0 \cap L^2_+(Q)$. With

$$G(x,t,\alpha_1;\alpha_0,\beta_0) \quad - \quad G(x,t,\beta_0;\alpha_0,\beta_0)$$
$$= \quad f_u(x,t,\alpha_0)(\alpha_1-\beta_0) + g_u(x,t,\beta_0)(\alpha_1-\beta_0)$$

we get by subtracting (6.3.20) from (??) that $p = \alpha_1 - \beta_0$ satisfies Corollary 6.3.1 which shows $\alpha_1 \leq \beta_0$. A similar argument proves that $\alpha_0 \leq \beta_1$. To prove $\alpha_1 \leq \beta_1$, we use (6.3.17) and (6.3.18) and the fact that $g_u(x,t,u)$ is monotone nonincreasing in u and $\alpha_1 \leq \beta_0$ to get

$$\left\langle \frac{\partial \alpha_1}{\partial t}, \varphi \right\rangle + B[\alpha_1, \varphi] \quad = \quad \int_Q F(x,t,\alpha_1;\alpha_0,\beta_0)\, \varphi \; dxdt \qquad (6.3.21)$$

$$\leq \quad \int_Q \Big[f(x,t,\alpha_1) + g(x,t,\alpha_1)$$
$$+ (g_u(x,t,\beta_0) - g_u(x,t,\alpha_1))$$
$$\times (\alpha_1 - \alpha_0) \Big]\, \varphi \; dxdt$$

$$\leq \quad \int_Q \Big[f(x,t,\alpha_1) + g(x,t,\alpha_1) \Big]\, \varphi \; dxdt$$

$$= \quad \int_Q F(x,t,\alpha_1;\alpha_1,\beta_1)\, \varphi \; dxdt,$$

for each $\varphi \in V_0 \cap L^2_+(Q)$. Similarly, for all $\varphi \in V_0 \cap L^2_+(Q)$, we have

$$\left\langle \frac{\partial \beta_1}{\partial t}, \varphi \right\rangle + B[\beta_1, \varphi] \quad = \quad \int_Q G(x,t,\beta_1;\alpha_0,\beta_0)\, \varphi \; dxdt \qquad (6.3.22)$$

$$\geq \quad \int_Q [f(x,t,\beta_1) + f_u(x,t,\beta_1)(\beta_0-\beta_1)$$
$$+ f_u(x,t,\alpha_0)(\beta_1-\beta_0) + g(x,t,\beta_1)]\, \varphi \; dxdt$$

$$= \quad \int_Q [f(x,t,\beta_1)$$
$$+ (-f_u(x,t,\beta_1) + f_u(x,t,\alpha_0))$$
$$\times (\beta_1-\beta_0) + g(x,t,\beta_1) \Big]\, \varphi \; dxdt$$

$$\geq \quad \int_Q \Big[f(x,t,\beta_1) + g(x,t,\beta_1) \Big]\, \varphi \; dxdt$$

$$= \quad \int_Q G(x,t,\beta_1;\alpha_1,\beta_1)\, \varphi \; dxdt,$$

because of the fact that $f_u(x,t,u)$ is monotone nondecreasing in u and $\alpha_0 \leq \beta_1$. The inequalities (6.3.21) and (6.3.22) imply that α_1 is a lower solution and β_1 is an upper solution of the IBVP (6.3.7). Since the nonlinearity $f(x,t,u) + g(x,t,u)$ of the right-hand side of (6.3.7) is Lipschitz continuous in $u \in [\alpha_0, \beta_0]$, it then follows from Theorem 6.3.1 that $\alpha_1 \leq \beta_1$ proving (6.3.16).

We shall next prove that if

$$\alpha_{k-1} \leq \alpha_k \leq \beta_k \leq \beta_{k-1} \qquad (6.3.23)$$

for some $k > 1$, then it follows that

$$\alpha_k \leq \alpha_{k+1} \leq \beta_{k+1} \leq \beta_k. \qquad (6.3.24)$$

Since α_k satisfies

$$\left\langle \frac{\partial \alpha_k}{\partial t}, \varphi \right\rangle + B[\alpha_k, \varphi] = \int_Q F(x, t, \alpha_k; \alpha_{k-1}, \beta_{k-1}) \, \varphi \, dxdt,$$

for all $\varphi \in V_0$, by following the arguments employed to obtain (6.3.21) and (6.3.22), we can prove, because of (6.3.23), that

$$\left\langle \frac{\partial \alpha_k}{\partial t}, \varphi \right\rangle + B[\alpha_k, \varphi] \leq \int_Q F(x, t, \alpha_k; \alpha_k, \beta_k) \, \varphi \, dxdt, \qquad (6.3.25)$$

and

$$\left\langle \frac{\partial \beta_k}{\partial t}, \varphi \right\rangle + B[\beta_k, \varphi] \geq \int_Q G(x, t, \beta_k; \alpha_k, \beta_k) \, \varphi \, dxdt, \qquad (6.3.26)$$

for each $\varphi \in V_0 \cap L_+^2(Q)$. Moreover, since α_{k+1} and β_{k+1} are the uniquely defined solutions of (6.3.12) and (6.3.11), respectively, we therefore obtain from Corollary 6.3.1 that $\alpha_k \leq \alpha_{k+1}$ and $\beta_{k+1} \leq \beta_k$. Next we show that $\alpha_{k+1} \leq \beta_k$ and $\alpha_k \leq \beta_{k+1}$. We find using (6.3.17) and (6.3.18) that

$$\left\langle \frac{\partial \alpha_{k+1}}{\partial t}, \varphi \right\rangle + B[\alpha_{k+1}, \varphi] \;=\; \int_Q F(x, t, \alpha_{k+1}; \alpha_k, \beta_k)\, \varphi \; dx dt$$

$$\leq \int_Q \Big[f(x, t, \beta_k) - f_u(x, t, \alpha_k)(\beta_k - \alpha_k)$$
$$+ g(x, t, \beta_k) - g_u(x, t, \beta_k)(\beta_k - \alpha_k)$$
$$+ f_u(x, t, \alpha_k)(\alpha_{k+1} - \alpha_k)$$
$$+ g_u(x, t, \beta_k)(\alpha_{k+1} - \alpha_k) \Big]\, \varphi \; dx dt$$

$$\leq \int_Q \Big[f(x, t, \beta_k) + g(x, t, \beta_k)$$
$$+ (f_u(x, t, \alpha_k)$$
$$+ g_u(x, t, \beta_k))(\alpha_{k+1} - \beta_k) \Big]\, \varphi \; dx dt$$

$$= \int_Q G(x, t, \alpha_{k+1}; \alpha_k, \beta_k)\, \varphi \; dx dt,$$

for each $\varphi \in V_0 \cap L^2_+(Q)$. This last inequality together with (6.3.26) yields by Corollary 6.3.1 that $\alpha_{k+1} \leq \beta_k$. Similarly, one can show that $\alpha_k \leq \beta_{k+1}$ using (6.3.25) and

$$\left\langle \frac{\partial \beta_{k+1}}{\partial t}, \varphi \right\rangle + B[\beta_{k+1}, \varphi] \;=\; \int_Q G(x, t, \beta_{k+1}; \alpha_k, \beta_k)\, \varphi \; dx dt$$

$$\geq \int_Q \Big[f(x, t, \alpha_k) + f_u(x, t, \alpha_k)(\beta_k - \alpha_k)$$
$$+ g(x, t, \alpha_k) + g_u(x, t, \beta_k)(\beta_k - \alpha_k)$$
$$+ f_u(x, t, \alpha_k)(\beta_{k+1} - \beta_k)$$
$$+ g_u(x, t, \beta_k)(\beta_{k+1} - \beta_k) \Big]\, \varphi \; dx dt$$

$$= \int_Q \Big[f(x, t, \alpha_k) + g(x, t, \alpha_k) +$$
$$+ (f_u(x, t, \alpha_k)$$
$$+ g_u(x, t, \beta_k))(\beta_{k+1} - \alpha_k) \Big]\, \varphi \; dx dt$$

$$= \int_Q F(x, t, \beta_{k+1}; \alpha_k, \beta_k)\, \varphi \; dx dt,$$

for each $\varphi \in V_0 \cap L^2_+(Q)$. Finally to prove $\alpha_{k+1} \leq \beta_{k+1}$, we show that

α_{k+1} and β_{k+1} satisfy

$$\left\langle \frac{\partial \alpha_{k+1}}{\partial t}, \varphi \right\rangle + B[\alpha_{k+1}, \varphi] \leq \int_Q F(x,t,\alpha_{k+1};\alpha_{k+1},\beta_{k+1})\, \varphi\, dxdt,$$

and

$$\left\langle \frac{\partial \beta_{k+1}}{\partial t}, \varphi \right\rangle + B[\beta_{k+1}, \varphi] \geq \int_Q G(x,t,\beta_{k+1};\alpha_{k+1},\beta_{k+1})\, \varphi\, dxdt,$$

for each $\varphi \in V_0 \cap L^2_+(Q)$. This precisely employs the arguments as we used in proving (6.3.21) and (6.3.22) replacing α_0, β_0, α_1, β_1 by α_k, β_k, α_{k+1}, β_{k+1} and using (6.3.17), (6.3.18) as well as the monotone character of f_u and g_u. Consequently, we get from Theorem 6.3.1 that $\alpha_{k+1} \leq \beta_{k+1}$, proving (6.3.24). Thus by induction, it follows that (6.3.15) is true for each $k = 0, 1, 2, \dots$.

(b) **Convergence of $\{\alpha_k\}$, $\{\beta_k\}$ to the unique solution in $[\alpha_0, \beta_0]$** By the monotonicity of the iterates (α_k) and (β_k) according to (6.3.15) there exist the a.e. pointwise limits

$$\varrho(x,t) = \lim_{k\to\infty} \alpha_k(x,t), \quad r(x,t) = \lim_{k\to\infty} \beta_k(x,t), \quad \text{for a.e. } (x,t) \in Q.$$

Moreover, since α_k, $\beta_k \in [\alpha_0, \beta_0]$, it follows by Lebesgue's dominated convergence theorem that

$$\alpha_k \to \varrho \text{ and } \beta_k \to r \text{ in } L^2(Q). \tag{6.3.27}$$

In what follows C denotes a generic positive constant whose value may be different at different places. Using (6.3.27) and (A3) we show that the right-hand sides of IBVP (6.3.11) and (6.3.12) defining β_{k+1} and α_{k+1}, respectively, converge in $L^2(Q)$. Consider

$$\begin{aligned} G(x,t,\beta_{k+1};\alpha_k,\beta_k) &= f(x,t,\beta_k) + g(x,t,\beta_k) \tag{6.3.28} \\ &\quad + (f_u(x,t,\alpha_k) + g_u(x,t,\beta_k))(\beta_{k+1} - \beta_k). \end{aligned}$$

By (A3) and due to $f(x,t,\eta) = f(x,t,\alpha_0) + f_u(x,t,\xi)(\eta - \alpha_0)$, where $\alpha_0 \leq \xi \leq \eta$, we obtain the estimate

$$|f(x,t,\eta)| \leq |f(x,t,\alpha_0)| + C\,|\beta_0 - \alpha_0|, \text{ for any } \eta \in [\alpha_0, \beta_0]$$

which implies that the Nemytski operator F related to f by $Fu(x,t) = f(x,t,u(x,t))$ is continuous and bounded as a mapping $F : [\alpha_0, \beta_0] \subset$

$L^2(Q) \to L^2(Q)$. The same arguments yield that $G : [\alpha_0, \beta_0] \subset L^2(Q)$ $\to L^2(Q)$ is continuous and bounded, where $Gu(x,t) = g(x,t,u(x,t))$ is the Nemytski operator related to g. By (6.3.27) we therefore get

$$F\beta_k \to Fr, \quad G\beta_k \to Gr, \quad \text{in } L^2(Q) \text{ as } k \to \infty. \tag{6.3.29}$$

Using (6.3.29) and (A3) we obtain for the right-hand sides $G(\cdot,\cdot,\beta_{k+1};\alpha_k,\beta_k)$ given by (6.3.28)

$$G(\cdot,\cdot,\beta_{k+1};\alpha_k,\beta_k) \to Fr + Gr, \quad \text{in } L^2(Q), \tag{6.3.30}$$

and by similar arguments

$$F(\cdot,\cdot,\alpha_{k+1};\alpha_k,\beta_k) \to F\varrho + G\varrho, \quad \text{in } L^2(Q). \tag{6.3.31}$$

Since β_k and α_k ($k \geq 1$) are solutions of the linear IBVP (6.3.11) and (6.3.12) with homogeneous initial and boundary values, respectively, we may apply the estimate of Theorem 6.3.2 which gives for some positive constant ν and for all $k \geq 1$ the estimates

$$\|\beta_k\|_W \leq \nu \, \|G(\cdot,\cdot,\beta_k;\alpha_{k-1},\beta_{k-1})\|_{L^2(Q)} \tag{6.3.32}$$

and

$$\|\alpha_k\|_W \leq \nu \, \|F(\cdot,\cdot,\alpha_k;\alpha_{k-1},\beta_{k-1})\|_{L^2(Q)}. \tag{6.3.33}$$

Due to (6.3.30) and (6.3.31), the estimates (6.3.32) and (6.3.33) imply the strong convergence of the sequences (α_k) and (β_k) in W whose limits must be ϱ and r, respectively, because of the (compact) embedding $W \subset L^2(Q)$. Since α_k and β_k belong for $k \geq 1$ to the subset $M := \{v \in W_0 \mid v(x,0) = 0\}$ which is closed in W it follows that also their limits $\varrho, r \in M$, that is, the limits also satisfy homogeneous initial and boundary conditions. The convergence result (6.3.30) and (6.3.31) together with

$$\alpha_k \to \varrho, \quad \beta_k \to r \quad \text{in } W \text{ as } k \to \infty,$$

allow us to pass to the limit in the corresponding variational forms (6.3.13) and (6.3.14) as $k \to \infty$ which yield

$$\left\langle \frac{\partial r}{\partial t}, \varphi \right\rangle + B[r,\varphi] = \int_Q (f(x,t,r) + g(x,t,r)) \, \varphi \, dxdt,$$

and

$$\left\langle \frac{\partial \varrho}{\partial t}, \varphi \right\rangle + B[\varrho, \varphi] = \int_Q (f(x,t,\varrho) + g(x,t,\varrho))\, \varphi \, dxdt,$$

for all $\varphi \in V_0$ showing that r and ϱ are solutions of (6.3.7). To show that $\varrho = r = u$ is the unique weak solution of (6.3.7), we note that, since $\alpha_0 \leq \varrho \leq r \leq \beta_0$, it is enough to show that $r \leq \varrho$. The latter, however, immediately follows by using Theorem 6.3.1, which proves the uniqueness.

(c) **Quadratic convergence of** $\{\alpha_k\}$, $\{\beta_k\}$ To prove quadratic convergence of the sequences $\{\alpha_k\}$, $\{\beta_k\}$ to the unique solution u, respectively, we set

$$p_{k+1} = u - \alpha_{k+1}, \quad q_{k+1} = \beta_{k+1} - u$$

so that $p_{k+1} \geq 0$ and $q_{k+1} \geq 0$ in Q as well as $p_{k+1}(x,t) = 0$ on Γ, $p_{k+1}(x,0) = 0$ in Ω, and $q_{k+1}(x,t) = 0$ on Γ, $q_{k+1}(x,0) = 0$ in Ω. We then have for $\varphi \in V_0 \cap L_+^2(Q)$ and using the fact that f_u is monotone nondecreasing and g_u is monotone nonincreasing

$$
\begin{aligned}
\left\langle \frac{\partial p_{k+1}}{\partial t}, \varphi \right\rangle + B[p_{k+1}, \varphi] \quad = \quad & \int_Q \Big[f(x,t,u) + g(x,t,u) \qquad (6.3.34) \\
& -f(x,t,\alpha_k) - f_u(x,t,\alpha_k) \\
& \times (\alpha_{k+1} - \alpha_k) - g(x,t,\alpha_k) \\
& -g_u(x,t,\beta_k)(\alpha_{k+1} - \alpha_k) \Big] \varphi \, dxdt \\
\leq \quad & \int_Q \Big[f_u(x,t,u)p_k \\
& +g_u(x,t,\alpha_k)p_k - f_u(x,t,\alpha_k) \\
& \times (p_k - p_{k+1}) - g_u(x,t,\beta_k) \\
& (p_k - p_{k+1}) \Big] \varphi \, dxdt \\
= \quad & \int_Q \Big[(f_u(x,t,u) - f_u(x,t,\alpha_k))p_k \\
& -(g_u(x,t,\beta_k) - g_u(x,t,\alpha_k))p_k \\
& +(f_u(x,t,\alpha_k) \\
& +g_u(x,t,\beta_k))p_{k+1} \Big] \varphi \, dxdt
\end{aligned}
$$

$$= \int_Q \Big[f_{uu}(x,t,\xi)p_k^2 - g_{uu}(x,t,\sigma)(\beta_k - \alpha_k)p_k$$
$$+ (f_u(x,t,\alpha_k)$$
$$+ g_u(x,t,\beta_k))p_{k+1} \Big] \, \varphi \, dxdt,$$

where $\alpha_k \le \xi \le u$, $\alpha_k \le \sigma \le \beta_k$. In the estimate (6.3.34) we have used the fact that, since by (A1) α_0, $\beta_0 \in L^4(Q)$, also α_k, $\beta_k \in L^4(Q)$ so that the integrals are defined. By (A3) and the definition of p_k, q_k we have

$$-g_{uu}(x,t,\sigma)(\beta_k - \alpha_k)p_k \le C \, (q_k + p_k)p_k \le C \, \Big(\frac{3}{2}p_k^2 + \frac{1}{2}q_k^2\Big), \quad (6.3.35)$$

thus, we get from (6.3.34), (6.3.35) and using $\|f_{uu}(\cdot,\cdot,\eta)\|_{L^\infty(Q)} \le C$ for all $\eta \in [\alpha_0, \beta_0]$

$$\Big\langle \frac{\partial p_{k+1}}{\partial t}, \varphi \Big\rangle + B[p_{k+1}, \varphi] \le \int_Q c_k(x,t)p_{k+1} \, \varphi \, dxdt \qquad (6.3.36)$$
$$+ \int_Q \Big(\frac{5}{2}Cp_k^2 + \frac{1}{2}Cq_k^2\Big) \, \varphi \, dxdt,$$

for all $\varphi \in V_0 \cap L_+^2(Q)$, where the coefficients $c_k \in L^\infty(Q)$ given by

$$c_k(x,t) = f_u(x,t,\alpha_k(x,t)) + g_u(x,t,\beta_k(x,t))$$

are uniformly bounded by some positive constant C, i.e., $\|c_k\|_{L^\infty(Q)} \le C$ for all k. Taking in (6.3.36), in particular, $\varphi = p_{k+1}$ we obtain for any subdomain $Q_t = \Omega \times (0,t) \subset Q$ with $t \in (0,T]$ (notice $p_{k+1}(x,0) = 0$, $k \ge 0$) the inequality

$$\frac{1}{2}\|p_{k+1}(\cdot,t)\|_{L^2(\Omega)}^2 + \mu\|\nabla p_{k+1}\|_{L^2(Q_t)}^2$$

$$\le C\|p_{k+1}\|_{L^2(Q_t)}^2 + \frac{1}{2}\|\frac{5}{2}Cp_k^2 + \frac{1}{2}Cq_k^2\|_{L^2(Q_t)}^2$$
$$+ \frac{1}{2}\|p_{k+1}\|_{L^2(Q_t)}^2,$$

and thus

$$\|p_{k+1}(\cdot,t)\|_{L^2(\Omega)}^2 \le c_1\|p_{k+1}\|_{L^2(Q_t)}^2 + c_2\|p_k^2 + q_k^2\|_{L^2(Q_t)}^2. \qquad (6.3.37)$$

Setting $y(t) = \| p_{k+1} \|^2_{L^2(Q_t)}$ and $h(t) = \| p_k^2 + q_k^2 \|^2_{L^2(Q_t)}$, (6.3.37) can equivalently be written as

$$y'(t) \le c_1 y(t) + c_2 h(t), \quad y(0) = 0,$$

which by applying Gronwall's inequality yields $y(t) \le c_2 e^{c_1 t} \int_0^t y(s) ds$, and thus there is a positive constant C_0 depending on c_1, c_2 and T such that

$$\| p_{k+1} \|^2_{L^2(Q)} \le C_0 \| p_k^2 + q_k^2 \|_{L^2(Q)}$$

which yields the desired estimate

$$\| p_{k+1} \|_{L^2(Q)} \le C_0 (\| p_k^2 \|_{L^2(Q)} + \| q_k^2 \|_{L^2(Q)}). \qquad (6.3.38)$$

One can get an estimate similar to (6.3.38) for q_{k+1} and we omit the details. This completes the proof.

Several remarks are now in order.

Remarks 6.3.1

(i) If $g(x, t, u) \equiv 0$ in (6.3.7), we get a result with $f(x, t, u)$ being convex in u.

(ii) If $f(x, t, u) \equiv 0$ in (6.3.7), one obtains the dual result where $g(x, t, u)$ is concave in u.

(iii) Consider the case when $g(x, t, u) \equiv 0$ and $f(x, t, u)$ is not convex in u. Suppose there is a function $k(x, t, u)$ which is convex in u such that $f(x, t, u) + k(x, t, u)$ is convex in u. Then (6.3.7) can be rewritten as

$$\frac{\partial u}{\partial t} + Au = \tilde{f}(x, t, u) + \tilde{g}(x, t, u)$$

where $\tilde{f}(x, t, u) = f(x, t, u) + k(x, t, u)$ and $\tilde{g}(x, t, u) = -k(x, t, u)$ so that the conditions of Theorem 6.3.1 are satisfied and we get the same conclusion when $f(x, t, u)$ is not convex also.

(iv) A dual situation of (iii) is also valid, that is, when $f(x, t, u) \equiv 0$ and $g(x, t, u)$ is not concave. Assume that there is a function $k(x, t, u)$ which is concave in u such that $g(x, t, u) + k(x, t, u)$ is concave in u. Then (6.3.7) can be written as before but now with $\tilde{f}(x, t, u) = -k(x, t, u)$ and $\tilde{g}(x, t, u) = g(x, t, u) + k(x, t, u)$. Then also the conditions of Theorem 6.3.1 are satisfied and we get the same conclusion when $g(x, t, u)$ is not concave.

(v) The semilinear parabolic problem (6.3.7) may be considered as a model for the Cauchy problem of a general semilinear evolution equation of the form

$$\frac{\partial u}{\partial t} + Au = f(u).$$

Therefore it should be possible to extend the quasilinearization method to abstract evolution equations in ordered Banach spaces.

6.4   Nonlocal Problems: Existence and Comparison Results

This and the next sections are devoted to the study of semilinear parabolic initial boundary value problems (IBVP for short) with nonlocal flux conditions. In this section, we introduce the problem, describe the needed notation and prove existence and comparison results, and in the next section, we shall consider the method of generalized quasilinearization for such problems.

Let $\Omega \subset \mathbb{R}^N$ be a bounded domain with boundary $\partial\Omega$ of class $C^{1,1}$, $Q = \Omega \times (0,T)$ and $\Gamma = \partial\Omega \times (0,T)$, $T > 0$. Consider the semilinear parabolic initial boundary value problem (IBVP)

$$\left[\begin{array}{rcll} \frac{\partial u(x,t)}{\partial t} + Au(x,t) &=& f(x,t,u(x,t)), & \text{in } Q\,, \\ u(x,0) &=& \psi(x), & \text{in } \Omega\,, \qquad (6.4.1) \\ \frac{\partial u(x,t)}{\partial \nu} + b(x,t)u(x,t) &=& \int_\Omega k(x,t,x',u(x',t))\,dx', & \text{on } \Gamma\,, \end{array} \right.$$

where A is a second-order strongly elliptic differential operator in the form

$$Au(x,t) = -\sum_{i,j=1}^N \frac{\partial}{\partial x_i}\left(a_{ij}(x,t)\frac{\partial u}{\partial x_j}\right),$$

with coefficients $a_{ij} \in L^\infty(Q)$ satisfying for all $\xi = (\xi_1,\ldots,\xi_N) \in \mathbb{R}^N$

$$\sum_{i,j=1}^N a_{ij}(x,t)\xi_i\xi_j \geq \mu|\xi|^2, \quad \text{for a.e.} \quad (x,t) \in Q \quad \text{with some constant } \mu > 0,$$

$\partial/\partial\nu$ denotes the exterior conormal derivative on Γ associated with the operator A, and $b \in L^\infty(\Gamma)$ satisfies $b(x,t) \geq 0$ a.e. on Γ. The nonlinearities $f : Q \times \mathbb{R} \to \mathbb{R}$ and $k : \Gamma \times \Omega \times \mathbb{R} \to \mathbb{R}$ are assumed to be Caratheodory functions, that is f (resp. k) is measurable in $(x,t) \in Q$ (resp. in $(x,t,x') \in$

$\Gamma \times \Omega$) for each $u \in \mathbb{R}$ and continuous in u for a.e. $(x,t) \in Q$ (resp. $(x,t,x') \in \Gamma \times \Omega$). The initial value ψ is assumed to be an element of $L^2(\Omega)$.

The following special case of problem (6.4.1) has been treated within the framework of classical solutions by means of the monotone iterative technique

$$\left[\begin{array}{rl} \frac{\partial u}{\partial t} + Au = f(x,u), & \text{in } Q, \\ u(x,0) = \psi(x), & \text{in } \Omega \\ \frac{\partial u}{\partial \nu} + bu = \int_\Omega K(x,x')\, u(x',t)\, dx', & \text{on } \Gamma, \end{array} \right. \tag{6.4.2}$$

where b is some nonnegative constant, $K(x,x') \geq 0$, A is a linear strongly elliptic operator and all data are assumed to be sufficiently smooth. The nonlocal IBVP (6.4.2) is taken from quasi-static thermoelasticity.

Let $H^1(\Omega)$ denote the usual Sobolev space of square integrable functions and let $(H^1(\Omega))^*$ denote its dual space. Then by identifying $L^2(\Omega)$ with its dual space, $H^1(\Omega) \subset L^2(\Omega) \subset (H^1(\Omega))^*$ forms an evolution triple with all the embeddings being continuous, dense and compact. We let $V = L^2(0,T;H^1(\Omega))$, denote its dual space by $V^* = L^2(0,T;(H^1(\Omega))^*)$, and define a function space W by

$$W = \left\{ w \in V \,\middle|\, \frac{\partial w}{\partial t} \in V^* \right\},$$

where the derivative $\partial/\partial t$ is understood in the sense of vector-valued distributions and characterized by

$$\int_0^T u'(t)\phi(t)\, dt = -\int_0^T u(t)\phi'(t)\, dt, \quad \text{for all } \phi \in C_0^\infty(0,T).$$

The space W endowed with the norm

$$\|w\|_W = \|w\|_V + \|\partial w/\partial t\|_{V^*}$$

is a Banach space which is separable and reflexive due to the separability and reflexivity of V and V^*, respectively. (Note that for any Banach space X the space $Y = L^2(0,T;X)$ of vector-valued functions consists of all measurable functions $u : (0,T) \to X$ for which $\|u\|_Y = (\int_0^T \|u(t)\|_X^2\, dt)^{1/2}$ is finite.) Furthermore it is well known that the embedding $W \subset C([0,T]; L^2(\Omega))$ is continuous. Finally, because $H^1(\Omega) \subset L^2(\Omega)$ is compactly embedded, we have a compact embedding of $W \subset L^2(Q) \equiv L^2(0,T;L^2(\Omega))$. We denote the duality pairing between the elements of V^* and V by $\langle \cdot, \cdot \rangle$, and define the bilinear form a associated with the operator A by

$$\langle Au, \varphi \rangle = a(u,\varphi) \equiv \sum_{i,j=1}^N \int_Q a_{ij}(x,t) \frac{\partial u}{\partial x_i} \frac{\partial \varphi}{\partial x_j}\, dx dt \tag{6.4.3}$$

which is well defined on $V \times V$, and which immediately implies that $A : V \to V^*$ is linear, continuous, and monotone. A partial ordering in $L^2(Q)$ is defined by $u \le w$ if and only if $w - u$ belongs to the cone $L^2_+(Q)$ of all nonnegative elements of $L^2(Q)$. This induces a corresponding partial ordering also in the subspace W of $L^2(Q)$, and if $\underline{u}, \bar{u} \in W$ with $\underline{u} \le \bar{u}$ then $[\underline{u}, \bar{u}] := \{u \in W \mid \underline{u} \le u \le \bar{u}\}$ denotes the order interval formed by $\underline{u}$ and $\bar{u}$. Furthermore, we introduce the Nemytski operator F related to the function $f : Q \times \mathbb{R} \to \mathbb{R}$ and an operator K related to the nonlocal boundary condition by

$$\begin{aligned} (Fu)(x,t) &:= f(x,t,u(x,t)), \quad (x,t) \in Q, \\ (Ku)(x,t) &:= \int_\Omega k(x,t,x',u(x',t))\,dx', \quad (x,t) \in \Gamma. \end{aligned} \qquad (6.4.4)$$

Let $\gamma : V \to L^2(\Gamma)$ denote the trace operator where $\gamma u \in L^2(\Gamma)$ are the generalized boundary values of $u \in V$ on Γ. It is well known that $\gamma : V \to L^2(\Gamma)$ is linear and continuous.

Due to the continuity of the trace operator γ and since $b(x,t) \ge 0$ a.e. on Γ, the bilinear form b given by

$$b(u,\varphi) := \int_\Gamma b\,\gamma u\,\gamma\varphi\,d\Gamma, \quad u, \varphi \in V,$$

defines a linear, continuous and monotone operator $B : V \to V^*$ by

$$\langle Bu, \varphi \rangle := b(u, \varphi). \qquad (6.4.5)$$

The notion of a weak solution (respectively, upper and lower solutions) of the IBVP (6.4.1) is obtained as usual by multiplying the equation by an appropriate test function, applying integration by parts, and taking into account the boundary condition. This yields the following definition.

Definition 6.4.1 *A function $u \in W$ is called a solution of the IBVP* (6.4.1) *if $Fu \in L^2(Q)$ and $Ku \in L^2(\Gamma)$ such that*

(i) $u(x,0) = \psi(x), \quad x \in \Omega;$

(ii) $\langle \frac{\partial u}{\partial t}, \varphi \rangle + a(u,\varphi) + b(u,\varphi) = \int_Q (Fu)\varphi\,dxdt + \int_\Gamma (Ku)\,\gamma\varphi\,d\Gamma, \; \forall \varphi \in V.$

In a natural way the notion of a lower solution for (6.4.1) is given by

Definition 6.4.2 *A function $\alpha \in W$ is called a lower solution of the IBVP* (6.4.1) *if $F\alpha \in L^2(Q)$ and $K\alpha \in L^2(\Gamma)$ such that*

(i) $\alpha(x,0) \leq \psi(x)$, $x \in \Omega$;

(ii) $\langle \frac{\partial \alpha}{\partial t}, \varphi \rangle + a(\alpha, \varphi) + b(\alpha, \varphi) \leq \int_Q (F\alpha)\varphi\,dx dt + \int_\Gamma (K\alpha)\gamma\varphi\,d\Gamma$, $\forall \varphi \in V \cap L_+^2(Q)$.

Similarly, a function $\beta \in W$ is an *upper solution* of (6.4.1) if the reversed inequalities hold in (i) and (ii) of Definition 6.4.2.

Without loss of generality, we may assume $\psi(x) \equiv 0$, since otherwise we can reduce the problem by translation $u \to u + v$ to zero initial values, such as for example by translation with $v \in W$, where v is the unique solution of the linear IBVP

$$\frac{\partial v}{\partial t} + Av = 0, \quad \frac{\partial v}{\partial \nu} + bv = 0, \quad v(x,0) = \psi.$$

Hence, in what follows we shall consider the IBVP (6.4.1) with zero initial values, i.e., with $\psi = 0$.

Let us make the following assumptions on the nonlinear right-hand side f and the nonlinear kernel function k.

(A1) $f : Q \times \mathbb{R} \to \mathbb{R}$ is a Caratheodory function satisfying $F0 = f(\cdot, \cdot, 0) \in L^2(Q)$ and $|f(x,t,u) - f(x,t,v)| \leq l(x,t)|u-v|$, $(x,t) \in Q$, $u, v \in \mathbb{R}$, where $l \in L_+^\infty(Q)$.

(A2) $k : \Gamma \times \Omega \times \mathbb{R} \to \mathbb{R}$ is a Caratheodory function satisfying $k(\cdot, \cdot, \cdot, 0) \in L^2(\Gamma \times \Omega)$, and $|k(x,t,x',u) - k(x,t,x',v)| \leq m(x,t,x')\,|u-v|$, $(x,t,x') \in \Gamma \times \Omega$, $u, v \in \mathbb{R}$, where $m \in L_+^\infty(\Gamma \times \Omega)$.

(A3) $u \mapsto k(x,t,x',u)$ is increasing.

Lemma 6.4.1 *Let (A1) and (A2) be satisfied. Then the operators $F : L^2(Q) \to L^2(Q)$ and $K : L^2(Q) \to L^2(\Gamma)$ are Lipschitz continuous.*

Proof   Let $u, v \in L^2(Q)$ be given. Then from (A1) we immediately get

$$\|Fu - Fv\|_{L^2(Q)} \leq \|l\|_{L^\infty(Q)} \|u - v\|_{L^2(Q)}.$$

To prove the Lipschitz continuity of K we apply Fubini's theorem to get the following estimate

$$\begin{aligned}
\|Ku - Kv\|_{L^2(\Gamma)}^2 &= \int_\Gamma \left| \int_\Omega (k(x,t,x',u(x',t)) - k(x,t,x',v(x',t)))\,dx' \right|^2 d\Gamma \\
&\leq \|m\|_{L^\infty(\Gamma \times \Omega)}^2 |\Omega| \int_\Gamma \left(\int_\Omega |u(x',t) - v(x',t)|^2\,dx' \right) d\Gamma \\
&\leq \|m\|_{L^\infty(\Gamma \times \Omega)}^2 |\Omega| |\partial \Omega| \|u - v\|_{L^2(Q)}^2,
\end{aligned}$$

where $|\Omega|$ and $|\partial\Omega|$ denote the Lebesgue measure and surface measure of Ω and $\partial\Omega$, respectively.

In the following lemma we provide a comparison result for weak upper and lower solutions of the IBVP (6.4.1).

Lemma 6.4.2 *Let α and β be weak lower and upper solutions of the IBVP (6.4.1), respectively. Then under assumptions (A1)–(A3) we have $\alpha \leq \beta$ in Q.*

Proof   Let $w = \alpha - \beta$, and $w^+ := \max(w, 0)$. Then $w^+ \in V \cap L^2_+(Q)$, and since $w(x, 0) \leq 0$, it follows that $w^+(x, 0) = 0$. Subtracting the defining inequalities for α and β from each other we obtain

$$\left\langle \frac{\partial w}{\partial t}, \varphi \right\rangle + a(w, \varphi) + b(w, \varphi) \leq \int_Q (F\alpha - F\beta)\,\varphi\,dx dt + \int_\Gamma (K\alpha - K\beta)\,\gamma\varphi\,d\Gamma,$$
$$(6.4.6)$$

for all $\varphi \in V \cap L^2_+(Q)$. Thus inequality (6.4.6) remains true for any subcylinder of the form $Q_\tau := \Omega \times (0, \tau) \subseteq Q$ and corresponding lateral boundary $\Gamma_\tau := \partial\Omega \times (0, \tau) \subseteq \Gamma$ with $\tau \in (0, T]$. Taking in (6.4.6) as a special test function w^+ we obtain by using Lemma 6.4.1 and the strong ellipticity of A for any $\tau \in (0, T]$

$$\frac{1}{2}\|w^+(\cdot, \tau)\|^2_{L^2(\Omega)} + \mu\|\nabla w^+\|^2_{L^2(Q_\tau)} + \|b\|_{L^\infty(\Gamma)}\|\gamma w^+\|^2_{L^2(\Gamma_\tau)} \qquad (6.4.7)$$

$$\leq \|l\|_{L^\infty(Q)}\|w^+\|^2_{L^2(Q_\tau)} + \int_{\Gamma_\tau} (K\alpha - K\beta)\,\gamma w^+\,d\Gamma.$$

The monotonicity of $u \mapsto k(x, t, x', u)$ according to (A3) yields

$$(K\alpha - K\beta)(x, t) \qquad\qquad\qquad\qquad\qquad\qquad\qquad\qquad\qquad (6.4.8)$$

$$= \int_{\{x':\alpha(x',t)<\beta(x',t)\}} (k(x, t, x', \alpha(x', t)) - k(x, t, x', \beta(x', t)))\,dx'$$

$$+ \int_{\{x':\alpha(x',t)\geq\beta(x',t)\}} (k(x, t, x', \alpha(x', t)) - k(x, t, x', \beta(x', t)))\,dx'$$

$$\leq \|m\|_{L^\infty(\Gamma\times\Omega)} \int_\Omega w^+(x', t)\,dx' \leq \|m\|_{L^\infty(\Gamma\times\Omega)}|\Omega|^{1/2}\|w^+(\cdot, t)\|_{L^2(\Omega)}.$$

By means of (6.4.8) and using Young's inequality the second term on the right-hand side of (6.4.7) can be estimated as follows

$$\int_{\Gamma_\tau} (K\alpha - K\beta)\,\gamma w^+ \, d\Gamma \tag{6.4.9}$$

$$\leq \|m\|_{L^\infty(\Gamma\times\Omega)}|\Omega|^{1/2} \int_{\Gamma_\tau} \|w^+(\cdot,t)\|_{L^2(\Omega)}\,\gamma w^+ \, d\Gamma$$

$$\leq \|m\|_{L^\infty(\Gamma\times\Omega)}|\Omega|^{1/2} \int_{\Gamma_\tau} \left(c(\varepsilon)\|w^+(\cdot,t)\|_{L^2(\Omega)}^2 + \varepsilon(\gamma w^+)^2 \right) d\Gamma$$

$$\leq \|m\|_{L^\infty(\Gamma\times\Omega)}|\Omega|^{1/2} \left(c(\varepsilon)|\partial\Omega|\|w^+\|_{L^2(Q_\tau)}^2 + \varepsilon\|\gamma w^+\|_{L^2(\Gamma_\tau)}^2 \right),$$

for any $\varepsilon > 0$, where $c(\varepsilon)$ denotes some positive constant depending on ε. The continuity of the trace operator $\gamma : V \to L^2(\Gamma)$ implies the following inequality (for some positive constant d)

$$\|\gamma w^+\|_{L^2(\Gamma_\tau)}^2 \leq d\,\|w^+\|_{V_\tau}^2 := d\left(\|\nabla w^+\|_{L^2(Q_\tau)}^2 + \|w^+\|_{L^2(Q_\tau)}^2\right). \tag{6.4.10}$$

Selecting ε small enough so that $d\,\|m\|_{L^\infty(\Gamma\times\Omega)}|\Omega|^{1/2}\varepsilon < \mu$ we get from (6.4.7) by taking (6.4.9) and (6.4.10) into account the inequality

$$\|w^+(\cdot,\tau)\|_{L^2(\Omega)}^2 \leq c\,\|w^+\|_{L^2(Q_\tau)}^2, \tag{6.4.11}$$

where c is some positive constant whose exact value can be derived from the above inequalities. Thus, by setting

$$y(\tau) = \|w^+(\cdot,\tau)\|_{L^2(\Omega)}^2$$

(6.4.11) implies

$$y(\tau) \leq c \int_0^\tau y(t)\,dt, \quad \text{for all } \tau \in [0,T],$$

which yields by applying the Gronwall lemma that $y(\tau) = 0$ for any $\tau \in [0,T]$, and hence $w^+ = 0$. This means $\alpha \leq \beta$ a.e. in Q.

Remarks 6.4.1

(i) It should be noted that the comparison result of Lemma 6.4.2 remains true if A is replaced by an extended linear operator of the form

$$Au = -\sum_{i,j=1}^{N} \frac{\partial}{\partial x_i}\left(a_{ij}(x,t)\frac{\partial u}{\partial x_j}\right) + \sum_{i=1}^{N} b_i(x,t)\frac{\partial u}{\partial x_i},$$

with bounded coefficients a_{ij} and b_i.

(ii) In the case that the monotonicity condition (A3) is dropped then the assertion of Lemma 6.4.2 may fail. However, one can prove a comparison result for a pair of so-called coupled upper and lower solutions whose definition is given as follows:

Definition 6.4.3 *A pair* α, $\beta \in W$ *is called a pair of coupled weak upper and lower solutions of* (6.4.1) *if* $F\alpha$, $F\beta \in L^2(Q)$ *and* $K\alpha$, $K\beta \in L^2(\Gamma)$ *such that for all* $\varphi \in V \cap L^2_+(Q)$ *the following inequalities hold*

$$\alpha(x,0) \leq \psi(x) \leq \beta(x,0), \quad x \in \Omega,$$

$$\left\langle \frac{\partial \alpha}{\partial t}, \varphi \right\rangle + a(\alpha, \varphi) + b(\alpha, \varphi)$$

$$\leq \int_Q (F\alpha)\varphi\, dx dt + \int_\Gamma (K\alpha)\, \gamma\varphi\, d\Gamma$$

$$+ \int_\Gamma \left(\int_\Omega m(x,t,x')(\alpha(x',t) - \beta(x',t))\, dx' \right) \gamma\varphi\, d\Gamma,$$

$$\left\langle \frac{\partial \beta}{\partial t}, \varphi \right\rangle + a(\beta, \varphi) + b(\beta, \varphi)$$

$$\geq \int_Q (F\beta)\varphi\, dx dt + \int_\Gamma (K\beta)\, \gamma\varphi\, d\Gamma$$

$$+ \int_\Gamma \left(\int_\Omega m(x,t,x')(\beta(x',t) - \alpha(x',t))\, dx' \right) \gamma\varphi\, d\Gamma,$$

Under the assumptions (A1) and (A2) one can show that a pair α, β of coupled weak upper and lower solutions satisfying the above inequalities must be ordered, i.e., $\alpha \leq \beta$ holds in Q. This can be proved in just the same way as in Lemma 6.4.2. Subtracting the above inequalities we obtain a similar inequality for $w = \alpha - \beta$ as in the proof of Lemma 6.4.2 with a nonlocal term in the form

$$\int_{\Gamma_\tau} (\tilde{K}\alpha - \tilde{K}\beta)\, \gamma w^+\, d\Gamma = \int_{\Gamma_\tau} (K\alpha - K\beta)\, \gamma w^+\, d\Gamma$$

$$+ \int_{\Gamma_\tau} \left(\int_\Omega 2m(x,t,x')w(x',t))\, dx' \right) \gamma w^+\, d\Gamma,$$

where $\tilde{K}u$ is given by

$$(\tilde{K}u)(x,t) = \int_\Omega (k(x,t,x',u(x',t) + 2m(x,t,x')u(x',t))\, dx'.$$

Observe that due to (A2) the modified kernel $k(x, t, x', u) + 2m(x, t, x')u$ is increasing with respect to u, and thus the proof of Lemma 6.4.2 applies.

In the next lemma we show that assumptions (A1) and (A2) are sufficient to get an existence and uniqueness result for the IBVP (6.4.1).

Lemma 6.4.3 *Let hypotheses (A1) and (A2) be satisfied. Then the IBVP (6.4.1) possesses a unique solution $u \in W$.*

Proof We first transform the IBVP (6.4.1) into an equivalent IBVP by performing the following exponential shift transformation

$$u(x, t) = e^{\lambda t} w(x, t), \qquad (6.4.12)$$

where $\lambda \geq 0$ is some constant to be specified later. Note that $u \in W$ if and only if $w \in W$. By definition u is a solution of the IBVP (6.4.1) (note $\psi = 0$) iff

$$u \in W, \ u(0) = 0 \quad \text{and} \quad u' + Au + Bu = Fu + Ku \quad \text{in} \quad V^*, \qquad (6.4.13)$$

where the operators $A : V \to V^*$ and $B : V \to V^*$ are defined by (6.4.3) and (6.4.4), respectively, and the operators F and K are given by (6.4.3) but considered as mappings from V into V^*, which means

$$\langle Fu, \varphi \rangle := \int_Q (Fu)\, \varphi \, dxdt, \quad \varphi \in V, \qquad (6.4.14)$$

$$\langle Ku, \varphi \rangle := \int_\Gamma (Ku)\, \gamma\varphi \, d\Gamma, \quad \varphi \in V. \qquad (6.4.15)$$

The continuous embedding $V \subset L^2(Q)$ and the continuity of the trace operator $\gamma : V \to L^2(\Gamma)$ show that $F : V \to V^*$ and $K : V \to V^*$ given by (6.4.14) and (6.4.15), respectively, are well defined. Moreover, by Lemma 6.4.1 we obtain

$$\left[\begin{array}{l} |\langle Fu - Fv, \varphi \rangle| \leq \|Fu - Fv\|_{L^2(Q)} \|\varphi\|_{L^2(Q)} \quad \leq c \|u - v\|_{L^2(Q)} \|\varphi\|_V, \\ |\langle Ku - Kv, \varphi \rangle| \leq \|Ku - Kv\|_{L^2(\Gamma)} \|\gamma\varphi\|_{L^2(\Gamma)} \quad \leq c \|u - v\|_{L^2(Q)} \|\varphi\|_V, \end{array} \right.$$
$$(6.4.16)$$

where c denotes some generic constant not depending on u and v, and whose value may be different at different places. The estimates (6.4.16) show that the operators $F, K : V \to V^*$ are even Lipschitz continuous. The exponential shift (6.4.12) transforms the IBVP (6.4.1) into the following equivalent one

$$w \in W, \ u(0) = 0 \quad \text{and}\ , \ w' + Aw + Bw + \lambda w = \hat{F}w + \hat{K}w \quad \text{in} \quad V^*, \quad (6.4.17)$$

where the operators $\hat{F}$ and $\hat{K}$ are generated by the Nemytski operator $\hat{F}$ and nonlocal term $\hat{K}$, respectively, given by

$$\left[\begin{array}{ll} (\hat{F}w)(x,t) = e^{-\lambda t}f(x,t,e^{\lambda t}w(x,t)), & (x,t) \in Q, \\ (\hat{K}w)(x,t) = \int_{\Omega} e^{-\lambda t}k(x,t,x',e^{\lambda t}w(x',t))\,dx', & (x,t) \in \Gamma. \end{array} \right. \quad (6.4.18)$$

One readily verifies that Lemma 6.4.1 holds likewise also for $\hat{F}$ and $\hat{K}$, and thus the operators $\hat{F}$, $\hat{K} : V \to V^*$ are Lipschitz continuous too. Define $P := A + B + \lambda I - \hat{F} - \hat{K}$, where I denotes the identity mapping. Then $P : V \to V^*$ is continuous and bounded, and there exists a unique solution of (6.4.17) provided P is, in addition, monotone and coercive. We shall prove next that for λ sufficiently large these last two properties are satisfied. Since (6.4.16) holds likewise also for $\hat{F}$ and $\hat{K}$, we have in view of the definition of A and B the following estimate:

$$\begin{aligned} \langle Pu - Pv, u - v \rangle &= \langle A(u-v), u-v \rangle + \langle B(u-v), u-v \rangle \quad (6.4.19) \\ &\quad + \langle \lambda(u-v) - (\hat{F}u - \hat{F}v), u-v \rangle + \langle \hat{K}u - \hat{K}v, u-v \rangle \\ &\geq \mu \|\nabla(u-v)\|_{L^2(Q)}^2 + (\lambda - c)\|u-v\|_{L^2(Q)}^2 \\ &\quad -c\|u-v\|_{L^2(Q)}\|\gamma(u-v)\|_{L^2(\Gamma)} \\ &\geq \mu \|u-v\|_V^2 + (\lambda - c - \mu)\|u-v\|_{L^2(Q)}^2 \\ &\quad -c(\varepsilon)\|u-v\|_{L^2(Q)}^2 - \varepsilon \|u-v\|_V^2, \end{aligned}$$

which is valid for any $\varepsilon > 0$ due to Young's inequality, where $c(\varepsilon)$ is some positive constant depending on ε. Selecting $\varepsilon < \frac{1}{2}\mu$ and choosing λ sufficiently large such that $\lambda - c - \mu - c(\varepsilon) > 0$ we get from (6.4.19)

$$\langle Pu - Pv, u - v \rangle \geq \frac{1}{2}\mu \|u-v\|_V^2,$$

which shows that $P : V \to V^*$ is even strongly monotone, and therefore also coercive. This completes the proof of the lemma.

Special Case Let us consider the following special case of the IBVP (6.4.1) with linear f and k in the form

$$\left[\begin{array}{ll} f(x,t,u) = l(x,t)u + h(x,t), & (x,t) \in Q, \\ k(x,t,x',u) = m(x,t,x')u + g(x,t,x'), & (x,t) \in \Gamma, \ x' \in \Omega, \end{array} \right. \quad (6.4.20)$$

where $l \in L^{\infty}(Q)$, $h \in L^2(Q)$, $m \in L^{\infty}(\Gamma \times \Omega)$, and $g \in L^2(\Gamma \times \Omega)$. Since assumptions (A1) and (A2) are trivially satisfied, Lemma 6.4.3 implies the following corollary for the IBVP

$$\left[\begin{array}{ll} \frac{\partial u(x,t)}{\partial t} + Au(x,t) = l(x,t)u(x,t) + h(x,t), & in\ Q, \\ u(x,0) = 0, & in\ \Omega, \\ \frac{\partial u(x,t)}{\partial \nu} + b(x,t)u(x,t) = \int_\Omega \Big(m(x,t,x')u(x',t) + g(x,t,x')\Big)\,dx', & on\ \Gamma. \end{array}\right.$$

$$(6.4.21)$$

Corollary 6.4.1 *The IBVP (6.4.21) possesses a uniquely defined solution $u \in W$ satisfying an estimate in the form*

$$\|u\|_W \le c\,(\|h\|_{L^2(Q)} + \|g\|_{L^2(\Gamma \times \Omega)}), \qquad (6.4.22)$$

where c is some positive constant not depending on u. Moreover, if the coefficient $m \in L^\infty(\Gamma \times \Omega)$ of the boundary condition is nonnegative, then the comparison result of Lemma 6.4.2 holds accordingly for the IBVP (6.4.21).

Proof   We only have to verify the estimate (6.4.4). Let F and K be the operators defined in (6.4.20) related to the affine functions f and k given by (6.4.20), which are Lipschitz continuous with respect to u with Lipschitz constants $\|l\|_{L^\infty(Q)}$ and $\|m\|_{L^\infty(\Gamma \times \Omega)}$, respectively. Denote by c a positive generic constant that may have different values at different places. Then by Lemma 6.4.1 we have

$$\left[\begin{array}{l} \|Fu\|_{L^2(Q)} \ \le \|Fu - F0\|_{L^2(Q)} + \|F0\|_{L^2(Q)} \le c\,\|u\|_{L^2(Q)} + \|F0\|_{L^2(Q)}, \\ \|Ku\|_{L^2(\Gamma)} \ \le \|Ku - K0\|_{L^2(\Gamma)} + \|K0\|_{L^2(\Gamma)} \le c\,\|u\|_{L^2(Q)} + \|K0\|_{L^2(\Gamma)}, \end{array}\right.$$

$$(6.4.23)$$

where

$$\|F0\|_{L^2(Q)} = \|h\|_{L^2(Q)}, \quad \|K0\|_{L^2(\Gamma)} \le c\,\|g\|_{L^2(\Gamma \times \Omega)}. \qquad (6.4.24)$$

Considering (6.4.21) on subcylinders $Q_\tau := \Omega \times (0,\tau) \subseteq Q$ with its corresponding lateral boundary $\Gamma_\tau := \partial\Omega \times (0,\tau) \subseteq \Gamma$, $\tau \in (0,T]$, we obtain from the weak formulation of (6.4.21) by taking as special test function the solution u itself the following estimate

$$\frac{1}{2}\|u(\cdot,\tau)\|^2_{L^2(\Omega)} + \mu\|\nabla u\|^2_{L^2(Q_\tau)} + \|b\|_{L^\infty(\Gamma)}\|\gamma u\|^2_{L^2(\Gamma_\tau)}$$

$$\le\ c\|u\|^2_{L^2(Q_\tau)} + \|F0\|_{L^2(Q_\tau)}\|u\|_{L^2(Q_\tau)}$$

$$+ \Big(c\|u\|_{L^2(Q_\tau)} + \|K0\|_{L^2(\Gamma_\tau)}\Big)\|\gamma u\|_{L^2(\Gamma_\tau)}$$

$$\le\ c(\varepsilon)\,\Big(\|u\|^2_{L^2(Q_\tau)} + \|F0\|^2_{L^2(Q_\tau)} + \|K0\|^2_{L^2(\Gamma_\tau)}\Big) + \varepsilon\|\gamma u\|^2_{L^2(\Gamma_\tau)}$$

$$\le\ c(\varepsilon)\,\Big(\|u\|^2_{L^2(Q_\tau)} + \|F0\|^2_{L^2(Q_\tau)} + \|K0\|^2_{L^2(\Gamma_\tau)}\Big) + \varepsilon\|\nabla u\|^2_{L^2(Q_\tau)},$$

which holds for any $\varepsilon > 0$ due to Young's inequality and the continuity of the trace operator $\gamma : V \to L^2(\Gamma)$. Selecting $\varepsilon < \frac{1}{2}\mu$ the last estimate yields

$$\|u(\cdot,\tau)\|_{L^2(\Omega)}^2 + \mu\|\nabla u\|_{L^2(Q_\tau)}^2 \tag{6.4.25}$$

$$\leq c\left(\|u\|_{L^2(Q_\tau)}^2 + \|F0\|_{L^2(Q_\tau)}^2 + \|K0\|_{L^2(\Gamma_\tau)}^2\right),$$

for any $\tau \in [0,T]$. Setting

$$y(\tau) = \|u(\cdot,\tau)\|_{L^2(\Omega)}^2, \quad p(\tau) = \|F0\|_{L^2(Q_\tau)}^2 + \|K0\|_{L^2(\Gamma_\tau)}^2,$$

and taking into account that $0 \leq p(\tau) \leq p(T)$ for all $\tau \in [0,T]$ we obtain from (6.4.25)

$$y(\tau) \leq c\int_0^\tau y(t)\,dt + c\,p(T). \tag{6.4.26}$$

Applying the Gronwall lemma to (6.4.26) we get

$$y(\tau) \leq c\,p(T)(1 + c\tau e^{c\tau}), \quad \tau \in [0,T],$$

which in view of (6.4.24) and $p(T) = \|F0\|_{L^2(Q)}^2 + \|K0\|_{L^2(\Gamma)}^2$ yields

$$y(\tau) \leq c(\|h\|_{L^2(Q)}^2 + \|g\|_{L^2(\Gamma\times\Omega)}^2), \quad \tau \in [0,T],$$

and thus by integrating the last inequality with respect to τ over the interval $[0,T]$ yields

$$\|u\|_{L^2(Q)} \leq c(\|h\|_{L^2(Q)} + \|g\|_{L^2(\Gamma\times\Omega)}). \tag{6.4.27}$$

Since (6.4.25) holds for any $\tau \in [0,T]$, we readily deduce for $\tau = T$ the estimate

$$\|\nabla u\|_{L^2(Q)}^2 \leq c\left(\|u\|_{L^2(Q)}^2 + \|F0\|_{L^2(Q)}^2 + \|K0\|_{L^2(\Gamma)}^2\right),$$

which by (6.4.24) and (6.4.27) implies

$$\|\nabla u\|_{L^2(Q)} \leq c(\|h\|_{L^2(Q)} + \|g\|_{L^2(\Gamma\times\Omega)}),$$

and thus due to (6.4.27) we have

$$\|u\|_V \leq c(\|h\|_{L^2(Q)} + \|g\|_{L^2(\Gamma\times\Omega)}). \tag{6.4.28}$$

To estimate $\|u'\|_{V^*}$, we consider the weak formulation of the IBVP (6.4.21), i.e.,

$$u \in W, \ u(0) = 0 \ \text{ and } \ u' + Au + Bu = Fu + Ku \ \text{ in } \ V^*, \tag{6.4.29}$$

where the norms $\|Fu\|_{V^*}$ and $\|Ku\|_{V^*}$ can be estimated as follows:

$$|\langle Fu, \varphi \rangle| = |\int_Q (Fu)\,\varphi\,dxdt| \leq \|Fu\|_{L^2(Q)} \|\varphi\|_{L^2(Q)} \leq \|Fu\|_{L^2(Q)} \|\varphi\|_V,$$

$$|\langle Ku, \varphi \rangle| = |\int_\Gamma (Ku)\,\gamma\varphi\,d\Gamma| \leq \|Ku\|_{L^2(\Gamma)} \|\gamma\varphi\|_{L^2(\Gamma)} \leq c\,\|Ku\|_{L^2(\Gamma)} \|\varphi\|_V,$$

and thus by using (6.4.23) and (6.4.24) it follows that

$$\left[\begin{array}{ll} \|Fu\|_{V^*} & \leq c\,(\|u\|_{L^2(Q)} + \|h\|_{L^2(Q)}), \\ \|Ku\|_{V^*} & \leq c\,(\|u\|_{L^2(Q)} + \|g\|_{L^2(\Gamma \times \Omega)}). \end{array} \right. \tag{6.4.30}$$

Since A and B are linear and continuous operators from V into V^*, i.e., we have $\|Au\|_{V^*} \leq c\,\|u\|_V$ and $\|Bu\|_{V^*} \leq c\,\|u\|_V$, we obtain from (6.4.29) and (6.4.30)

$$\|u'\|_{V^*} \leq c\,(\|u\|_V + \|h\|_{L^2(Q)} + \|g\|_{L^2(\Gamma \times \Omega)}),$$

which in view of (6.4.28) gives

$$\|u'\|_{V^*} \leq c\,(\|h\|_{L^2(Q)} + \|g\|_{L^2(\Gamma \times \Omega)}), \tag{6.4.31}$$

and thus by (6.4.28) and (6.4.31) the assertion, i.e.,

$$\|u\|_W = \|u\|_V + \|u'\|_{V^*} \leq c\,(\|h\|_{L^2(Q)} + \|g\|_{L^2(\Gamma \times \Omega)}).$$

By inspection of the proof of Corollary 6.4.1 we readily obtain the following result for the linear IBVP

$$\left[\begin{array}{rl} \frac{\partial u(x,t)}{\partial t} + Au(x,t) = \hat{h}(x,t), & \text{in } Q, \\ u(x,0) = 0, & \text{in } \Omega, \\ \frac{\partial u(x,t)}{\partial \nu} + b(x,t)u(x,t) = \hat{g}(x,t), & \text{on } \Gamma. \end{array} \right. \tag{6.4.32}$$

Corollary 6.4.2 *Let $\hat{h} \in L^2(Q)$ and $\hat{g} \in L^2(\Gamma)$ be given. Then the IBVP (6.4.32) possesses a uniquely defined solution $u \in W$ satisfying an estimate of the form*

$$\|u\|_W \leq c\,(\|\hat{h}\|_{L^2(Q)} + \|\hat{g}\|_{L^2(\Gamma)}),$$

where c is some positive constant not depending on u. Moreover, a comparison principle holds.

Remark 6.4.2 The comparison result of Lemma 6.4.2 and the existence and uniqueness result for the associated linear IBVP given in Lemma 6.4.3 enable us to apply the monotone iterative technique for the nonlinear IBVP (6.4.1). However, our main goal is to establish the generalized quasilinearization method which is a refined monotone iteration that, in addition, provides a measure of the convergence rate. This we consider in the next section.

6.5   Generalized Quasilinearization (GQ): Nonlocal Problems

We shall continue to consider the nonlocal IBVP (6.4.1) and establish the method of generalized quasilinearization, when the nonlinear term f and the nonlinear kernel k admit a decomposition in the form

$$f(x,t,u) = f^{(1)}(x,t,u) + f^{(2)}(x,t,u),$$

$$k(x,t,x',u) = k^{(1)}(x,t,x',u) + k^{(2)}(x,t,x',u),$$

for $u \in [\underline{u}, \bar{u}]$ formed by lower and upper solutions. Here $f^{(1)}$, $k^{(1)}$ are convex functions and $f^{(2)}$, $k^{(2)}$ are concave functions in u. As we have seen, such a decomposition covers several special cases of interest.

Let us consider the IBVP (6.4.1) with a right-hand side f and a nonlinear kernel function k admitting a splitting as given below

$$
\left[
\begin{array}{ll}
\frac{\partial u(x,t)}{\partial t} + Au(x,t) = f^{(1)}(x,t,u(x,t)) + f^{(2)}(x,t,u(x,t)), & \text{in } Q, \\
u(x,0) = 0, & \text{in } \Omega, \\
\frac{\partial u(x,t)}{\partial \nu} + b(x,t)u(x,t) = \int_\Omega \Big(k^{(1)}(x,t,x',u(x',t)) \\
\qquad\qquad + k^{(2)}(x,t,x',u(x',t)) \Big)\, dx', & \text{on } \Gamma,
\end{array}
\right.
$$

$$(6.5.1)$$

where we have assumed without loss of generality the initial value $\psi(x) = 0$.

Our aim is to establish the generalized quasilinearization method. For this purpose we assume the following hypotheses:

($H1$) There exist lower and upper solutions α_0 and β_0 of the IBVP (6.5.1), respectively, such that $\alpha_0 \le \beta_0$ and α_0, $\beta_0 \in L^\infty(Q)$.

($H2$) The first and second derivatives of $f^{(i)}(x,t,u)$, $i = 1,2$, with respect to u exist and $f^{(i)}$, $f_u^{(i)}$, $f_{uu}^{(i)} : Q \times \mathbb{R} \to \mathbb{R}$ are Caratheodory functions satisfying

(i) $f_{uu}^{(1)}(x,t,u) \ge 0$, $f_{uu}^{(2)}(x,t,u) \le 0$ for all $u \in [\text{ess inf}_Q\, \alpha_0(x,t),$ $\text{ess sup}_Q\, \beta_0(x,t)]$ and for a.e. $(x,t) \in Q$.

(ii) There is some positive constant c such that

$$\|f_u^{(i)}(\cdot,\cdot,\eta),\ f_{uu}^{(i)}(\cdot,\cdot,\eta)\|_{L^\infty(Q)} \le c, \quad \forall \eta \in [\alpha_0, \beta_0].$$

($H3$) The first and second derivatives of $k^{(i)}(x,t,x',u)$, $i = 1,2$, with respect to u exist and $k^{(i)}$, $k_u^{(i)}$, $k_{uu}^{(i)} : \Gamma \times \Omega \times \mathbb{R} \to \mathbb{R}$ are Caratheodory functions satisfying

(i) $k_{uu}^{(1)}(x,t,x',u) \geq 0$, $k_{uu}^{(2)}(x,t,x',u) \leq 0$ for all $u \in [\text{ess inf}_Q \, \alpha_0(x,t),$ ess $\sup_Q \beta_0(x,t)]$ and for a.e. $(x,t,x') \in \Gamma \times \Omega$.

(ii) There is some positive constant c such that

$$\|k_u^{(i)}(\cdot,\cdot,\cdot,\eta), \ k_{uu}^{(i)}(\cdot,\cdot,\cdot,\eta)\|_{L^\infty(\Gamma \times \Omega)} \leq c, \quad \forall \eta \in [\alpha_0, \beta_0].$$

(iii) $k_u^{(1)}(\cdot,\cdot,\cdot,\eta) + k_u^{(2)}(\cdot,\cdot,\cdot,\xi) \geq 0$ for all $\eta, \ \xi \in [\text{essinf}_Q \, \alpha_0(x,t),$ esssup$_Q \beta_0(x,t)]$.

With $f = f^{(1)} + f^{(2)}$ and $k = k^{(1)} + k^{(2)}$ the weak formulation of (6.5.1) reads as

$$u \in W, \ u(0) = 0: \quad u' + Au + Bu = F(u) + K(u) \quad \text{in} \quad V^*,$$

where F and K are the operators related to f and k, respectively.

Remark 6.5.1 Note that the convexity of $f^{(1)}$, $k^{(1)}$ and the concavity of $f^{(2)}$, $k^{(2)}$ as well as the boundedness condition on the derivatives according to (H2) and (H3) are only required to hold in the ordered interval $[\alpha_0, \beta_0]$ formed by the lower and upper solutions α_0 and β_0, respectively.

The main result of this section is given by the following theorem.

Theorem 6.5.1 *Let hypotheses (H1)–(H3) be satisfied. Then there exist monotone sequences* $(\alpha_n)_{n=1}^\infty$, $(\beta_n)_{n=1}^\infty \subset W$ *satisfying*

$$\alpha_0 \leq \alpha_1 \leq \cdots \leq \alpha_n \leq \alpha_{n+1} \leq \cdots \leq \beta_{n+1} \leq \beta_n \leq \cdots \leq \beta_1 \leq \beta_0, \quad (6.5.2)$$

which converge to the unique solution u of IBVP (6.5.1) within the interval $[\alpha_0, \beta_0]$. The convergence rate of these monotone sequences is quadratic and the following estimate holds:

$$\|\beta_{n+1} - u\|_W + \|u - \alpha_{n+1}\|_W \leq c \, \|(\beta_n - u)^2 + (u - \alpha_n)^2\|_{L^2(Q)}. \quad (6.5.3)$$

Proof The proof will be given in three steps, (a), (b) and (c).

(a) **Generalized quasilinearization scheme**

For any pair of functions α, $\beta \in [\alpha_0, \beta_0]$, we introduce the following linearizations of $f^{(1)} + f^{(2)}$ and $k^{(1)} + k^{(2)}$ at β and α which play a key role in establishing the generalized quasilinearization method:

$$
\begin{aligned}
\bar{f}(x,t,u;\alpha,\beta) &:= f^{(1)}(x,t,\beta) + f^{(2)}(x,t,\beta) & (6.5.4)\\
&\quad + \Big(f_u^{(1)}(x,t,\alpha) + f_u^{(2)}(x,t,\beta)\Big)(u-\beta),\\
\underline{f}(x,t,u;\alpha,\beta) &:= f^{(1)}(x,t,\alpha) + f^{(2)}(x,t,\alpha)\\
&\quad + \Big(f_u^{(1)}(x,t,\alpha) + f_u^{(2)}(x,t,\beta)\Big)(u-\alpha),\\
\bar{k}(x,t,x',u;\alpha,\beta) &:= k^{(1)}(x,t,x',\beta) + k^{(2)}(x,t,x',\beta) & (6.5.5)\\
&\quad + \Big(k_u^{(1)}(x,t,x',\alpha) + k_u^{(2)}(x,t,x',\beta)\Big)(u-\beta),\\
\underline{k}(x,t,x',u;\alpha,\beta) &:= k^{(1)}(x,t,x',\alpha) + k^{(2)}(x,t,x',\alpha)\\
&\quad + \Big(k_u^{(1)}(x,t,x',\alpha) + k_u^{(2)}(x,t,x',\beta)\Big)(u-\alpha),
\end{aligned}
$$

and associated with them the corresponding operators

$$
\bar{F}(u;\alpha,\beta)(x,t) = \bar{f}(x,t,u(x,t);\alpha(x,t),\beta(x,t)), \quad (x,t)\in Q
$$

and analogously $\underline{F}(u;\alpha,\beta)$, and

$$
\bar{K}(u;\alpha,\beta)(x,t) := \int_\Omega \bar{k}(x,t,x',u(x',t);\alpha(x',t),\beta(x',t))\,dx', \;\; (x,t)\in\Gamma,
$$

and analogously $\underline{K}(u;\alpha,\beta)$. With the above linearizations, we form the following iteration schemes $(n=0,1,2,\dots)$

$$
\beta_{n+1}\in W,\;\beta_{n+1}(0)=0:\;\;\beta'_{n+1} + A\beta_{n+1} + B\beta_{n+1} \qquad (6.5.6)
$$
$$
= \bar{F}(\beta_{n+1};\alpha_n,\beta_n) + \bar{K}(\beta_{n+1};\alpha_n,\beta_n) \;\;\text{in}\;\; V^*,
$$
$$
\alpha_{n+1}\in W,\;\alpha_{n+1}(0)=0:\;\;\alpha'_{n+1} + A\alpha_{n+1} + B\alpha_{n+1} \qquad (6.5.7)
$$
$$
= \underline{F}(\alpha_{n+1};\alpha_n,\beta_n) + \underline{K}(\alpha_{n+1};\alpha_n,\beta_n) \;\;\text{in}\;\; V^*.
$$

Starting the iterations (6.5.6) and (6.5.7) with the given upper and lower solutions β_0 and α_0 we shall show that these iterations yield well-defined monotone sequences (β_n) and (α_n) satisfying (6.5.2).

Consider $n=0$: In view of (6.5.4) and (6.5.5) problems (6.5.6) and (6.5.7) are the corresponding weak formulation of the linear IBVP for β_1 and α_1 which are of the form (6.4.21). Thus by Corollary 6.4.1 there is a unique solution β_1 of (6.5.6) and a unique solution α_1 of (6.5.7). Let us prove the following order relation:

$$
\alpha_0 \le \alpha_1 \le \beta_1 \le \beta_0. \qquad (6.5.8)
$$

From definition (6.5.4) and (6.5.5) it follows that

$$\left[\begin{array}{ll} \bar{F}(\beta;\alpha,\beta) = F\beta, & \underline{F}(\alpha;\alpha,\beta) = F\alpha, \\ \bar{K}(\beta;\alpha,\beta) = K\beta, & \underline{K}(\alpha;\alpha,\beta) = K\alpha, \end{array}\right. \tag{6.5.9}$$

and thus the given upper solution β_0 is an upper solution of the IBVP

$$u \in W,\ u(0) = 0:\ u' + Au + Bu = \bar{F}(u;\alpha_0,\beta_0) + \bar{K}(u;\alpha_0,\beta_0) \quad \text{in } V^*, \tag{6.5.10}$$

and β_1 is the unique solution of (6.5.10). By hypothesis (H3)(iii), the coefficient of the linear term in $\bar{k}$ is nonnegative, so that the comparison result of Lemma 6.4.2 can be applied which proves that $\beta_1 \le \beta_0$. By similar arguments it follows that $\alpha_0 \le \alpha_1$. To complete the proof of the inequalities (6.5.8) we have to show that $\alpha_1 \le \beta_1$. For this purpose we make use of the convexity and concavity assumptions of (H2) and (H3), from which the following inequalities are immediate consequences:

$$\left[\begin{array}{l} f^{(1)}(x,t,u) \ge f^{(1)}(x,t,v) + f_u^{(1)}(x,t,v)(u-v), \\ f^{(2)}(x,t,u) \ge f^{(2)}(x,t,v) + f_u^{(2)}(x,t,u)(u-v), \end{array}\right. \tag{6.5.11}$$

$$\left[\begin{array}{l} k^{(1)}(x,t,x',u) \ge k^{(1)}(x,t,x',v) + k_u^{(1)}(x,t,x',v)(u-v), \\ k^{(2)}(x,t,x',u) \ge k^{(2)}(x,t,x',v) + k_u^{(2)}(x,t,x',u)(u-v) \end{array}\right. \tag{6.5.12}$$

for all $u,v \in [\alpha_0,\beta_0]$. Inequalities (6.5.11) imply

$$f^{(1)}(\cdot,\cdot,\alpha_0) \le f^{(1)}(\cdot,\cdot,\beta_0) - f_u^{(1)}(\cdot,\cdot,\alpha_0)(\beta_0-\alpha_0),$$

$$f^{(2)}(\cdot,\cdot,\alpha_0) \le f^{(2)}(\cdot,\cdot,\beta_0) - f_u^{(2)}(\cdot,\cdot,\beta_0)(\beta_0-\alpha_0),$$

and in view of definition (6.5.4) we have the estimate

$$\begin{aligned} \underline{F}(\alpha_1;\alpha_0,\beta_0) &= \underline{f}(\cdot,\cdot,\alpha_1;\alpha_0,\beta_0) \le f^{(1)}(\cdot,\cdot,\beta_0) + f^{(2)}(\cdot,\cdot,\beta_0) \\ &\quad + \left(f_u^{(1)}(\cdot,\cdot,\alpha_0) + f_u^{(2)}(\cdot,\cdot,\beta_0)\right)(\alpha_1 - \beta_0) \\ &= \bar{f}(\cdot,\cdot,\alpha_1;\alpha_0,\beta_0) = \bar{F}(\alpha_1;\alpha_0,\beta_0). \end{aligned}$$

Similarly we obtain from (6.5.12) and in view of (6.5.5) that the following inequality holds

$$\underline{K}(\alpha_1;\alpha_0,\beta_0) \le \bar{K}(\alpha_1;\alpha_0,\beta_0),$$

and thus we have

$$\underline{F}(\alpha_1;\alpha_0,\beta_0) + \underline{K}(\alpha_1;\alpha_0,\beta_0) \le \bar{F}(\alpha_1;\alpha_0,\beta_0) + \bar{K}(\alpha_1;\alpha_0,\beta_0). \tag{6.5.13}$$

Consider the IBVP (6.5.10) whose unique solution is β_1. Due to (6.5.13) α_1 is a lower solution of (6.5.10), and hence by the comparison result of Lemma 6.4.2 it follows that $\alpha_1 \leq \beta_1$, which completes the proof of (6.5.8).

We shall next prove that if

$$\alpha_{n-1} \leq \alpha_n \leq \beta_n \leq \beta_{n-1} \qquad (6.5.14)$$

for some $n > 1$, then it follows that

$$\alpha_n \leq \alpha_{n+1} \leq \beta_{n+1} \leq \beta_n. \qquad (6.5.15)$$

To this end the following inequalities will be shown:

$$\begin{bmatrix} \underline{F}(\alpha_n;\alpha_{n-1},\beta_{n-1}) & \leq \underline{F}(\alpha_n;\alpha_n,\beta_n) = F\alpha_n \\ \overline{F}(\beta_n;\alpha_{n-1},\beta_{n-1}) & \geq \overline{F}(\beta_n;\alpha_n,\beta_n) = F\beta_n \\ \underline{K}(\alpha_n;\alpha_{n-1},\beta_{n-1}) & \leq \underline{K}(\alpha_n;\alpha_n,\beta_n) = K\alpha_n \\ \overline{K}(\beta_n;\alpha_{n-1},\beta_{n-1}) & \geq \overline{K}(\beta_n;\alpha_n,\beta_n) = K\beta_n. \end{bmatrix} \qquad (6.5.16)$$

We verify the first inequality of (6.5.16) only, since all the others can be shown in a similar way. From inequalities (6.5.11) we get

$$f^{(1)}(\cdot,\cdot,\alpha_{n-1}) \leq f^{(1)}(\cdot,\cdot,\alpha_n) - f_u^{(1)}(\cdot,\cdot,\alpha_{n-1})(\alpha_n - \alpha_{n-1}),$$

$$f^{(2)}(\cdot,\cdot,\alpha_{n-1}) \leq f^{(2)}(\cdot,\cdot,\alpha_n) - f_u^{(2)}(\cdot,\cdot,\alpha_n)(\alpha_n - \alpha_{n-1}),$$

which gives the following estimate of $\underline{F}(\alpha_n;\alpha_{n-1},\beta_{n-1})$

$$\underline{F}(\alpha_n;\alpha_{n-1},\beta_{n-1}) \leq f^{(1)}(\cdot,\cdot,\alpha_n) + f^{(2)}(\cdot,\cdot,\alpha_n) \qquad (6.5.17)$$
$$+ \left(f_u^{(2)}(\cdot,\cdot,\beta_{n-1}) - f_u^{(2)}(\cdot,\cdot,\alpha_n) \right)(\alpha_n - \alpha_{n-1}).$$

Because $f^{(2)}$ is concave with respect to u it follows that $f_u^{(2)}$ is decreasing, so that in view of $\alpha_n \leq \beta_{n-1}$ and $\alpha_n - \alpha_{n-1} \geq 0$ we get from (6.5.17)

$$\underline{F}(\alpha_n;\alpha_{n-1},\beta_{n-1}) \leq f^{(1)}(\cdot,\cdot,\alpha_n) + f^{(2)}(\cdot,\cdot,\alpha_n) = F\alpha_n,$$

which proves the first inequality of (6.5.16). Consider the IBVP

$$u \in W,\ u(0) = 0:\ u'+Au+Bu = \underline{F}(u;\alpha_n,\beta_n)+\underline{K}(u;\alpha_n,\beta_n),\ \text{in }V^*, \qquad (6.5.18)$$

whose unique solution is α_{n+1}. From (6.5.16) it follows that

$$\underline{F}(\alpha_n; \alpha_{n-1}, \beta_{n-1}) + \underline{K}(\alpha_n; \alpha_{n-1}, \beta_{n-1})$$

$$\leq \underline{F}(\alpha_n; \alpha_n, \beta_n) + \underline{K}(\alpha_n; \alpha_n, \beta_n),$$

which shows that α_n must be a lower solution of (6.5.18), and thus $\alpha_n \leq \alpha_{n+1}$ due to the comparison result of Lemma 6.4.2. Similarly, by using the inequality

$$\bar{F}(\beta_n; \alpha_{n-1}, \beta_{n-1}) + \bar{K}(\beta_n; \alpha_{n-1}, \beta_{n-1}) \geq \bar{F}(\beta_n; \alpha_n, \beta_n) + \bar{K}(\beta_n; \alpha_n, \beta_n),$$

it follows that β_n is an upper solution of the IBVP

$$u \in W,\ u(0) = 0\ :\ u' + Au + Bu = \bar{F}(u; \alpha_n, \beta_n) + \bar{K}(u; \alpha_n, \beta_n),\quad \text{in}\ \ V^*,$$
(6.5.19)

whose unique solution is β_{n+1}, and thus again by comparison we have $\beta_{n+1} \leq \beta_n$. In just the same way as (6.5.13) has been obtained by using (6.5.11) and (6.5.12), we get the inequality

$$\underline{F}(\alpha_{n+1}; \alpha_n, \beta_n)\ +\ \underline{K}(\alpha_{n+1}; \alpha_n, \beta_n) \qquad (6.5.20)$$
$$\leq\ \bar{F}(\alpha_{n+1}; \alpha_n, \beta_n) + \bar{K}(\alpha_{n+1}; \alpha_n, \beta_n).$$

Consider again the IBVP (6.5.19) whose unique solution is β_{n+1}. Then due to (6.5.20) α_{n+1} is a lower solution of (6.5.19), and hence by comparison it follows that $\alpha_{n+1} \leq \beta_{n+1}$, which completes the proof of (6.5.15), and thus inequality (6.5.2) holds.

(b) **Convergence of the monotone sequences $\{\alpha_n\}$, $\{\beta_n\}$**

The monotonicity of the iterates (α_n), $(\beta_n) \in [\alpha_0, \beta_0]$ imply the existence of their a.e. pointwise limits

$$\varrho(x,t) = \lim_{n \to \infty} \alpha_n(x,t),\quad r(x,t) = \lim_{n \to \infty} \beta_n(x,t),\quad \text{for a.e.}\ (x,t) \in Q.$$

In view of α_0, $\beta_0 \in L^\infty(Q)$ by Lebesgue's dominated convergence theorem we have

$$\alpha_n \to \varrho \quad \text{and} \quad \beta_n \to r \quad \text{in} \quad L^2(Q). \qquad (6.5.21)$$

The boundedness of the derivatives $f_u^{(i)}$ and $k_u^{(i)}$ within $[\alpha_0, \beta_0]$ and the convergence (6.5.21) imply as $n \to \infty$:

$$\left[\begin{array}{l} \| \bar{F}(\beta_{n+1}; \alpha_n, \beta_n)\ - Fr \|_{L^2(Q)} \to 0, \\ \| \bar{K}(\beta_{n+1}; \alpha_n, \beta_n)\ - Kr \|_{L^2(\Gamma)} \to 0, \end{array} \right. \qquad (6.5.22)$$

$$\left[\begin{array}{ll} \|\underline{F}(\alpha_{n+1};\alpha_n,\beta_n) \ -F\varrho\|_{L^2(Q)} \to 0, \\ \|\underline{K}(\alpha_{n+1};\alpha_n,\beta_n) \ -K\varrho\|_{L^2(\Gamma)} \to 0, \end{array}\right. \qquad (6.5.23)$$

where F and K are generated by $f = f^{(1)} + f^{(2)}$ and $k = k^{(1)} + k^{(2)}$, respectively. Since β_n and α_n $(n \geq 1)$ are solutions of the linear IBVP (6.5.6) and (6.5.7) with homogeneous initial value, respectively, we may apply the estimate of Corollary 6.4.2 which gives for some positive constant c and for all $n \geq 1$ the estimates

$$\|\alpha_{n+1}\|_W \leq c\left(\|\underline{F}(\alpha_{n+1};\alpha_n,\beta_n)\|_{L^2(Q)} + \|\underline{K}(\alpha_{n+1};\alpha_n,\beta_n)\|_{L^2(\Gamma)}\right),$$
$$(6.5.24)$$

$$\|\beta_{n+1}\|_W \leq c\left(\|\bar{F}(\beta_{n+1};\alpha_n,\beta_n)\|_{L^2(Q)} + \|\bar{K}(\beta_{n+1};\alpha_n,\beta_n)\|_{L^2(\Gamma)}\right).$$
$$(6.5.25)$$

Due to (6.5.22) and (6.5.23) the estimates (6.5.24) and (6.5.25) imply the strong convergence of the sequences (α_n) and (β_n) in W whose limits must be ϱ and r, respectively, because of the (compact) embedding $W \subset L^2(Q)$. Since α_n and β_n belong for $n \geq 1$ to the subset $M := \{v \in W \mid v(x,0) = 0\}$ which is closed in W it follows that also their limits ϱ, $r \in M$, that is, the limits also satisfy the homogeneous initial condition. The convergence result (6.5.22) and (6.5.23) together with

$$\alpha_n \to \varrho, \quad \beta_n \to r \text{ in } W \text{ as } n \to \infty,$$

allow us to pass to the limit in the corresponding weak formulation of (6.5.6) and (6.5.7) as $n \to \infty$ which yield

$$\varrho \in W, \ \varrho(0) = 0 : \quad \varrho' + A\varrho + B\varrho = F\varrho + K\varrho \text{ in } V^*,$$

$$r \in W, \ r(0) = 0 : \quad r' + Ar + Br = Fr + Kr \text{ in } V^*,$$

which shows that the limits ϱ and r are solutions of the IBVP (6.5.1) within $[\alpha_0, \beta_0]$. By the comparison result of Lemma 6.4.2 it follows that $\varrho = r = u$ is the unique solution of (6.5.1) within the interval $[\alpha_0, \beta_0]$.

(c) **Quadratic convergence of $\{\alpha_n\}$, $\{\beta_n\}$**

To prove quadratic convergence of the monotone sequences (α_n), (β_n) to the unique solution u, respectively, we set

$$p_n = u - \alpha_n, \quad q_n = \beta_n - u, \quad n \geq 1.$$

Since $\alpha_n \leq u \leq \beta_n$ for all n, we have $p_n \geq 0$ and $q_n \geq 0$. By means of (6.5.11) we deduce

$$f^{(1)}(\cdot, \cdot, u) \leq f^{(1)}(\cdot, \cdot, \alpha_n) + f_u^{(1)}(\cdot, \cdot, u)(u - \alpha_n),$$

$$f^{(2)}(\cdot, \cdot, u) \leq f^{(2)}(\cdot, \cdot, \alpha_n) + f_u^{(2)}(\cdot, \cdot, \alpha_n)(u - \alpha_n),$$

which will be used to estimate

$$f(\cdot, \cdot, u) - \underline{f}(\cdot, \cdot, \alpha_{n+1}; \alpha_n, \beta_n) \qquad (6.5.26)$$

$$
\begin{aligned}
=\ & f^{(1)}(\cdot, \cdot, u) + f^{(2)}(\cdot, \cdot, u) - f^{(1)}(\cdot, \cdot, \alpha_n) - f^{(2)}(\cdot, \cdot, \alpha_n) \\
& - f_u^{(1)}(\cdot, \cdot, \alpha_n)(\alpha_{n+1} - \alpha_n) - f_u^{(2)}(\cdot, \cdot, \beta_n)(\alpha_{n+1} - \alpha_n) \\
\leq\ & \left(f_u^{(1)}(\cdot, \cdot, u) - f_u^{(1)}(\cdot, \cdot, \alpha_n) \right) p_n - \left(f_u^{(2)}(\cdot, \cdot, \beta_n) - f_u^{(2)}(\cdot, \cdot, \alpha_n) \right) p_n \\
& + \left(f_u^{(1)}(\cdot, \cdot, \alpha_n) + f_u^{(2)}(\cdot, \cdot, \beta_n) \right) p_{n+1} \\
=\ & f_{uu}^{(1)}(\cdot, \cdot, \eta)\, p_n^2 - f_{uu}^{(2)}(\cdot, \cdot, \xi)(\beta_n - \alpha_n)\, p_n \\
& + \left(f_u^{(1)}(\cdot, \cdot, \alpha_n) + f_u^{(2)}(\cdot, \cdot, \beta_n) \right) p_{n+1},
\end{aligned}
$$

where η, $\xi \in [\alpha_0, \beta_0]$. Since the derivatives of $f^{(i)}$ with respect to u are bounded within the interval $[\alpha_0, \beta_0]$ we get

$$-f_{uu}^{(2)}(\cdot, \cdot, \xi)(\beta_n - \alpha_n)\, p_n \leq c\,(q_n + p_n)p_n \leq c\,(p_n^2 + q_n^2),$$

and thus from (6.5.26) we obtain

$$f(\cdot, \cdot, u) - \underline{f}(\cdot, \cdot, \alpha_{n+1}; \alpha_n, \beta_n) \leq c\,(p_{n+1} + p_n^2 + q_n^2). \qquad (6.5.27)$$

In just the same way, we obtain

$$k(\cdot, \cdot, \cdot, u) - \underline{k}(\cdot, \cdot, \cdot, \alpha_{n+1}; \alpha_n, \beta_n) \leq c\,(p_{n+1} + p_n^2 + q_n^2). \qquad (6.5.28)$$

In view of (6.5.27) we can estimate

$$\langle Fu - \underline{F}(\alpha_{n+1}; \alpha_n, \beta_n), p_{n+1} \rangle \qquad (6.5.29)$$

$$
\begin{aligned}
=\ & \int_Q \left(f(x, t, u) - \underline{f}(x, t, \alpha_{n+1}; \alpha_n, \beta_n) \right) p_{n+1}\, dx\, dt \\
\leq\ & c\,(\|p_{n+1}\|_{L^2(Q)}^2 + \|p_n^2 + q_n^2\|_{L^2(Q)}),
\end{aligned}
$$

and analogously by means of (6.5.28) one gets

$$\langle Ku - \underline{K}(\alpha_{n+1}; \alpha_n, \beta_n), p_{n+1}\rangle \qquad (6.5.30)$$

$$
\begin{aligned}
&= \int_\Gamma \left(\int_\Omega (k(x,t,x',u) - \underline{k}(x,t,x',\alpha_{n+1};\alpha_n,\beta_n))\, dx' \right) \gamma p_{n+1}\, d\Gamma \\
&\le c(\varepsilon)\left(\|p_{n+1}\|^2_{L^2(Q)} + \|p_n^2 + q_n^2\|^2_{L^2(Q)} \right) + \varepsilon \|\gamma p_{n+1}\|^2_{L^2(\Gamma)} \\
&\le c(\varepsilon)\left(\|p_{n+1}\|^2_{L^2(Q)} + \|p_n^2 + q_n^2\|^2_{L^2(Q)} \right) + \varepsilon \|\nabla p_{n+1}\|^2_{L^2(Q)},
\end{aligned}
$$

which holds for any $\varepsilon > 0$. In (6.5.30) Young's inequality and the continuity of the trace operator $\gamma : V \to L^2(\Gamma)$ has been applied. The difference $p_{n+1} = u - \alpha_{n+1} \in W$ satisfies $p_{n+1} \ge 0$, $p_{n+1}(0) = 0$ and the equation

$$p'_{n+1} + Ap_{n+1} + Bp_{n+1} \qquad (6.5.31)$$

$$= Fu - \underline{F}(\alpha_{n+1};\alpha_n,\beta_n) + Ku - \underline{K}(\alpha_{n+1};\alpha_n,\beta_n), \quad \text{in } V^*.$$

Testing equation (6.5.31) with the special nonnegative test function p_{n+1}, and using estimates (6.5.29), (6.5.30) with ε sufficiently small we finally obtain for some positive constant c an estimate in the form

$$\|p_{n+1}\|_W \le c\|p_n^2 + q_n^2\|_{L^2(Q)}, \qquad (6.5.32)$$

which can be derived in just the same way as, for instance, the estimate (6.4.22) of Corollary 6.4.1. Analogously one gets the estimate for q_{n+1} in the form

$$\|q_{n+1}\|_W \le c\|p_n^2 + q_n^2\|_{L^2(Q)}, \qquad (6.5.33)$$

so that (6.5.32) and (6.5.33) complete the proof of the theorem.

Due to the continuous embeddings $W \subset C([0,T]; L^2(\Omega)) \subset L^2(Q)$ one readily obtains from (6.5.2) and (6.5.3) of Theorem 6.5.1 the following corollary.

Corollary 6.5.1 *The iterates (α_n) and (β_n) of the generalized quasilinearization method satisfy an estimate in the form:*

$$
\begin{aligned}
\|\beta_{n+1} - \alpha_{n+1}\|_{L^2(Q)} &\le c \max_{[0,T]} \|(\beta_{n+1} - \alpha_{n+1})(\cdot,t)\|_{L^2(\Omega)} \qquad (6.5.34) \\
&\le c\|(\beta_n - \alpha_n)^2\|_{L^2(Q)} \le c\|\beta_n - \alpha_n\|^2_{L^\infty(Q)}.
\end{aligned}
$$

The convergence estimate (6.5.34) can be improved under additional hypotheses, such as, for instance, for the special case of the IBVP (6.5.11) where $k^{(2)} = 0$ and $k^{(1)}$ is given by the linear affine function

$$k(x, t, x', u) \equiv k^{(1)}(x, t, x', u) = m(x, t, x')u + g(x, t, x'), \qquad (6.5.35)$$

where $m \in L^\infty(\Gamma \times \Omega)$ with $m \geq 0$ on $\Gamma \times \Omega$, and $g \in L^2(\Gamma \times \Omega)$. Then the following stronger convergence estimate holds.

Corollary 6.5.2 *Let the kernel function k of the IBVP (6.5.1) be given by (6.5.35) with m and g as above, and assume that the coefficient b in the boundary condition of (6.5.1) satisfies*

$$b(x, t) \geq \int_\Omega m(x, t, x')\, dx', \quad \text{for a.e. } (x, t) \in \Gamma. \qquad (6.5.36)$$

Then there is some constant $c > 0$ such that the following estimate holds:

$$\|p_{n+1}\|_{L^\infty(Q)} + \|q_{n+1}\|_{L^\infty(Q)} \leq c\, (\|p_n\|^2_{L^\infty(Q)} + \|q_n\|^2_{L^\infty(Q)}). \qquad (6.5.37)$$

Proof   The special form (6.5.35) of the kernel function implies according to (6.5.5) that

$$k(x, t, x', u) = \underline{k}(x, t, x', u; \alpha, \beta) = \bar{k}(x, t, x', u; \alpha, \beta),$$

and thus $Ku = \underline{K}(u; \alpha, \beta) = \bar{K}(u; \alpha, \beta)$, which yields

$$(Ku)(x, t) - \underline{K}(\alpha_{n+1}; \alpha_n, \beta_n)(x, t) = \int_\Omega m(x, t, x')\, p_{n+1}(x', t)\, dx'. \qquad (6.5.38)$$

We have p_n, $q_n \in L^\infty(Q)$, since α_0, $\beta_0 \in L^\infty(Q)$. Set $S = \|p_n\|^2_{L^\infty(Q)} + \|q_n\|^2_{L^\infty(Q)}$, and consider the IBVP

$$\left[\begin{array}{rl} u' + Au = c\,(u + p_n^2 + q_n^2), & \text{in } Q, \\ u = 0, & \text{in } \Omega, \\ \frac{\partial u}{\partial \nu} + bu = \int_\Omega m(\cdot, \cdot, x')\, u(x', \cdot)\, dx', & \text{on } \Gamma. \end{array}\right. \qquad (6.5.39)$$

By using (6.5.27) and (6.5.31) one can show that $\underline{u} := p_{n+1}$ is a lower solution of the IBVP (6.5.39). Define $\bar{u}(x, t) := \delta S\, e^{\lambda t}$. Then under assumption (6.5.36) one verifies that $\bar{u}$ is an upper solution of (6.5.39) for $\lambda > 0$ and $\delta > 0$ sufficiently large. Due to Corollary 6.4.2 a comparison result holds for the IBVP (6.5.39), and thus it follows that

$$0 \leq p_{n+1}(x, t) \leq \delta S\, e^{\lambda t} \leq \delta S\, e^{\lambda T} = cS.$$

Similarly one can also show that $0 \leq q_{n+1}(x, t) \leq cS$, which proves (6.5.37) and thus the assertion of the corollary.

6.6   Quasilinear Problems: Existence and Comparison Results

In this and the next sections, we consider the initial boundary value problems for a class of quasilinear parabolic equations whose lower-order nonlinearities are of d.e. function type relative to the dependent variable. The second order quasilinear differential operator in divergence form that is employed shall be of Leray–Lions type and consequently we need to prove existence as well as comparison results that are required for the development of the method of generalized quasilinearization in the next section.

Let $\Omega \subset R^N$ be a bounded domain with Lipschitz boundary $\partial\Omega$; let $Q = \Omega \times (0, T)$ and $\Gamma = \partial\Omega \times (0, T)$, $T > 0$. Consider the initial boundary value problem (IBVP)

$$
\left[
\begin{aligned}
\frac{\partial u(x,t)}{\partial t} + Au(x,t) &= f(x,t,u(x,t)) + h, \quad \text{in } Q, \\
u(x,t) &= 0, \quad \text{on } \Gamma, \\
u(x,0) &= 0, \quad \text{in } \Omega
\end{aligned}
\right.
\tag{6.6.1}
$$

where A is a second-order quasilinear differential operator in divergence form of Leray–Lions type given by

$$
Au(x,t) = - \sum_{i=1}^{N} \left(\frac{\partial}{\partial x_i} \right) a_i(x,t,\nabla u(x,t)).
$$

Let $W^{1,p}(\Omega)$ denote the usual Sobolev space, and let $(W^{1,p}(\Omega))^*$ denote its dual space. We assume $p \geq 2$, and let $q \in R$ be the dual real satisfying

$$
\frac{1}{p} + \frac{1}{q} = 1,
$$

Then, by identifying $L^2(\Omega)$ with its dual space,

$$
W^{1,p}(\Omega) \subset L^2(\Omega) \subset (W^{1,p}(\Omega))^*
$$

forms an evolution triple with all the embeddings being continuous, dense, and compact. Let

$$
V = L^p(0, T; W^{1,p}(\Omega))
$$

be the L^p-space of vector-valued functions $u : (0, T) \rightarrow W^{1,p}(\Omega)$; then, its dual space is given by

$$
V^* = L^q(0, T; (W^{1,p}(\Omega))^*).
$$

Recall that, for any Banach space X, the space $Y = L^p(0,T;X)$ consists of all measurable functions $u : (0,T) \to X$ for which

$$\| u \|_Y := \left(\int_0^T \| u(t) \|_X^p \, dt \right)^{1/p}$$

is finite. We define a function space W by

$$W = \left\{ w \in V : \frac{\partial w}{\partial t} \in V^* \right\},$$

where the derivative $\frac{\partial}{\partial t}$ is understood in the sense of vector-valued distributions and is characterized by

$$\int_0^T u'(t)\phi(t)dt = - \int_0^T u(t)\phi'(t)dt, \quad \text{for all } \phi \in C_0^\infty(0,T).$$

The space W, endowed with the norm

$$\| w \|_W = \| w \|_V + \left\| \frac{\partial w}{\partial t} \right\|_{V^*},$$

is a Banach space which is separable and reflexive due to the separability and reflexivity of V and V^*, respectively. Furthermore, it is well known that the embedding $W \subset C([0,T]; L^2(\Omega))$ is continuous, and because of the compact embedding $W^{1,p}(\Omega) \subset L(\Omega)$, we have a compact embedding of $W \subset L^p(Q) \equiv L^p(0,T; L^p(\Omega))$. Denote by $W_0^{1,p}(\Omega)$ the subspace of $W^{1,p}(\Omega)$ whose elements have generalized homogeneous boundary values, and let $W^{-1,q}(\Omega)$ be its dual space. Then obviously,

$$W_0^{1,p}(\Omega) \subset L^2(\Omega) \subset W^{-1,q}(\Omega)$$

forms an evolution triple too, and all statements made above remain true also in the situation when setting

$$V_0 = L^p(0,T; W_0^{1,p}(\Omega)),$$

$$V_0^* = L^q(0,T; W^{-1,q}(\Omega)),$$

$$W_0 = \left\{ w \in V_0 : \frac{\partial w}{\partial t} \in V_0^* \right\}.$$

We impose the following conditions of Leray–Lions type on the coefficients a_i, $i = 1, \ldots, N$, of the operator A:

(A1) Each $a_k : Q \times R^N \to R$ satisfies the Caratheodory conditions; i.e., $a_i(x,t,\xi)$ is measurable in $(x,t) \in Q$, for all $\xi \in R^N$, and continuous in ξ for almost every (a.e.) $(x,t) \in Q$. There exist a constant $c_0 \geq 0$ and a function $k_0 \in L^q_+(Q)$ such that

$$| a_i(x,t,\xi) |\leq k_0(x,t) + c_0 \mid \xi \mid^{p-1},$$

for a.e. $(x,t) \in Q$ and for all $\xi \in R^N$.

(A2) $\sum_{i=1}^N (a_i(x,t,\xi) - a_i(x,t,\xi'))(\xi_i - \xi_i') > 0$, for a.e. $(x,t) \in Q$ and for all $\xi, \xi' \in R^N$ with $\xi \neq \xi'$.

(A3) There exist a positive constant μ and a function $k_1 \in L^1(Q)$ such that

$$\sum_{i=1}^N a_i(x,t,\xi)\xi_i \geq \mu \mid \xi \mid^p + k_1(x,t),$$

for a.e. $(x,t) \in Q$ and for all $\xi \in R^N$.

Denote by $\langle \cdot, \cdot \rangle$ the duality pairing in $V_0^* \times V_0$. Then, as a consequence of (A1),

$$\langle Au, \phi \rangle := a(u,\phi) = \sum_{i=1}^N \int_Q a_i(x,t,\nabla u)\left(\frac{\partial \phi}{\partial x_i}\right) dx dt, \quad \phi \in V_0,$$

is well defined for any $u \in V$ and gives rise to a continuous and bounded operator $A : V \to V_0^*$. Note that $A = -\Delta_p$, where Δ_p is the p-Laplacian given by

$$\Delta_p u = \sum_{i=1}^N \left(\frac{\partial}{\partial x_i}\right)\left[\mid \nabla u \mid^{p-2}\left(\frac{\partial u}{\partial x_i}\right)\right],$$

is a special case of the operator satisfying (A1)–(A3) with

$$a_i(x,t,\xi) =\mid \xi \mid^{p-2} \xi_i.$$

A partial ordering in $L^p(Q)$ is defined by $u \leq w$ if and only if $w - u$ belongs to the set $L^p_+(Q)$ of all nonnegative elements of $L^p(Q)$. This induces a corresponding partial ordering also in the subspace W of $L^p(Q)$; and, if $\underline{u}, \bar{u} \in W$ with $\underline{u} \leq \bar{u}$, then

$$[\underline{u}, \bar{u}] := \{u \in W \mid \underline{u} \leq u \leq \bar{u}\}$$

denotes the order interval formed by $\underline{u}$ and $\bar{u}$. Let F denote the Nemytski operator related to the Caratheodory function f by

$$Fu(x,t) = f(x,t,u(x,t)),$$

and let $h \in V_0^*$ be given. Then, we introduce the notion of a weak solution (respectively upper and lower solutions) of the IBVP (6.6.1) as follows.

Definition 6.6.1 *A function $u \in W_0$ is called a weak solution of the IBVP (6.6.1) if $Fu \in L^q(Q)$ and*

(i) $u(x,0) = 0$, *in* Ω,

(ii) $\langle \frac{\partial u}{\partial t}, \phi \rangle = \int_Q (Fu)\phi \, dx dt + \langle h, \phi \rangle$, *for all* $\phi \in V_0$.

Definition 6.6.2 *A function $\alpha \in W$ is said to be a lower solution of the IBVP (6.6.1) if $F\alpha \in L^q(Q)$ and*

(i) $\alpha(x,t) \le 0$, *on* Γ, $\alpha(x,0) \le 0$, *in* Ω,

(ii) $\langle \frac{\partial \alpha}{\partial t}, \phi \rangle + a(\alpha, \phi) \le \int_Q (F\alpha)\phi \, dx dt + \langle h, \phi \rangle$, *for all* $\phi \in V_0 \cap L_+^p(Q)$.

An upper solution of (6.6.1) is defined similarly by reversing the inequalities in Definition 6.6.2.

Remark 6.6.1 The structure of the IBVP (6.6.1) and the conditions imposed on the coefficients of the operator A allow us to assume homogeneous initial and boundary conditions without loss of generality. Inhomogeneous initial and boundary conditions which are the initial values and traces, respectively, of some function $w \in W$ can be reduced to homogeneous ones by a simple translation preserving the structure of the equation and the conditions on the coefficients.

The notion of d.c function, which is used in optimization theory, is adapted properly to nonlinearities that, in addition, depend on the space-time variable, as follows.

Definition 6.6.3 *Let $J \subset \mathbb{R}$ be an interval. The function $f : Q \times J \to \mathbb{R}$ is of d.c. type on J if, for all $u \in J$, f can be expressed in the form*

$$f(x,t,u) = f^1(x,t,u) - f^2(x,t,u), \text{ for a.e. } (x,t) \in Q, \qquad (6.6.2)$$

where the functions $f^i : Q \times J \to \mathbb{R}$, $i = 1,2$, are convex in $u \in J$ for a.e. $(x,t) \in Q$.

We now prove existence and comparison results for the IBVP (6.6.1) which are used to establish the generalized quasilinearization method to be presented in Section 6.7.

Throughout this section, we shall assume that hypotheses $(A1)$–$(A3)$ are satisfied. The following comparison result will be used frequently later.

Lemma 6.6.1 *Let α and β be lower and upper solutions of the IBVP (6.6.1), respectively. Further, suppose that there is a nonnegative function $l \in L^\infty_+(Q)$ such that f satisfies*

$$f(x,t,s_1) - f(x,t,s_2) \le l(x,t)(s_1 - s_2), \qquad (6.6.3)$$

for a.e. $(x,t) \in Q$, whenever $s_1 \ge s_2$. Then we have that $\alpha \le \beta$ in Q.

Proof   By definition of lower and upper solutions of the IBVP (6.6.1), we get

$$\alpha(x,t) - \beta(x,t) \le 0, \text{ on } \Gamma,$$

$$\alpha(x,t) - \beta(x,0) \le 0, \text{ in } \Omega,$$

and subtracting the corresponding differential inequalities for α and β yields

$$\left\langle \frac{\partial(\alpha - \beta)}{\partial t}, \phi \right\rangle + a(\alpha,\phi) - a(\beta,\phi) \le \int_Q (F\alpha - F\beta)\phi\,dxdt, \qquad (6.6.4)$$

for all $\phi \in V_0 \cap L^p_+(Q)$. Taking in (6.6.4) as special test function

$$\phi = (\alpha - \beta)^+ := \max(\alpha - \beta, 0) \in V_0 \cap L^p_+(Q),$$

then

$$(\alpha - \beta)^+(x,0) = 0,$$

and for any $\tau \in (0,T]$, we obtain the following inequality

$$\frac{1}{2} \| (\alpha - \beta)^+(\cdot,\tau) \|^2_{L^2(\Omega)} \qquad (6.6.5)$$

$$+ \int_{Q_\tau} \sum_{i=1}^N [a_i(x,t,\nabla\alpha) - a_i(x,t,\nabla\beta)]\left[\frac{\partial(\alpha - \beta)^+}{\partial x_i}\right] dxdt$$

$$\le \int_{Q_\tau} (F\alpha - F\beta)(\alpha - \beta)^+ dxdt,$$

where $Q_\tau := \Omega \times (0, \tau) \subset Q$. Using hypothesis $(A2)$, the second term on the left-hand side of (6.6.5) yields the estimate

$$\int_{Q_\tau} \sum_{i=1}^N [a_i(x, t, \nabla\alpha) - a_i(x, t, \nabla\beta)] \left[\frac{\partial(\alpha - \beta)^+}{\partial x_i} \right] dxdt \qquad (6.6.6)$$

$$= \int_{\{\alpha > \beta\}} \sum_{i=1}^N [a_i(x, t, \nabla\alpha) - a_i(x, t, \nabla\beta)] \left[\frac{\partial(\alpha - \beta)}{\partial x_i} \right] dxdt$$

$$\geq 0,$$

and the right-hand side of (6.6.5) can be estimated by using (6.6.3) as follows:

$$\int_{Q_\tau} (F\alpha - F\beta)(\alpha - \beta)^+ dxdt \qquad (6.6.7)$$

$$\leq \int_{\{\alpha > \beta\}} l(\alpha - \beta)^2 dxdt$$

$$\leq \|l\|_{L^\infty(Q)} \int_{Q_\tau} ((\alpha - \beta)^+)^2 dxdt.$$

Note that $\alpha, \beta \in W \subset L^p(Q) \subset L^2(Q)$, since $p \geq 2$. Thus, by setting

$$y(\tau) = \|(\alpha - \beta)^+(\cdot, \tau)\|_{L^2(\Omega)}^2, \text{ for any } \tau \in [0, T],$$

we obtain from (6.6.5)–(6.6.7) the inequality

$$y(\tau) \leq 2\|l\|_{L^\infty(Q)} \int_0^\tau y(t)dt;$$

by applying the Gronwall lemma and taking into account the continuous embedding $W \subset C([0, T]; L^2(\Omega))$, the above inequality yields that

$$y(\tau) = 0, \text{ for a.e. } (x, t) \in Q,$$

i.e., $\alpha \leq \beta$, proving the claim.

In the discussion of generalized quasilinearization, we also need the following existence and uniqueness result.

Lemma 6.6.2 *Let $f : Q \times \mathbb{R} \to \mathbb{R}$ be a Caratheodory function satisfying $F_0 \in L^q(Q)$ and a Lipschitz condition in the form*

$$| f(x, t, s_1) - f(x, t, s_2) | \leq l(x, t) | s_1 - s_2 |, \quad s_1, s_2 \in \mathbb{R}, \qquad (6.6.8)$$

for some nonnegative function $l \in L_+^\infty(Q)$. Then the IBVP (6.6.1) has a unique solution $u \in W_0$.

Proof   Since any solution of (6.6.1) is in particular an upper or a lower solution of (6.6.1), the uniqueness follows immediately from Lemma 6.6.1. Thus, we need to show only the existence of solutions.

We first derive an IBVP which is equivalent to (6.6.1) by performing the following simple transformation (also called an exponential shift):

$$u(x,t) = e^{\lambda,t} w(x,t), \tag{6.6.9}$$

with a constant $\lambda \geq 0$ to be specified later. Note that $u \in W$ if and only if $w \in W$. By definition, u is a solution of the IBVP (6.6.1) iff

$$u \in W_0, u(0) = 0; \quad u' + Au = Fu + h, \text{ in } V_0^*, \tag{6.6.10}$$

where u is considered as a vector-valued function. Applying the transformation (6.6.9), we obtain the following problem which is equivalent to (6.6.10)

$$w \in W_0, w(0) = 0; \quad w' + \tilde{A}w + \lambda w = \tilde{F}w + \tilde{h}, \text{ in } V_0^*, \tag{6.6.11}$$

where the operators $\tilde{A}$, $\tilde{F}$ and the element $\tilde{h} \in V_0^*$ are given by

$$\langle \tilde{A}w, \phi \rangle = \int_Q \sum_{i=1}^N e^{\lambda t} a_i(x,t,e^{\lambda t}\nabla w)\left(\frac{\partial \phi}{\partial x_i}\right) dxdt,$$

$$\langle \tilde{F}w, \phi \rangle = \int_Q e^{\lambda t} f(x,t,e^{\lambda t}w)\phi dxdt,$$

$$\tilde{h}(t) = e^{-\lambda t}h(t).$$

Denoting

$$\tilde{a}_i(x,t,\xi) = e^{-\lambda t}a_i(x,t,e^{\lambda t}\xi), \quad \tilde{f}(x,t,s) = e^{-\lambda t}f(x,t,e^{\lambda t}s),$$

one verifies readily that the transformed coefficients $\tilde{a}_i : Q \times \mathbb{R}^N \to \mathbb{R}$ preserve the structure conditions $(A1)$–$(A3)$. For example, to verify $(A3)$, we have

$$\sum_{i=1}^N \tilde{a}_i(x,t,\xi)\xi_i \;=\; \sum_{i=1}^N e^{-\lambda t}a_i(x,t,e^{\lambda t}\xi)\xi_i \tag{6.6.12}$$

$$=\; e^{-2\lambda t}\sum_{i=1}^N a_i(x,t,e^{\lambda t}\xi)(e^{\lambda t}\xi_i)$$

$$\geq\; e^{-2\lambda t}(\mu \mid e^{\lambda t}\xi \mid^p + k_1(x,t))$$

$$\geq\; \mu e^{(p-2)\lambda t} \mid \xi \mid^p + e^{-2\lambda t}k_1(x,t)$$

$$\geq\; \mu \mid \xi \mid^p + \tilde{k}_1(x,t),$$

where

$$\tilde{k}_1(x,t) = e^{-2\lambda t}k_1(x,t) \text{ and } \tilde{k}_1 \in L^1(Q);$$

note that $p \geq 2$ and $\lambda \geq 0$. Obviously $\tilde{f}$ satisfies the Lipschitz condition (6.6.8) and $\tilde{F}_0 \in L^q(Q)$. Problem (6.6.11) can be rewritten in the form

$$w \in W_0, w(0) = 0; \ w' + \tilde{A}w + (\lambda I - \tilde{F})w = \tilde{h}, \text{ in } V_0^*, \qquad (6.6.13)$$

where I denotes the identity mapping. An existence result for (6.6.13) can be obtained by means of the main theorem on monotone first-order evolution equations as given. To this end, we shall show that the operator $\tilde{A} + \lambda I - \tilde{F}$: $V_0 \to V_0^*$ is monotone, continuous, coercive, and bounded provided that λ is sufficiently large. Since the coefficients $\tilde{a}_i$ satisfy the structure conditions $(A1)$–$(A3)$, the operator $\tilde{A} : V_0 \to V_0^*$ is in particular continuous, bounded, and monotone. The function $(x,t,s) \to \lambda s - \tilde{f}(x,t,s)$ obviously satisfies the Caratheodory conditions and the following estimate:

$$\mid \lambda s - \tilde{f}(x,t,s) \mid \leq \mid \tilde{f}(x,t,0) \mid + (\lambda + \parallel l \parallel_{L^\infty(Q)}) \mid s \mid \leq k(x,t) + c \mid s \mid^{p-1},$$

where c is some positive constant and $k \in L_+^q(Q)$, which shows that the operator $\lambda I - \tilde{f} : L^p(Q) \to L^q(Q)$ is continuous and bounded. Thus, in view of the continuous embeddings $V_0 \subset L^p(Q)$ and $L^q(Q) \subset V_0^*$, it follows that $\lambda I - \tilde{F} : V_0 \to V_0^*$ is also continuous and bounded. Furthermore, we easily get the estimate

$$[\lambda(s - s') - (\tilde{f}(x,t,s) - \tilde{f}(x,t,s'))](s - s') \qquad (6.6.14)$$

$$\geq (\lambda - \|l\|_{L^\infty(Q)})(s - s')^2, \ s, s' \in \mathbb{R},$$

which implies that the operator $\lambda I - \tilde{F} : V_0 \to V_0^*$ is monotone for $\lambda \geq \|l\|_{L^\infty(Q)}$; thus, we have shown that $\tilde{A} + \lambda I - \tilde{F} : V_0 \to V_0^*$ is continuous, bounded, and monotone provided that $\lambda \geq \|l\|_{L^\infty(Q)}$. Let the latter be satisfied. We use (6.6.12) and (6.6.14) to obtain the following estimate:

$$\langle (\tilde{A} + \lambda I - \tilde{F})u, u \rangle$$

$$\geq \ \mu\|\nabla u\|_{L^p(Q)}^p - \|\tilde{k}_1\|_{L^1(Q)} + [\lambda - \|l\|_{L^\infty(Q)}]$$

$$\times \int_Q u^2 dx dt - \|\tilde{F}_0\|_{L^q(Q)}\|u\|_{L^p(Q)}$$

$$\geq \ \mu\|\nabla u\|_{L^p(Q)}^p - \|\tilde{k}\|_{L^1(Q)} - \|\tilde{F}_0\|_{L^q(Q)}\|u\|_{L^q(Q)},$$

which implies that $\tilde{A} + \lambda I - \tilde{F} : V_0 \to V_0^*$ is also coercive for $\lambda \geq \|l\|_{L^\infty(Q)}$. Note that $\|\nabla u\|_{L^p(Q)}$ and $\|u\|_V$ are equivalent norms in V_0. Now, applying a known result ensures the existence and this completes the proof.

As an immediate consequence of Lemmas 6.4.1 and 6.4.2, we obtain the following result.

Corollary 6.6.1 *Let $d \in L^\infty(Q)$ and $h \in V_0^*$ be given. Then, the IBVP*

$$u \in W_0, u(0) = 0; \quad u' + u = du + h, \quad in\ V_0^*, \qquad (6.6.15)$$

has a unique solution. Moreover, if α and β are lower and upper solutions of (6.6.15) respectively, then $\alpha \leq \beta$ in Q.

6.7 Generalized Quasilinearization (GQ): Quasilinear Problems

Having the necessary ingredients prepared in Section 6.6, we are now ready to discuss the method of generalized quasilinearization for quasilinear problems.

Instead of (6.6.1), we shall investigate the following IBVP

$$\left[\begin{array}{l} \frac{\partial u(x,t)}{\partial t} + Au(x,t) \\ = f(x,t,u(x,t)) + g(x,t,u(x,t)), \quad in\ Q, \\ \qquad\qquad\qquad u(x,t) = 0, \quad on\ \Gamma, \\ \qquad\qquad\qquad u(x,0) = 0, \quad in\ \Omega. \end{array}\right. \qquad (6.7.1)$$

where the Caratheodory functions f and g are assumed to be convex and concave in u respectively for a.e. $(x,t) \in Q$. Thus, the right-hand side of (6.7.1) is of d.c. function type according to Definition 6.6.3 with the convex functions f^i of the d.c. decomposition (6.6.2) given by

$$f^1(x,t,u) = f(x,t,u), \quad f^2(x,t,u) = -g(x,t,u).$$

It should be noted that an additional term $h \in V_0^*$ in (6.7.1) can be taken into account as well without any difficulties and has been omitted only for the sake of simplicity.

More precisely, we assume the following hypotheses:

(*H*1) There exist lower and upper solutions α_0 and β_0 of the IBVP (6.7.1) respectively such that $\alpha_0 \leq \beta_0$ and $\alpha_0, \beta_0 \in L^\infty(Q)$.

($H2$) The first and second derivatives of $f(x, t, u)$ and $g(x, t, u)$ with respect to u exist; f, g, f_u, g_u, f_{uu}, $g_{uu} : Q \times \mathbb{R} \to \mathbb{R}$ are Caratheodory functions, and $f_{uu}(x, t, u) \geq 0$, $g_{uu}(x, t, u) \leq 0$ for all $u \in [\text{ess } inf_Q \alpha_0(x, t),$ $\text{esssup}_Q \beta_0(x, t)]$ and for a.e. $(x, t) \in Q$.

($H3$) There is some positive constant C such that

$$\|f_u(\cdot, \cdot, \eta), g_u(\cdot, \cdot, \eta), f_{uu}(\cdot, \cdot, \eta), g_{uu}(\cdot, \cdot, \eta)\|_{L^\infty(Q)} \leq C, \qquad (6.7.2)$$

for any $\eta \in [\alpha_0, \beta_0]$.

Note that the convexity of f and the concavity of g as well as the boundedness condition (6.7.2) on the derivatives according to ($H2$)–($H3$) are required to hold only with respect to the order interval $[\alpha_0, \beta_0]$ formed by the lower and upper solutions α_0 and β_0 respectively. In particular, the right-hand side of (6.7.1) is, by hypothesis ($H2$), a function of d.c. type on the interval $[\text{ess } inf_Q \alpha_0(x, t), \text{ ess } sup_Q \beta_0(x, t)]$. The main result of this section is given by the following theorem.

Theorem 6.7.1 *Let hypotheses ($A1$)–($A3$) and ($H1$)–($H3$) be satisfied. Then there exist monotone sequences $(\alpha_k)_{k=1}^\infty$, $(\beta_k)_{k=1}^\infty \subset W_0$ such that $\alpha_k \to \varrho$, $\beta_k \to r$ in W_0, where $\varrho = r = u$, is the unique weak solution of (6.7.1) satisfying $\alpha_0 \leq u \leq \beta_0$. Furthermore, the convergence of these monotone sequences is quadratic with respect to the $L^2(Q)$-norm.*

Proof   The proof is provided in several steps.

Step 1 The generalized quasilinearization and related iterative schemes For any pair of functions $\alpha, \beta \in [\alpha_0, \beta_0]$, we introduce linearizations of the right-hand side $f + g$ at β and α in the form

$$\begin{aligned}
G(x, t, u; \alpha, \beta) & := f(x, t, \beta) + g(x, t, \beta) + f_u(x, t, \alpha)(u - \beta \quad (6.7.3) \\
& \quad + g_u(x, t, \beta)(u - \beta),
\end{aligned}$$

$$\begin{aligned}
F(x, t, u; \alpha, \beta) & := f(x, t, \alpha) + g(x, t, \alpha) + f_u(x, t, \alpha)(u - \alpha) \quad (6.7.4) \\
& \quad + g_u(x, t, \beta)(u - \alpha),
\end{aligned}$$

respectively, and associate with them the following iteration schemes for $k = 0, 1, 2, \ldots$;

$$\left[\begin{array}{rl}
\frac{\partial \beta_{k+1}}{\partial t} + A\beta_{k+1} = G(x, t, \beta_{k+1}; \alpha_k, \beta_k), & \text{in } Q, \\
\beta_{k+1}(x, t) = 0, & \text{on } \Gamma, \\
\beta_{k+1}(x, 0) = 0, & \text{in } \Omega,
\end{array} \right. \qquad (6.7.5)$$

and

$$\left[\begin{array}{l} \frac{\partial \alpha_{k+1}}{\partial t} + A\alpha_{k+1} = F(x,t,\alpha_{k+1};\alpha_k,\beta_k), \quad \text{in } Q, \\ \qquad\qquad\qquad \alpha_{k+1}(x,t) = 0, \quad \text{on } \Gamma, \\ \qquad\qquad\qquad \alpha_{k+1}(x,0) = 0, \quad \text{in } \Omega. \end{array}\right. \tag{6.7.6}$$

The corresponding weak formulations of (6.7.5), (6.7.6) are given by: Find $\beta_{k+1} \in W_0$ and $\alpha_{k+1} \in W_0$ satisfying homogeneous initial conditions such that

$$\left\langle \frac{\partial \beta_{k+1}}{\partial t}, \phi \right\rangle + a(\beta_{k+1}, \phi) \tag{6.7.7}$$

$$= \int_Q G(x,t,\beta_{k+1};\alpha_k,\beta_k)\phi\,dxdt, \quad \text{for all } \phi \in V_0,$$

$$\left\langle \frac{\partial \alpha_{k+1}}{\partial t}, \phi \right\rangle + a(\alpha_{k+1}, \phi) \tag{6.7.8}$$

$$= \int_Q F(x,t,\alpha_{k+1};\alpha_k,\beta_k)\phi\,dxdt, \quad \text{for all } \phi \in V_0,$$

respectively. We shall show that the weak solutions β_{k+1} and α_{k+1} of (6.7.5), (6.7.6) respectively are uniquely defined and satisfy

$$\alpha_0 \le \alpha_1 \le \alpha_2 \le \ldots \le \alpha_k \le \alpha_{k+1} \ldots \tag{6.7.9}$$

$$\le \beta_{k+1} \le \beta_k \le \ldots \le \beta_2 \le \beta_1 \le \beta_0.$$

Consider the case $k = 0$. According to (6.7.3), $G(x,t,u;\alpha_0,\beta_0)$ is of the form

$$G(x,t,u;\alpha_0,\beta_0) = d(x,t)u + h(x,t),$$

where

$$d(x,t) = f_u(x,t,\alpha_0) + g_u(x,t,\beta_0),$$

$$h(x,t) = f(x,t,\beta_0) + g(x,t,\beta_0) - f_u(x,t,\alpha_0)\beta_0 - g_u(x,t,\beta_0)\beta_0.$$

Since α_0 and β_0 are essentially bounded lower and upper solutions of (6.7.1), respectively, it follows by $(H3)$ that $d \in L^\infty(Q)$ and $h \in L^q(Q) \subset V_0^*$. Thus, by Corollary 6.6.1, there is a unique solution $\beta_1 \in W_0$ of (6.7.5). In just the same way, one obtains the existence of a unique solution of $\alpha_1 \in W_0$ of (6.7.6). We prove next that

$$\alpha_0 \le \alpha_1 \le \beta_1 \le \beta_0. \tag{6.7.10}$$

From (6.7.4), we get

$$F(x,t,\alpha_0;\alpha_0,\beta_0) = f(x,t,\alpha_0) + g(x,t,\alpha_0),$$

and thus the lower solution α_0 of the IBVP (6.7.1) can be considered as a lower solution of the following IBVP

$$u \in W_0, u(0) = 0; u' + Au = F(\cdot, \cdot, u; \alpha_0, \beta_0), \text{ in } V_0^*, \qquad (6.7.11)$$

with the linear affine right-hand side $F(\cdot, \cdot, u; \alpha_0, \beta_0)$, which in particular is Lipschitz continuous. By definition, α_1 is the uniquely defined solution of (6.7.8) for $k = 0$ which corresponds with (6.7.11). Now $\alpha_0 \leq \alpha_1$ follows readily from Lemma 6.6.1. Similarly, we can show that $\beta_1 \leq \beta_0$.

Next, we prove that

$$\alpha_1 \leq \beta_0 \text{ and } \alpha_0 \leq \beta_1.$$

To this end, we make use of the following inequalities which are immediate consequences of $f_{uu}(x, t, u) \geq 0$ (f convex in u) and $g_{uu}(x, t, u) \leq 0$ (g concave in u) imposed by $(H2)$:

$$f(x, t, u) \geq f(x, t, v) + f_u(x, t, v)(u - v), \qquad (6.7.12)$$

$$g(x, t, u) \geq g(x, t, v) + g_u(x, t, u)(u - v), \qquad (6.7.13)$$

for all $u, v \in [\alpha_0, \beta_0]$. To show that $\alpha_1 \leq \beta_0$, we employ (6.7.12), (6.7.13) to obtain, for all $\phi \in V_0 \cap L_+^p(Q)$, the estimate

$$\left\langle \frac{\partial \alpha_1}{\partial t}, \phi \right\rangle + a(\alpha_1, \phi) \qquad (6.7.14)$$

$$= \int_Q F(x, t, \alpha_1; \alpha_0, \beta_0)\phi dx dt$$

$$\leq \int_Q [f(x, t, \beta_0) - f_u(x, t, \alpha_0)(\beta_0 - \alpha_0) + g(x, t, \beta_0)$$

$$-g_u(x, t, \beta_0)(\beta_0 - \alpha_0) + f_u(x, t, \alpha_0)(\alpha_1 - \alpha_0) + g_u(x, t, \beta_0)(\alpha_1 - \alpha_0)]\phi dx dt$$

$$\leq \int_Q [f(x, t, \beta_0) + g(x, t, \beta_0) + f_u(x, t, \alpha_0)(\alpha_1 - \beta_0)$$

$$+g_u(x, t, \beta_0)(\alpha_1 - \beta_0)]\phi dx dt$$

$$= \int_Q G(x, t, \alpha_1; \alpha_0, \beta_0)\phi dx dt;$$

that is, α_1 is a lower solution of the IBVP

$$u \in W_0, u(0) = 0; \ u' + Au = G(\cdot, \cdot, u; \alpha_0, \beta_0), \text{ in } V_0^*, \qquad (6.7.15)$$

with the linear affine function $G(x, t, u; \alpha_0, \beta_0)$. Furthermore, since

$$G(x, t, \beta_0; \alpha_0, \beta_0) = f(x, t, \beta_0) + g(x, t, \beta_0),$$

it implies readily that the upper solution β_0 of the IBVP (6.7.1) is an upper solution of the IBVP (6.7.15); thus, $\alpha_1 \leq \beta_0$ follows by applying Lemma 6.6.1. By similar arguments, we prove that $\alpha_0 \leq \beta_1$. To prove that $\alpha_1 \leq \beta_1$, we use (6.7.12), (6.7.13) and the fact that $g_u(x, t, u)$ is monotone nonincreasing in u as well as the obtained inequality $\alpha_1 \leq \beta_0$ to get

$$\left\langle \frac{\partial \alpha_1}{\partial t}, \phi \right\rangle + a(\alpha_1, \phi) \tag{6.7.16}$$

$$= \int_Q F(x, t, \alpha_1; \alpha_0, \beta_0)\phi dx dt$$

$$\leq \int_Q [f(x, t, \alpha_1) + g(x, t, \alpha_1) + (g_u(x, t, \beta_0) - g_u(x, t, \alpha_1))(\alpha_1 - \alpha_0)]\phi dx dt$$

$$\leq \int_Q [f(x, t, \alpha_1) + g(x, t, \alpha_1)]\phi dx dt$$

$$= \int_Q F(x, t, \alpha_1; \alpha_1, \beta_1)\phi dx dt,$$

for each $\phi \in V_0 \cap L_+^p(Q)$. Similarly, because of the fact that $f_u(x, t, u)$ is monotone nondecreasing in u and $\alpha_0 \leq \beta_1$, we obtain, for all $\phi \in V_0 \cap L_+^p(Q)$, the estimate

$$\left\langle \frac{\partial \beta_1}{\partial t}, \phi \right\rangle + a(\beta_1, \phi) \tag{6.7.17}$$

$$= \int_Q G(x, t, \beta_1; \alpha_0, \beta_0)\phi dx dt$$

$$\geq \int_Q [f(x, t, \beta_1) + f_u(x, t, \beta_1)(\beta_0 - \beta_1)$$
$$+ f_u(x, t, \alpha_0)(\beta_1 - \beta_0)$$
$$+ g(x, t, \beta_1)]\phi dx dt$$

$$= \int_Q [f(x, t, \beta_1) + (-f_u(x, t, \beta_1) + f_u(x, t, \alpha_0))(\beta_1 - \beta_0) + g(x, t, \beta_1)]\phi dx dt$$

$$\geq \int_Q [f(x, t, \beta_1)$$
$$+ g(x, t, \beta_1)]\phi dx dt$$

$$= \int_Q G(x, t, \beta_1; \alpha_1, \beta_1)\phi dx dt.$$

Since

$$F(x,t,\alpha_1;\alpha_1,\beta_1) = f(x,t,\alpha_1) + g(x,t,\alpha_1)$$

$$G(x,t,\beta_1;\alpha_1,\beta_1) = f(x,t,\beta_1) + g(x,t,\beta_1),$$

the inequalities (6.7.16), (6.7.17) imply that α_1 is a lower solution and β_1 is an upper solution of the IBVP (6.7.1) within the interval $[\alpha_0,\beta_0]$. Finally, in view of $(H3)$, the nonlinearity $f(x,t,u)+g(x,t,u)$ is Lipschitz continuous in u with respect to the order interval $[\alpha_0,\beta_0]$; hence, from Lemma 6.6.1, it follows that $\alpha_1 \le \beta_1$, which proves the inequality (6.7.10).

Next, we shall prove that, if

$$\alpha_{k-1} \le \alpha_k \le \beta_k \le \beta_{k-1}, \tag{6.7.18}$$

for some $k > 1$, then it follows that

$$\alpha_k \le \alpha_{k+1} \le \beta_{k+1} \le \beta_k. \tag{6.7.19}$$

By definition, $\alpha_k \in W_0$ satisfies homogeneous initial conditions and

$$\left\langle \frac{\partial \alpha_k}{\partial t}, \phi \right\rangle + a(\alpha_k, \phi) = \int_Q F(x,t,\alpha_k;\alpha_{k-1},\beta_{k-1})\phi dx dt,$$

for all $\phi \in V_0$. Following the arguments employed to obtain (6.7.16), (6.7.17), by using the assumption (6.7.18), we get the following inequalities:

$$\left\langle \frac{\partial \alpha_k}{\partial t}, \phi \right\rangle + a(\alpha_k, \phi) \le \int_Q F(x,t,\alpha_k;\alpha_k,\beta_k)\phi dx dt, \tag{6.7.20}$$

$$\left\langle \frac{\partial \beta_k}{\partial t}, \phi \right\rangle + a(\beta_k, \phi) \ge \int_Q G(x,t,\beta_k;\alpha_k,\beta_k)\phi dx dt, \tag{6.7.21}$$

for each $\phi \in V_0 \cap L_+^p(Q)$, which shows that α_k is a lower solution of the IBVP

$$u \in W_0, u(0) = 0; u' + Au = F(\cdot,\cdot,u;\alpha_k,\beta_k), \text{ in } V_0^*, \tag{6.7.22}$$

and β_k is an upper solution of

$$u \in W_0, u(0) = 0; u' + Au = G(\cdot,\cdot,u;\alpha_k,\beta_k), \text{ in } V_0^*. \tag{6.7.23}$$

Since α_{k+1} and β_{k+1} are the uniquely defined solutions (6.7.22), (6.7.23) respectively, whose right-hand sides are linear affine functions, from Lemma 6.6.1, it follows that

$$\alpha_k \le \alpha_{k+1} \text{ and } \beta_{k+1} \le \beta_k.$$

Next, we show that

$$\alpha_{k+1} \le \beta_k \text{ and } \alpha_k \le \beta_{k+1}.$$

Using (6.7.12), (6.7.13), we find that

$$\left\langle \frac{\partial \alpha_{k+1}}{\partial t}, \phi \right\rangle + a(\alpha_{k+1}, \phi)$$

$$= \int_Q F(x, t, \alpha_{k+1}; \alpha_k, \beta_k) \phi \, dx dt$$

$$\le \int_Q [f(x, t, \beta_k) - f_u(x, t, \alpha_k)(\beta_k - \alpha_k) + g(x, t, \beta_k)$$
$$- g_u(x, t, \beta_k)(\beta_k - \alpha_k)$$
$$+ f_u(x, t, \alpha_k)(\alpha_{k+1} - \alpha_k) + g_u(x, t, \beta_k)(\alpha_{k+1} - \alpha_k)] \phi \, dx dt$$

$$\le \int_Q [f(x, t, \beta_k) + g(x, t, \beta_k)$$
$$+ (f_u(x, t, \alpha_k) + g_u(x, t, \beta_k))(\alpha_{k+1} - \beta_k)] \phi \, dx dt$$

$$= \int_Q G(x, t, \alpha_{k+1}; \alpha_k, \beta_k) \phi \, dx dt,$$

for each $\phi \in V_0 \cap L^p_+(Q)$, which shows that α_{k+1} is a lower solution of (6.7.23); thus, in view of (6.7.21), by applying Lemma 6.6.1, we obtain that $\alpha_{k+1} \le \beta_k$.

Similarly, we get the estimate

$$\left\langle \frac{\partial \beta_{k+1}}{\partial t}, \phi \right\rangle + a(\beta_{k+1}, \phi)$$

$$= \int_Q G(x, t, \beta_{k+1}; \alpha_k, \beta_k) \phi \, dx dt$$

$$\ge \int_Q [f(x, t, \alpha_k) + f_u(x, t, \alpha_k)(\beta_k - \alpha_k) + g(x, t, \alpha_k)$$
$$+ g_u(x, t, \beta_k)(\beta_k - \alpha_k)$$
$$+ f_u(x, t, \alpha_k)(\beta_{k+1} - \beta_k) + g_u(x, t, \beta_k)(\beta_{k+1} - \beta_k)] \phi \, dx dt$$

$$= \int_Q [f(x, t, \alpha_k) + g(x, t, \alpha_k) + (f_u(x, t, \alpha_k)$$
$$+ g_u(x, t, \beta_k))(\beta_{k+1} - \alpha_k)] \phi \, dx dt$$

$$= \int_Q F(x, t, \beta_{k+1}; \alpha_k, \beta_k) \phi \, dx dt,$$

for each $\phi \in V_0 \cap L_+^p(Q)$, which proves that β_{k+1} is an upper solution of (6.7.22). By (6.7.20), α_k is a lower solution of (6.7.22); thus, $\alpha_k \leq \beta_{k+1}$.

Finally, to prove that $\alpha_{k+1} \leq \beta_{k+1}$, we use the inequalities $\alpha_{k+1} \leq \beta_k$ and $\alpha_k \leq \beta_{k+1}$ to show that α_{k+1} and β_{k+1} satisfy

$$\left\langle \frac{\partial \alpha_{k+1}}{\partial t}, \phi \right\rangle + a(\alpha_{k+1}, \phi) \leq \int_Q F(x, t, \alpha_{k+1}; \alpha_{k+1}, \beta_{k+1}) \phi \, dx dt, \quad (6.7.24)$$

$$\left\langle \frac{\partial \beta_{k+1}}{\partial t}, \phi \right\rangle + a(\beta_{k+1}, \phi) \leq \int_Q G(x, t, \beta_{k+1}; \alpha_{k+1}, \beta_{k+1}) \phi \, dx dt, \quad (6.7.25)$$

for each $\phi \in V_0 \cap L_+^p(Q)$. The last inequalities (6.7.24), (6.7.25) can be obtained in the same way as (6.7.16), (6.7.17) replacing α_0, β_0, α_1, β_1 by α_k, β_k, α_{k+1}, β_{k+1} and using (6.7.12), (6.7.13) as well as the monotone character of f_u and g_u. Consequently, we get from Lemma 6.6.1 that $\alpha_{k+1} \leq \beta_{k+1}$, which proves (6.7.19); thus, by induction, it follows that (6.7.9) is true for each $k = 0, 1, 2, \ldots$.

Step 2 Convergence of $\{\alpha_k\}$ $\{\beta_k\}$ to the unique solution in $[\alpha_0, \beta_0]$
According to (6.7.9), the monotonicity of the iterates (α_k), $(\beta_k) \in [\alpha_0, \beta_0]$ imply the existence of their a.e. pointwise limits,

$$\varrho(x, t) = \lim_{k \to \infty} \alpha_k(x, t), \quad r(x, t) = \lim_{k \to \infty} \beta_k(x, t), \quad \text{for a.e. } (x, t) \in Q.$$

In view of $\alpha_0, \beta_0 \in L^\infty(Q)$, the Lebesgue dominated convergence theorem yields

$$\alpha_k \to \varrho \text{ and } \beta_k \to r, \text{ in } L^\sigma(Q), \quad (6.7.26)$$

for any $\sigma \in [1, \infty)$, and thus, in particular, also for $\sigma = p$. Let C denote a generic positive constant whose value may be different at different places. By $(H3)$, for any $\eta \in [\alpha_0, \beta_0]$, we get an estimate in the form

$$\mid f(x, t, \eta) + g(x, t, \eta) \mid \leq \mid f(x, t, \alpha_0) + g(x, t, \alpha_0) \mid + C \mid \beta_0 - \alpha_0 \mid. \quad (6.7.27)$$

Since the right-hand side of (6.7.27) is in $L^q(Q)$ [note that $f(\cdot, \cdot, \alpha_0) + g(\cdot, \cdot, \alpha_0) \in L^q(Q)$], it follows that the Nemytski operator $F + G$ related to $f + g$ is continuous and bounded as a mapping $F + G : [\alpha_0, \beta_0] \subset L^\sigma(Q) \to L^q(Q)$ for any $\sigma \in [1, \infty)$. Thus, in view of (6.7.3), (6.7.4) and $(H3)$, from (6.7.26) it follows that the right-hand sides of (6.7.5) and (6.7.6) satisfy

$$\begin{bmatrix} G(\cdot, \cdot, \beta_{k+1}; \alpha_k, \beta_k) \to (F + G)r, \\ F(\cdot, \cdot, \alpha_{k+1}; \alpha_k, \beta_k) \to (F + G)\varrho \quad \text{in } L^q(Q). \end{bmatrix} \quad (6.7.28)$$

Next, we show that the sequences $(\alpha_k)_{k=1}^{\infty}$ and $(\beta_k)_{k=1}^{\infty}$ defined by (6.7.5), (6.7.6), respectively are uniformly bounded in W_0. Taking $\phi = \beta_{k+1}$ as special test function in the weak formulation (6.7.7), using $(A3)$ and taking into account that $\beta_{k+1}(x,0) = 0$, we obtain the following estimate:

$$\frac{1}{2}\|\beta_{k+1}(\cdot,T)\|_{L^2(\Omega)}^2 + \mu\|\nabla\beta_{k+1}\|_{L^P(Q)}^P - \|k_1\|_{L^1(Q)} \qquad (6.7.29)$$

$$\leq \|G(\cdot,\cdot,\beta_{k+1};\alpha_k,\beta_k)\|_{L^q(Q)}\|\beta_{k+1}\|_{L^P(Q)},$$

for all $k = 1,2,\dots$. In view of (6.7.26) and (6.7.28), the right-hand side of (6.7.29) is uniformly bounded; thus, it follows that

$$\|\beta_k\|_{V_0} \leq C, \text{ for all } k = 1,2,\dots. \qquad (6.7.30)$$

Further, from the weak formulation (6.7.7), which corresponds with

$$\left[\begin{array}{l} \beta_{k+1} \in W_0, \quad \beta_{k+1}(x,0) = 0; \\ \beta_{k+1}' + A\beta_{k+1} = G(\cdot,\cdot,\beta_{k+1};\alpha_k,\beta_k), \quad \text{in } V_0^+, \end{array} \right. \qquad (6.7.31)$$

we get immediately, by means of (6.7.28) and (6.7.30)

$$\left\|\frac{\partial\beta_{k+1}}{\partial t}\right\|_{V_0^+} \leq \left\|A\beta_{k+1}\right\|_{V_0^+} + \|G(\cdot,\cdot,\beta_{k+1};\alpha_k,\beta_k)\|_{V_0^+} \leq C, \qquad (6.7.32)$$

where we have employed the boundedness of the operator $A : V_0 \to V_0^*$ and the continuous embedding $L^q(Q) \subset V_0^*$. Thus, by (6.7.30) and (6.7.32), we obtain

$$\|\beta_k\|_{W_0} \leq C, \text{ for all } k = 1,2,\dots, \qquad (6.7.33)$$

and by similar arguments,

$$\|\alpha_k\|_{W_0} \leq C, \text{ for all } k = 1,2,\dots. \qquad (6.7.34)$$

Since W_0 is reflexive, there exists a weakly convergent subsequence of $(\beta_k)_{k=1}^{\infty}$ whose weak limit must be r in view of (6.7.26) and the compact embedding $W_0 \subset L^P(Q)$. Since any weakly convergent subsequence $(\beta_k)_{k=1}^{\infty}$ has the same limit r, it follows that the whole sequence must be weakly convergent to $r \in W_0$. Let $L = \frac{\partial}{\partial t}$, and let its domain be given by

$$D(L) = \{u \in W_0 \mid u(x,0) = 0\};$$

then, $L : D(L) \subset V_0 \to V_0^*$ can be shown to be a closed, densely defined, and maximal monotone operator. Since $\beta_k \in D(L)$ and $D(L)$ is closed and convex, and thus weakly closed, it follows that the weak limit $r \in D(L)$. Hypotheses $(A1)$–$(A3)$ imply that the operator A possesses the so-called (S_+)-property with respect to the graph norm topology of $D(L)$ which means that, for any sequence $(u_n) \subset D(L)$, with $u_n \mapsto u$ in V_0, $Lu_n \mapsto Lu$ in V_0^*, and

$$\limsup_{n \to \infty} \langle Au_n, u_n - u \rangle \leq 0, \qquad (6.7.35)$$

it follows that $u_n \mapsto u$ (strongly) in V_0. This property allows one to ensure the strong convergence in V_0 of the sequence (β_k) to r. To this end, we have to verify only (6.7.35). From (6.7.9), we get, for $k = 1, 2, \ldots$,

$$
\begin{aligned}
\langle A\beta_k, \beta_k - r \rangle &= -\left\langle \frac{\partial \beta_k}{\partial t}, \beta_k - r \right\rangle \\
&\quad + \int_Q G(\cdot, \cdot \beta_k; \alpha_{k-1}, \beta_{k-1})(\beta_k - r) dx dt \\
&= -\left\langle \frac{\partial(\beta_k - r)}{\partial t}, \beta_k - r \right\rangle - \left\langle \frac{\partial r}{\partial t}, \beta_k - r \right\rangle \\
&\quad + \int_Q G(\cdot, \cdot, \beta_k; \alpha_{k-1}, \beta_{k-1})(\beta_k - r) dx dt \\
&\leq -\left\langle \frac{\partial r}{\partial t}, \beta_k - r \right\rangle \\
&\quad + \int_Q G(\cdot, \cdot, \beta_k; \alpha_{k-1}, \beta_{k-1})(\beta_k - r) dx dt \to 0, \\
&\text{as } k \to \infty,
\end{aligned}
$$

which shows that

$$\limsup_{k \to \infty} \langle A\beta_k, \beta_k - r \rangle \leq 0,$$

and thus the strong convergence $\beta_k \to r$ in V_0. In the last estimate, we have used the fact that

$$(\beta_k - r)(x, 0) = 0,$$

which implies that

$$\left\langle \frac{\partial(\beta_k - r)}{\partial t}, \beta_k - r \right\rangle \geq 0.$$

From (6.7.28) and the continuous embedding $L^q(Q) \subset V_0^*$, it follows that

$$G(\cdot, \cdot, \beta_{k+1}; \alpha_k, \beta_k) \to (F + G)r, \text{ in } V_0^*. \qquad (6.7.36)$$

and the continuity of $A : V_0 \to V_0^*$ together with $\beta_k \to r$ in V_0 implies

$$A\beta_k \to Ar, \text{ in } V_0^*. \qquad (6.7.37)$$

Hence, by (6.7.36) and (6.7.37), in view of (6.7.31) we get the strong convergence of the sequence $(\frac{\partial \beta_k}{\partial t})$ in V_0^* to $\frac{\partial r}{\partial t}$, which due to $\beta_k \to r$ in V_0 yields

$$\beta_k \to r, \quad \text{in } W_0, \text{ as } k \to \infty. \tag{6.7.38}$$

Finally, the convergence (6.7.38) allows us to pass to the limit in the defining equation (6.7.31) of the β_k as $k \to \infty$, which shows that the limit r is a solution of the IBVP (6.7.1). Similar arguments can be applied to prove that the sequence (α_k) satisfies

$$\alpha_k \to \varrho, \quad \text{in } W_0, \text{ as } k \to \infty, \tag{6.7.39}$$

where ϱ is a solution of the IBVP (6.7.1). By $(H3)$, the right-hand side $f(x,t,u) + g(x,t,u)$ of (6.7.1) is in particular Lipschitz continuous in u with respect to the order interval $[\alpha_0, \beta_0]$, which by Lemma 6.6.1 yields uniqueness with respect to this interval, and thus $r = \varrho = u$ is the unique solution of (6.7.1) in $[\alpha_0, \beta_0]$.

Step 3 Quadratic convergence of $\{\alpha_k\}, \{\beta_k\}$ For $k = 1, 2, \ldots$, we set

$$p_{k+1} = u - \alpha_{k+1}, \quad q_{k+1} = \beta_{k+1} - u,$$

where u is the unique solution of (6.7.1) within $[\alpha_0, \beta_0]$ so that

$$p_{k+1} \geq 0, \quad q_{k+1} \geq 0, \quad \text{in } Q,$$

as well as

$$p_{k+1}(x,t) = 0, \quad \text{on } \Gamma,$$

$$p_{k+1}(x,0) = 0, \quad \text{in } \Omega,$$

and

$$q_{k+1}(x,t) = 0, \quad \text{on } \Gamma$$

$$q_{k+1}(x,0) = 0, \quad \text{in } \Omega.$$

We then obtain, for $\phi \in V_0 \cap L_+^p(Q)$ and using the fact that f_u is monotone nondecreasing and g_u is monotone nonincreasing, the following estimate:

$$\left\langle \frac{\partial p_{k+1}}{\partial t}, \phi \right\rangle + a(u, \phi) - a(\alpha_{k+1}, \phi) \tag{6.7.40}$$

$$= \int_Q [f(x,t,u) + g(x,t,u) - f(x,t,\alpha_k) - f_u(x,t,\alpha_k)(\alpha_{k+1} - \alpha_k)$$

$$-g(x,t,\alpha_k) - g_u(x,t,\beta_k)(\alpha_{k+1} - \alpha_k)]\phi dx dt$$

$$\leq \int_Q [f_u(x,t,u)p_k + g_u(x,t,\alpha_k)p_k - f_u(x,t,\alpha_k)(p_k - p_{k+1})$$

$$-g_u(x,t,\beta_k)(p_k - p_{k+1})]\phi dx dt$$

$$= \int_Q [(f_u(x,t,u) - f_u(x,t,\alpha_k))p_k - (g_u(x,t,\beta_k) - g_u(x,t,\alpha_k))p_k$$

$$+(f_u(x,t,\alpha_k) + g_u(x,t,\beta_k))p_{k+1}]\phi dx dt$$

$$= \int_Q [f_{uu}(x,t,\eta_1)p_k^2 - g_{uu}(x,t,\eta_2)(\beta_k - \alpha_k)p_k$$

$$+(f_u(x,t,\alpha_k) + g_u(x,t,\beta_k))p_{k+1}]\phi dx dt,$$

for some η_1, η_2 with $\alpha_k \leq \eta_1 \leq u$, $\alpha_k \leq \eta_2 \leq \beta_k$. By using $(H3)$ and the definitions of p_k, q_k, we have

$$-g_{uu}(x,t,\sigma)(\beta_k - \alpha_k)p_k \leq C(q_k + p_k)p_k \qquad (6.7.41)$$

$$\leq C\left[\frac{3}{2}p_k^2 + \frac{1}{2}q_k^2\right];$$

thus, from (6.7.40) and (6.7.41), we get

$$\left\langle \frac{\partial p_{k+1}}{\partial t}, \phi \right\rangle + a(u,\phi) - a(\alpha_{k+1},\phi) \qquad (6.7.42)$$

$$\leq \int_Q c_k(x,t)p_{k+1}\phi dx dt + \int_Q \left[\frac{5}{2}Cp_k^2 + \frac{1}{2}Cq_k^2\right]\phi dx dt,$$

for all $\phi \in V_0 \cap L_+^p(Q)$, where the coefficients $c_k \in L^\infty(Q)$, given by

$$c_k(x,t) = f_u(x,t,\alpha_k(x,t)) + g_u(x,t,\beta_k(x,t)),$$

are uniformly bounded by some positive constant C due to $(H3)$; i.e.,

$$\|c_k\|_{L^\infty(Q)} \leq C, \text{ for all } k.$$

Taking in (6.7.42) the special test function $\phi = p_{k+1}$, we obtain by $(A2)$ and in view of $p_{k+1}(x,0) = 0$, for any subdomain $Q_t = \Omega \times (0,t) \subset Q$ with $t \in (0,T]$, the inequality

$$\frac{1}{2}\|p_{k+1}(\cdot,t)\|_{L^2(\Omega)}^2 \leq C\|p_{k+1}\|_{L^2(Q_t)}^2 + \frac{1}{2}\|\frac{5}{2}Cp_k^2\|$$

$$+\frac{1}{2}Cq_k^2\|_{L^2(Q_t)}^2 + \frac{1}{2}\|p_{k+1}\|_{L^2(Q_t)}^2,$$

and thus,

$$\|p_{k+1}(\cdot, t)\|^2_{L^2(\Omega)} \le c_1 \|p_{k+1}\|^2_{L^2(Q_t)} + c_2 \|p_k^2 + q_k^2\|^2_{L^2(Q_t)}.$$

Setting
$$y(t) = \|p_{k+1}\|^2_{L^2(Q_t)}, \ h(t) = \|p_k^2 + q_k^2\|^2_{L^2(Q_t)},$$

the last inequality implies

$$y'(t) \le c_1 y(t) + c_2 h(t), \ y(0) = 0,$$

which by applying the Gronwall inequality yields

$$y(t) \le c_2 e^{c_1 t} \int_0^t h(s) ds;$$

thus, there is a positive constant (again denoted by C) depending on c_1, c_2 and T such that

$$\|p_{k+1}\|^2_{L^2(Q)} \le C \|p_k^2 + q_k^2\|^2_{L^2(Q)}. \tag{6.7.43}$$

An estimate similar to (6.7.43) also holds for q_{k+1}, which completes the proof.

Remark 6.7.1 The convergence of the iterates improves if we impose a stronger monotonicity condition on the operator A. Let the coefficients a_i satisfy the following p-ellipticity condition:

$(A4)$
$$\sum_{i=1}^N (a_i(x, t, \xi) - a_i(x, t, \xi'))(\xi_i - \xi_i') \ge \mu \mid \xi - \xi' \mid^p,$$

for a.e. $(x, t) \in Q$ and for all $\xi, \xi' \in \mathbb{R}^N$;

then obviously, $(A2)$–$(A3)$ are satisfied, and the operator $A : V_0 \to V_0^*$ is uniformly monotone. An important example of an operator satisfying $(A4)$ is the p-Laplacian for $p \ge 2$. By means of $(A4)$ and (6.7.43), one obtains readily from (6.7.42) with the special test function $\phi = p_{k+1}$, the following gradient estimate:

$$\|\nabla p_{k+1}\|^p_{L^p(Q)} \le C \|p_k^2 + q_k^2\|^2_{L^2(Q)},$$

which yields

$$\|p_{k+1}\|_{V_0} \le C \|p_k^2 + q_k^2\|^2_{L^{\frac{2}{p}}(Q)}.$$

Further improvements of the convergence can be achieved for example if the coefficients $a_i(x,t,\xi)$ satisfy in addition a Lipschitz condition with respect to ξ. Then, one can also get an estimate of $\left\|\frac{\partial p_{k+1}}{\partial t}\right\|_{V_0^+}$ in terms of $\|p_k^2 + q_k^2\|_{L^2(Q)}^2$, and thus also of the norm $\|p_{k+1}\|_{W_0}$, which due to the continuous embedding $W_0 \subset C([0,T]; L^2(\Omega))$ gives an estimate of $\max_{0 \le t \le T} \|p_{k+1}(t)\|_{L^2(\Omega)}$ in terms of $\|p_k^2 + q_k^2\|_{L^2(Q)}^2$.

Remark 6.7.2

(i) As special cases, we get results for the right-hand sides of (6.7.1) that are convex with respect to u if $g(x,t,u) \equiv 0$, and for concave right-hand sides if $f(x,t,u) \equiv 0$ in (6.7.1).

(ii) Consider the case when $g(x,t,u) \equiv 0$ and $f(x,t,u)$ is not convex in u. Suppose there is a function $k(x,t,u)$ which is convex in u such that $f(x,t,u) + k(x,t,u)$ is convex in u. Then, the initial condition in (6.7.1) can be rewritten as

$$\frac{\partial u}{\partial t} + Au = \tilde{f}(x,t,u) + \tilde{g}(x,t,u),$$

where

$$\tilde{f}(x,t,u) = f(x,t,u) + k(x,t,u)$$

and

$$\tilde{g}(x,t,u) = -k(x,t,u),$$

so that the conditions of Theorem 6.7.1 are satisfied, and we get the same conclusion also when $f(x,t,u)$ is not convex.

(iii) A dual situation of (ii) is also valid, that is, when $f(x,t,u) \equiv 0$ and $g(x,t,u)$ is not concave. However, assume that there is a function $k(x,t,u)$ which is concave in u such that $g(x,t,u) + k(x,t,u)$ is concave in u. Then, the conditions in (6.7.1) can be written as before, but now with

$$\tilde{f}(x,t,u) = -k(x,t,u), \quad \tilde{g}(x,t,u) = g(x,t,u) + k(x,t,u).$$

Then, also the conditions of Theorem 6.7.1 are satisfied, and we get the same conclusion when $g(x,t,u)$ is not concave.

(iv) There is a large class of nonlinearities f that admit a d.c. decomposition in the form (6.6.2). Note that the d.c. decomposition is needed only on the interval formed by the upper and lower solutions. Thus, this method can be applied to any nonlinearity $f(x, t, u)$ whose second derivative $f_{uu}(x, t, u)$ exists and is essentially bounded on the interval

$$J = [\operatorname*{ess\,inf}_Q \alpha_0, \operatorname*{ess\,sup}_Q \beta_0],$$

i.e.,

$$\|f_{uu}(\cdot, \cdot, \eta)\|_{L^\infty(Q)} \le C, \text{ for all } \eta \in J.$$

This is because f can readily be d.c. decomposed in the following way:

$$f(x, t, u) = (f(x, t, u) + Cu^2) - Cu^2,$$

with the convex functions f^i given by

$$f^1(x, t, u) = f(x, t, u) + Cu^2, \ f^2(x, t, u) = Cu^2.$$

6.8   Notes and Comments

The results on the monotone iterative technique in a unified setting discussed in Section 6.2 for semilinear parabolic IBVPs are taken from Köksal and Lakshmikantham [42]. For the existence and uniqueness Theorem 6.2.2, see Ladyzhenskaya [48], Lions [65], Lions and Magenes [66], Wloka [82], and Evans [32]. Section 6.3, consisting of the methodology of generalized quasilinearization for semilinear parabolic IBVPs, is adapted from the work of Carl and Lakshmikantham [16]. See Carl and Lakshmikantham [18], [17] for the material in Sections 6.4 and 6.5, which deals with the formulation, existence, comparison results and the method of generalized quasilinearization for nonlocal parabolic IBVPs. The development starting with posing the problem, proving existence, the comparison principle and describing the method of generalized quasilinearization for quasilinear parabolic IBVPs of Leray–Lions type, can be found in Carl and Lakshmikantham [17].

For the discussion of the monotone method for discontinuous quasilinear parabolic IBVPs, see Carl and Heikkilä [15].

Chapter 7

Hyperbolic Equations

7.1 Introduction

We shall consider, in this chapter, second-order hyperbolic differential equations which are natural generalizations of the equations studied in Chapter 4. As we are aware, the study of hyperbolic equations is of a different character compared with parabolic equations. Consequently, we shall provide, in Section 7.2, the appropriate notions and functional analytic framework that would facilitate the desired investigation of the methodology of monotone iterates and fast convergence. We shall also prove the required comparison principle using different ideas and tools. We shall extend, in Section 7.3 the monotone iterative technique for nonlinear hyperbolic IBVPs and in Section 7.4, the method of generalized quasilinearization. Both of these techniques are presented in a unified set-up so that several important special cases can be derived from them.

7.2 Notation and Comparison Results

Second-order hyperbolic differential equations are a natural generalization of the wave equation. Although parabolic and hyperbolic partial differential equations are physically of different character, we shall provide, in this section, a similar functional analytic framework for such hyperbolic equations. We will describe appropriate notation and then consider comparison results that are necessary to develop the monotone iterative technique and the method of generalized quasilinearization in a unified setup.

Let $\Omega \subset R^n$ be an open, bounded domain with boundary $\partial\Omega$ and let $Q = \Omega \times (0, T]$, $\Gamma = \partial\Omega \times [0, T]$ for some $T > 0$. We shall consider the second-order hyperbolic IBVP

$$\left[\begin{array}{rl} u_{tt} + Lu = f(x,t,u) & \text{in } Q, \\ u = 0 & \text{on } \Gamma, \\ u = g, u_t = h, & \text{on } \Omega \times \{t = 0\}, \end{array} \right. \tag{7.2.1}$$

where the operator L, for each t, denotes a second-order partial differential operator in the divergence form

$$Lu = -\sum_{i,j=1}^{n} (a_{ij}(x,t)u_{x_i})_{x_j} + \sum_{i=1}^{n} b_i(x,t)u_{x_i} + c(x,t)u \tag{7.2.2}$$

with coefficients a_{ij}, b_i, $c \in L^\infty(\Omega) \times C^1[0,T]$, $g \in H_0^1(\Omega)$ and $h \in L^2(\Omega)$, $f : Q \times R \to R$ is a Caratheodory function and for $(x,t) \in Q$,

$$\sum_{i,j=1}^{n} a_{ij}(x,t)\xi_i\xi_j \geq \theta \mid \xi \mid^2, \xi \in R^n, \theta > 0. \tag{7.2.3}$$

The operator $\frac{\partial^2}{\partial t^2} + L$ is said to be uniformly hyperbolic if (7.2.3) holds. Moreover, we assume that $a_{ij} = a_{ji}$, $i = 1, 2, \ldots, n$.

As before, we introduce the time-dependent bilinear form

$$B[u,v;t] = \int_\Omega \left[\sum_{i,j=1}^{n} a_{ij}(\cdot,t)u_{x_i}v_{x_j} + \sum_{i=1}^{n} b_i(\cdot,t)u_{x_i}v + c(\cdot,t)u, v \right] dx \tag{7.2.4}$$

for $u, v \in H_0^1(\Omega)$ and $0 \leq t \leq T$.

We let, as in Chapter 6, $H^1(\Omega)$ be the Sobolev space of all square integrable functions and $(H^1(\Omega))^*$ denote its dual space. Then identifying $L^2(\Omega)$ with its dual space, $H^1(\Omega) \subset L^2(\Omega) \subset (H^1(\Omega)^*)$ forms an evolution triple with all the embeddings being continuous, dense and compact. Let $V = L^2(0, T, H^1(\Omega))$, its dual space, $V^* = L^2(0, T; (H^1(\Omega)^*))$, and we define

$$W = \left[w \in V : w' \in L^2(Q), w'' \in V^* \right].$$

The space W endowed with the norm

$$\|u\|_W = \left[\|u\|_V + \|u'\|_{L^2(Q)} + \|u''\|_{V^*} \right]$$

is a Banach space which is separable and reflexive. Also the embedding $W \subset C[0,T]$ is continuous. Since $H^1(\Omega) \subset L^2(\Omega)$ is compactly embedded, the embedding $W \subset L^2(Q)$ is compact, where $L^2(Q) = L^2(0, T; L^2(\Omega))$.

Let $H_0^1(\Omega)$ be the subspace of $H^1(\Omega)$ whose elements have generalized homogeneous boundary values and let $H^{-1}(\Omega)$ be its dual. Then
$H_0^1(\Omega) \subset L^2(\Omega) \subset H^{-1}(\Omega)$ forms an evolution triple, and all the statements made in the foregoing paragraph remain valid, in this situation, when
$V_0 = L^2(0, T; H_0^1(\Omega))$, $V_0^* = L^2(0, T; H^{-1}(\Omega))$ and

$$W_0 = [u \in V_0 : u' \in L^2(Q), u'' \in V_0^*].$$

We denote the duality pairing between the elements of V^* and V (respectively V_0^* and V_0) by $\langle \cdot, \cdot \rangle$.

If $u = u(t, x)$ is a solution of (7.2.1) that is smooth enough, we define
the associated mapping

$$\mathbf{u} : [0, T] \to H_0^1(\Omega)$$

by $[\mathbf{u}(t)](x) = u(x, t)$, $x \in \Omega$, $t \in [0, T]$. Similarly $[(\mathbf{fu})(t)](x)$
$= f(x, t, u(x, t))$. For any $v \in H_0^1(\Omega)$ multiplying $u_{tt} + Lu = f$ by v and
integrating by parts, we get

$$(\mathbf{u}'', v) + B[\mathbf{u}, v; t] = (\mathbf{fu}, v) \tag{7.2.5}$$

for $t \in [0, T]$, where $(\cdot, \cdot)$ denotes the inner product in $L^2(\Omega)$. As usual, we
can reinterpret the first term in (7.2.5) as $\langle \mathbf{u}'', v \rangle$ where $\langle \cdot, \cdot \rangle$ denotes the
pairing between $H^{-1}(\Omega)$ and $H_0^1(\Omega)$. Also, the bilinear form B is related to
the operator $\mathbf{L} : V_0 \to V_0^*$ by

$$\langle \mathbf{Lu}, \phi \rangle = \int_0^T \mathbf{L}u(t)\phi(t)dt = B[\mathbf{u}, \phi].$$

Definition 7.2.1 *A function* $\mathbf{u} \in W_0$ *is called a weak solution of (7.2.1) if*
$\mathbf{fu} \in L^2(Q)$ *and*

(i) $\mathbf{u}(0) = g$, $\mathbf{u}'(0) = h$;

(ii) $\langle \mathbf{u}'', \phi \rangle + B[\mathbf{u}, \phi; t] = (\mathbf{fu}, \phi)$

for all $\phi \in V_0$.

We note that since $\mathbf{u} \in C[0, T; L^2(\Omega)]$ and $\mathbf{u}' \in C[0, T; H^{-1}(\Omega)]$, the
equalities in (i) make sense.

A partial ordering in $L^2(Q)$ is defined by $u \leq v$ iff $v - u \in L_+^2(Q)$ of
all nonnegative elements of $L^2(Q)$. This induces a corresponding partial
ordering also in the subset W of $L^2(Q)$ and if $\alpha, \beta \in W$ with $\alpha \leq \beta$ then

$$[\alpha, \beta] = [\mathbf{u} \in W : \alpha \leq \mathbf{u} \leq \beta]$$

denotes the order interval formed by α, β.

We shall next define weak lower and upper solutions of (7.2.1).

Definition 7.2.2 *A function $\alpha \in W$ is said to be a weak lower solution of*
(7.2.1) if $\mathbf{f}\alpha \in L^2(Q)$ and

(i) $\alpha(0) \leq g$, $\alpha'(0) \leq h$;

(ii) $\langle \alpha'', \phi \rangle + B[\alpha, \phi; t] \leq (\mathbf{f}\alpha, \phi)$

for all $\phi \in V_0 \cap L^2_+(Q)$.

A weak upper solution of (7.2.1) is defined by reversing the inequalities
in Definition 7.2.2.
We are now in a position to prove the comparison result.

Theorem 7.2.1 *Let α, β be weak lower and upper solutions of (7.2.1) re-*
spectively. Suppose further that

$$f(x,t,u_1) - f(x,t,u_2) \leq k(x,t)(u_1 - u_2), u_1 \geq u_2, \qquad (7.2.6)$$

for $u_1, u_2 \in R$, for some nonnegative $k \in L^\infty(Q)$. Then $\alpha \leq \beta$.

Proof   The definition of weak lower and upper solutions of (7.2.1) yields

$$\alpha(0) - \beta(0) \leq 0, \quad \alpha'(0) - \beta'(0) \leq 0,$$

and

$$\langle (\alpha - \beta)'', \phi \rangle + B[\alpha - \beta, \phi; t] \leq (\mathbf{f}\alpha - \mathbf{f}\beta, \phi)$$

for all $\phi \in V_0 \cap L^2_+(Q)$. Define $\mathbf{m}(t) = [\alpha(t) - \beta(t)]^+ = \max(\alpha(t) - \beta(t), 0)$,
so that $\mathbf{m}(0) = 0$ and $\mathbf{m} \in V_0 \cap L^2_+(Q)$. For fixed $0 \leq s \leq T$, we let

$$\phi(t) = \begin{bmatrix} \int_t^s \mathbf{m}(\sigma)d\sigma, & 0 \leq t \leq s \\ 0, & s \leq t \leq T. \end{bmatrix}$$

Then we have $\phi \in V_0 \cap L^2_+(Q)$ and as a result it follows that

$$\int_0^s \{\langle (\alpha - \beta)'', \phi \rangle + B[\alpha - \beta, \phi; t]\}dt \leq \int_0^s (\mathbf{f}\alpha - \mathbf{f}\beta, \phi)dt.$$

Since $\alpha'(0) - \beta'(0) \leq 0$, $\phi(s) = 0$ and $\phi(0) \geq 0$, we get, integrating by parts,
the first term,

$$\int_0^s \{-\langle (\alpha - \beta)', \phi' \rangle + B[\alpha - \beta, \phi; t]\}dt \leq \int_0^s (\mathbf{f}\alpha - \mathbf{f}\beta, \phi)dt.$$

Observe that $\phi'(t) = -\mathbf{m}(t)$ and $\mathbf{m}'(t) = \begin{bmatrix} (\alpha - \beta)', & \text{a.e.} & \text{if } \alpha - \beta > 0, \\ 0, & & \text{otherwise.} \end{bmatrix}$

We therefore obtain, using (7.2.6) and letting $k_0 = \|k\|_{L^\infty(Q)}$,

$$\int_0^s \{\langle \mathbf{m}', \mathbf{m} \rangle + B[\mathbf{m}, \phi; t]\} dt \leq k_0 \int_0^s (\mathbf{m}, \phi) dt,$$

which can be rewritten as

$$\int_0^s \{\langle \mathbf{m}', \mathbf{m} \rangle + B_0[\mathbf{m}, \phi; t]\} dt \leq 0,$$

where

$$B_0[\mathbf{m}, \phi; t] = B[\mathbf{m}, \phi; t] - k_0(\mathbf{m}, \phi).$$

Since $\phi'(t) = -\mathbf{m}(t)$, the foregoing inequality becomes

$$\int_0^s \{\langle \mathbf{m}', \mathbf{m} \rangle - B_0[\phi', \phi; t]\} dt \leq 0.$$

We see that this is equivalent to

$$\int_0^s \frac{d}{dt} \left(\frac{1}{2} \|\mathbf{m}\|_{L^2(\Omega)}^2 - \frac{1}{2} B_0[\phi, \phi; t] \right) dt \leq - \int_0^s \{ B_1[\mathbf{m}, \phi; t] + B_2[\phi, \phi; t]\} dt,$$

where

$$B_1[\mathbf{m}, \phi; t] = \int_\Omega \left[\sum_{i,j=1}^n b_i \phi_{x_i} \mathbf{m} + \frac{1}{2} b_i \mathbf{m}_{x_i} \phi \right] dx$$

and

$$B_2[\phi, \phi; t] = \frac{1}{2} \int_\Omega \left[\sum_{i,j=1}^n (a_{ij})_t \phi_{x_i} \phi_{x_j} + \sum_{i=1}^n (b_i)_t \phi_{x_i} \phi_{x_j} + c_t \phi^2 \right] dx.$$

It then follows that if $0 \leq s \leq T$,

$$\frac{1}{2} \|\mathbf{m}(s)\|_{L^2(\Omega)}^2 + \frac{1}{2} B_0[\phi(0), \phi(0); t] \leq - \int_0^s \{ B_1[\mathbf{m}, \phi; t] + B_2[\phi, \phi; t]\} dt$$

and consequently, we arrive at, using the standard estimates for B_1, B_2,

$$\|\mathbf{m}(s)\|_{L^2(\Omega)}^2 + \|\phi(0)\|_{H_0^1(\Omega)}^2 \qquad (7.2.7)$$

$$\leq C \int_0^s (\|\phi\|^2_{H^1_0(\Omega)} + \|\mathbf{m}\|^2_{L^2(\Omega)})dt + \|\phi(0)\|^2_{L^2(\Omega)},$$

for some constant $c > 0$. Let us now set $\mathbf{w}(t) = \int_0^t \mathbf{m}(\sigma)d\sigma$, $0 \leq t \leq T$, so that (7.2.7) reduces to

$$\|\mathbf{m}(s)\|^2_{L^2(\Omega)} + \|\mathbf{w}(s)\|^2_{H^1_0(\Omega)}$$

$$\leq c \int_0^s (\|\mathbf{w}(t) - \mathbf{w}(s)\|^2_{H^1_0(\Omega)} + \|\mathbf{m}(t)\|^2_{L^2(\Omega)})dt + \|\mathbf{w}(s)\|^2_{L^2(\Omega)}.$$

Since $\|\mathbf{w}(t) - \mathbf{w}(s)\|^2_{H^1_0(\Omega)} \leq 2[\|\mathbf{w}(t)\|^2_{H^1_0(\Omega)} + \|\mathbf{w}(s)\|^2_{H^1_0(\Omega)}]$ and $\|\mathbf{w}(s)\|_{L^2(\Omega)}$ $\leq \int_0^s \|\mathbf{m}(t)\|_{L^2(\Omega)}ds$, we are led to

$$\|\mathbf{m}(s)\|^2_{L^2(\Omega)} + (1 - 2sc_1)\|\mathbf{w}(s)\|^2_{H^1_0(\Omega)} \leq c_1 \int_0^s (\|\mathbf{m}\|^2_{L^2(\Omega)} + \|\mathbf{w}\|^2_{H^1_0(\Omega)})dt,$$

for some constant $c_1 > 0$. Choosing T_1 small enough to satisfy $1 - 2T_1 c_1 \geq \frac{1}{2}$, we find for $0 \leq s \leq T_1$

$$\|\mathbf{m}(s)\|^2_{L^2(\Omega)} + \|\mathbf{w}(s)\|^2_{H^1_0(\Omega)} \leq c \int_0^s (\|\mathbf{m}^2_{L^2(\Omega)} + \|\mathbf{w}\|^2_{H^1_0(\Omega)})dt. \qquad (7.2.8)$$

An application of Gronwall's inequality now yields, because of the fact that $\mathbf{m}(0) = 0$, $\mathbf{m}(t) = 0$ on $[0, T_1]$. This implies that $\alpha \leq \beta$ on $[0, T_1]$. A repeated application of the foregoing arguments on the intervals $[T_1, T_2]$, $[2T_1, 3T_1]$ and so on, gives finally the desired conclusion $\alpha \leq \beta$, completing the proof.

The following corollary of Theorem 7.2.1 is also useful.

Corollary 7.2.1 *For any* $\mathbf{p} \in W$ *such that* $\mathbf{p}(0) \leq 0$, $\mathbf{p}'(0) \leq 0$ *and* $\langle \mathbf{p}'', \phi \rangle$ $+ B[\mathbf{p}, \phi; t] \leq 0$ *for all* $\phi \in V_0 \cap L^2_+(Q)$ *implies* $\mathbf{p} \leq 0$.

We also require the following existence and uniqueness theorem for linear IBVPs

$$\left[\begin{array}{l} u_{tt} + Lu = \tilde{h} \quad \text{in } Q, \\ u = 0 \text{ on } \Gamma, \quad u = h_1, u_t = h_2 \text{ on } \Omega \times \{t = 0\}, \end{array} \right. \qquad (7.2.9)$$

where $\tilde{h} : Q \to R$ is such that $\tilde{h} \in L^2(Q)$.

Theorem 7.2.2 *There exists a unique solution* $\mathbf{u} \in W_0$ *to the IBVP (7.2.9) that satisfies the estimate*

$$\|\mathbf{u}\|_{W_0} \leq c[\|h_1\|_{H^1_0(\Omega)} + \|h_2\|_{L^2(\Omega)} + \|\tilde{h}\|_{L^2(Q)}] \qquad (7.2.10)$$

for some $c > 0$.

7.3 Monotone Iterative Technique

We shall extend the monotone iterative technique for the following second-order hyperbolic IBVP

$$\left[\begin{array}{rl} u_{tt} + Lu = f(x,t,u) + g(x,t,u) & \text{in } Q, \\ u = 0 & \text{on } \Gamma, \\ u = 0, u_t = 0 & \text{on } \Omega \times \{t = 0\}, \end{array}\right. \tag{7.3.1}$$

where $f, g : Q \times R \to R$ are Caratheodory functions, the other assumptions being the same as in Section 7.2, including the notation. We follow the unified approach, considered in other types of equations, since it includes several special cases of interest.

Definition 7.3.1 *The functions $\alpha_0, \beta_0 \in W$ are said to be coupled weak solutions of (7.3.1) if $\mathbf{f}\alpha_0 + \mathbf{g}\beta_0, \mathbf{f}\beta_0 + \mathbf{g}\alpha_0 \in L^2(Q)$ and*

$$\alpha_0(0) \leq \beta_0(0), \alpha_0'(0) \leq \beta_0'(0),$$

$$\langle \alpha_0'', \phi \rangle + B[\alpha_0, \phi] \leq (\mathbf{f}\alpha_0 + \mathbf{g}\beta_0, \phi),$$

$$\langle \beta_0'', \phi \rangle + B[\beta_0, \alpha] \geq (\mathbf{f}\beta_0 + \mathbf{g}\alpha_0, \phi),$$

for all $\phi \in V_0 \cap L_+^2(Q)$.

We can now prove the following result.

Theorem 7.3.1 *Assume that*

(A_1) *α_0, β_0 are coupled weak lower and upper solutions of (7.3.1) such that $\alpha_0 \leq \beta_0$;*

(A_2) *$f, g : Q \times R \to R$ are Caratheodory functions such that $f(x,t,u)$ is nondecreasing in u and $g(x,t,u)$ is nonincreasing in u for $(x,t) \in Q$ a.e.;*

(A_3) *for any $\mu, \eta \in [\alpha_0, \beta_0]$, $\eta, \mu \in W$, $h \in L^2(Q)$ where $h(x,t) = f(x,t,\eta(x,t)) + g(x,t,\mu(x,t))$.*

Then there exist monotone sequences $\{\alpha_k\}, \{\beta_k\} \in W_0$ such that $\alpha_k \to \rho$, $\beta_k \to \mathbf{r}$ in W_0 and $\rho, \mathbf{r}$ are the coupled weak solutions of (7.3.1).

Proof   Consider the following linear IBVPs for each $k = 1, 2, 3, \ldots$,

$$\left[\begin{array}{ll} (\alpha_k)_{tt} + L\alpha_k = f(x, t, \alpha_{k-1}) + g(x, t, \beta_{k-1}) & \text{in } Q, \\ \hspace{3cm} \alpha_k = 0 & \text{on } \Gamma, \\ \hspace{1.5cm} \alpha_k = 0, (\alpha_k)_t = 0 & \text{on } \Omega \times \{t = 0\}, \end{array} \right. \tag{7.3.2}$$

$$\left[\begin{array}{ll} (\beta_k)_{tt} + L\beta_k = f(x, t, \beta_{k-1}) + g(x, t, \alpha_{k-1}) & \text{in } Q, \\ \beta_k = 0 \text{ on } \Gamma, \; \beta_k = 0, (\beta_k)_t = 0 & \text{on } \Omega \times \{t = 0\}, \end{array} \right. \tag{7.3.3}$$

whose variational forms associated with (7.3.2) and (7.3.3) are

$$\left[\begin{array}{l} \hspace{1cm} \alpha_k = 0, \alpha_k'(0) = 0, \\ \langle \alpha_k'', \phi \rangle + B[\alpha_k, \phi] = (\mathbf{f}\alpha_{k-1} + \mathbf{g}\beta_{k-1}, \phi), \end{array} \right. \tag{7.3.4}$$

and

$$\left[\begin{array}{l} \hspace{1cm} \beta_k(0) = 0, \beta_k'(0) = 0, \\ \langle \beta_k'', \phi \rangle + B[\beta_k, \phi] = (\mathbf{f}\beta_{k-1} + \mathbf{g}\alpha_{k-1}, \phi) \end{array} \right. \tag{7.3.5}$$

for all $\phi \in V_0 \cap L_+^2(Q)$. Because of (A_3), $h \in L^2(Q)$ where $h(x, t) = f(x, t, \alpha_0)$ $+ g(x, t, \beta_0)$. Hence, by Theorem 7.2.2 there exists a unique weak solution $\alpha_1 \in W_0$ of (7.3.2) for $k = 1$. Similarly, we have the existence of unique weak solution β_1 of (7.3.3) for $k = 1$. We shall first prove that

$$\alpha_0 \leq \alpha_1 \leq \beta_1 \leq \beta_0. \tag{7.3.6}$$

Note that α_1 satisfies the relation

$$\left[\begin{array}{l} \langle \alpha_1'', \phi \rangle + B[\alpha_1, \phi] = (\mathbf{f}\alpha_0 + \mathbf{g}\beta_0, \phi), \\ \hspace{0.5cm} \alpha_1(0) = 0, \alpha_1'(0) = 0, \end{array} \right. \tag{7.3.7}$$

and α_0 satisfies Definition 7.3.1, for all $\phi \in V_0 \cap L_+^2(Q)$. Let $\mathbf{p} = \alpha_0 - \alpha_1$. Then $\mathbf{p}$ fulfills $\mathbf{p} \leq 0$, $\mathbf{p}' \leq 0$ and

$$\langle \mathbf{p}'', \phi \rangle + B[\mathbf{p}, \phi] \leq 0 \text{ for all } \phi \in V_0 \cap L_+^2(Q).$$

Hence by Corollary 7.2.1, we get $\mathbf{p} \leq 0$, that is, $\alpha_0 \leq \alpha_1$. A similar argument proves $\beta_1 \leq \beta_0$. To show $\alpha_0 \leq \beta_1$, we find from (7.3.5),

$$\left[\begin{array}{l} \langle \beta_1'', \phi \rangle + B[\beta_1, \phi] = (\mathbf{f}\beta_0 + \mathbf{g}\alpha_0, \phi) \geq (\mathbf{f}\alpha_0 + \mathbf{g}\beta_0, \phi), \\ \hspace{3cm} \beta_1(0) = 0, \beta_1' = 0, \\ \hspace{2.5cm} \text{for all } \phi \in V_0 \cap L_+^2(Q); \end{array} \right. \tag{7.3.8}$$

using the monotone nature of $\mathbf{f}, \mathbf{g}$. By Corollary 7.2.1, we get, in view of (7.3.7), (7.3.8), $\alpha_1 \leq \beta_1$, proving (7.3.6).

We shall prove next that if for some $j \geq 1$,

$$\alpha_0 \leq \alpha_{j-1} \leq \alpha_j \leq \beta_j \leq \beta_{j-1} \leq \beta_0, \tag{7.3.9}$$

then we have

$$\alpha_j \leq \alpha_{j+1} \leq \beta_{j+1} \leq \beta_j. \tag{7.3.10}$$

Assumption (A_3), together with the relation (7.3.9) yields by Theorem 7.2.2 the existence of a unique weak solution α_j, α_{j+1}, β_j, $\beta_{j+1} \in W_0$ of (7.3.4), (7.3.5) with $k = j$, $k = j+1$. To show that $\alpha_j \leq \alpha_{j+1}$, we consider

$$\langle \alpha''_{j+1}, \phi \rangle + B[\alpha_j, \phi] = (\mathbf{f}\alpha_j + \mathbf{g}\beta_j, \phi) \tag{7.3.11}$$

and

$$\langle \alpha''_j, \phi \rangle + B[\alpha_j, \phi] = (\mathbf{f}\alpha_{j-1} + \mathbf{g}\beta_{j-1}, \phi) \leq (\mathbf{f}\alpha_j + \mathbf{g}\beta_j, \phi), \tag{7.3.12}$$

for each $\phi \in V_0 \cap L_+^2(Q)$. In (7.3.12), we have utilized the monotone character of $\mathbf{f}$, $\mathbf{g}$ and (7.3.9). Corollary 7.2.1 then yields $\alpha_j \leq \alpha_{j+1}$. Similarly, we can prove that $\beta_{j+1} \leq \beta_j$. To prove $\alpha_{j+1} \leq \beta_{j+1}$, we see, using the monotone nature of $\mathbf{f}$, $\mathbf{g}$ and (7.3.9), that

$$\langle \beta''_{j+1}, \phi \rangle + B[\beta_{j+1}, \phi] = (\mathbf{f}\beta_j + \mathbf{g}\alpha_j, \phi) \geq (\mathbf{f}\alpha_j + \mathbf{g}\beta_j, \phi) \tag{7.3.13}$$

for each $\phi \in V_0 \cap L_+^2(Q)$. Then the relations (7.3.11) and (7.3.13) prove, by Corollary 7.2.1, that $\alpha_{j+1} \leq \beta_{j+1}$, showing that (7.3.10) is true. Hence by induction, we arrive at

$$\alpha_0 \leq \alpha_1 \leq \ldots \leq \alpha_k \leq \beta_k \leq \ldots \leq \beta_1 \leq \beta_0, \tag{7.3.14}$$

for all $k = 1, 2, \ldots$. By monotonicity of the iterates $\{\alpha_k\}$, $\{\beta_k\}$, there exist pointwise limits

$$\lim_{k \to \infty} \alpha_k(x, t) = \rho(x, t), \ \lim_{k \to \infty} \beta_k(x, t) = r(x, t), \ \text{a.e. in} \ Q. \tag{7.3.15}$$

Moreover, since α_k, $\beta_k \in [\alpha_0, \beta_0]$, it follows by Lebesgue's dominated convergence theorem that

$$\alpha_k \to \rho, \quad \beta_k \to r \text{ in } L^2(Q). \tag{7.3.16}$$

By (A_3), for any η, $\mu \in [\alpha_0, \beta_0]$, $\mathbf{f}\eta + \mathbf{g}\mu$ is continuous and bounded as a mapping $[\alpha_0, \beta_0] \subset L^2(Q) \to L^2(Q)$ and consequently, in view of (7.3.16), we find that

$$\mathbf{f}\alpha_k \to \mathbf{f}\rho, \quad \mathbf{g}\beta_k \to \mathbf{g}r, \quad \mathbf{f}\beta_k \to \mathbf{f}r, \quad \mathbf{g}\alpha_k \to \mathbf{g}\rho \text{ in } L^2(Q) \text{ as } k \to \infty.$$
$$\tag{7.3.17}$$

Since α_k, β_k are weak solutions of linear IBVPs (7.3.2), (7.3.3) with homogeneous initial boundary conditions, the estimate (7.2.10) of Theorem 7.2.2 gives

$$\|\alpha_k\|_W \le c\|\mathbf{h}_1\|_{L^2(Q)}, \text{ where } \mathbf{h}_1 = \mathbf{f}\alpha_{k-1} + \mathbf{g}\beta_{k-1},$$

$$\|\beta_k\|_W \le c\|\mathbf{h}_2\|^2_{L^2(Q)}, \text{ where } \mathbf{h}_2 = \mathbf{f}\beta_{k-1} + \mathbf{g}\alpha_{k-1}.$$

Due to (7.3.4), (7.3.5), the foregoing estimates for α_k, β_k imply the strong convergence of the sequences $\{\alpha_k\}$, $\{\beta_k\}$ in W, whose limits must be ρ, $\mathbf{r}$ respectively, because of the compact embedding $W \subset L^2(Q)$. Since α_k, β_k, $k \ge 1$, belong to the subset $M = [\mathbf{u} \in W : \mathbf{u}(0) = 0, \mathbf{u}'(0) = 0]$ which is closed in W, it follows that the limits ρ, $\mathbf{r} \in M$, that is, the limits satisfy homogeneous initial and boundary conditions. The convergence result (7.3.17) together with

$$\alpha_k \to \rho, \quad \beta_k \to \mathbf{r} \text{ in } W \text{ as } k \to \infty,$$

allow us to pass to the limits in the corresponding variational forms (7.3.4), (7.3.5) as $k \to \infty$. This results in

$$\langle \rho'', \phi \rangle + B[\rho, \phi; t] = (\mathbf{f}\rho + \mathbf{g}\mathbf{r}, \phi),$$

$$\langle \mathbf{r}'', \phi \rangle + B[\mathbf{r}, \phi; t] = (\mathbf{f}\mathbf{r} + \mathbf{g}\rho, \phi),$$

for all $\phi \in V_0 \cap L^2_+(Q)$, showing that $(\rho, \mathbf{r})$ are coupled weak solutions of (7.2.1).

Let $\mathbf{u} \in [\alpha_0, \beta_0]$ be any weak solution of (7.2.1). Then we shall show that $\alpha_1 \le \mathbf{u} \le \beta_1$. Since α_1 satisfies (7.3.7), we obtain using the monotone character of f, g

$$\langle \alpha_1'', \phi \rangle + B[\alpha_1, \phi; t] = (\mathbf{f}\mathbf{u} + \mathbf{g}\mathbf{u}, \phi).$$

Consequently, Corollary 7.2.1 yields $\alpha_1 \le \mathbf{u}$. Similarly, we can prove $\mathbf{u} \le \beta_1$. Next we suppose that for some $j > 1$,

$$\alpha_0 \le \alpha_j \le \mathbf{u} \le \beta_j \le \beta_0.$$

Then utilizing the monotone nature of $\mathbf{f}$, $\mathbf{g}$, we get

$$
\begin{aligned}
\langle \alpha_{j+1}'', \phi \rangle + B[\alpha_{j+1}, \phi; t] &= (\mathbf{f}\alpha_j + \mathbf{g}\beta_j, \phi) \\
&\le (\mathbf{f}\mathbf{u} + \mathbf{g}\mathbf{u}, \phi)
\end{aligned}
$$

for each $\phi \in V_0 \cap L^2_+(Q)$. It then follows from Corollary 7.2.1 that $\alpha_{j+1} \leq \mathbf{u}$. In the same way, one can show that $\mathbf{u} \leq \beta_{j+1}$. Thus we arrive at

$$\alpha_0 \leq \alpha_j \leq \alpha_{j+1} \leq \beta_{j+1} \leq \beta_j \leq \beta_0.$$

Hence by induction $\alpha_k \leq \mathbf{u} \leq \beta_k$ for all k and as a result, taking the limit as $k \to \infty$, we get $\rho \leq \mathbf{u} \leq \mathbf{r}$, proving that $(\rho, \mathbf{r})$ are coupled weak minimal and maximal solutions of (7.2.1). The proof is now complete.

Corollary 7.3.1 *In addition to the assumptions of Theorem 7.3.1, suppose that f satisfies condition (7.2.6) and g satisfies*

$$g(x, t, u_1) - g(x, t, u_2) \geq -d(x, t)(u_1 - u_2), u_1 \geq u_2, \qquad (7.3.18)$$

for $u_1, u_2 \in R$, for some nonnegative $d \in L^\infty(Q)$. Then $\rho = \mathbf{r} = \mathbf{u}$ is the unique weak solution of (7.3.1).

Proof Since $\rho \leq \mathbf{r}$, we need only show $\mathbf{r} \leq \rho$. Setting $\mathbf{p} = \mathbf{r} - \rho$, we find using (7.2.6) and (7.3.18), $\mathbf{p}(0) = \mathbf{p}'(0) = 0$ and

$$\langle \mathbf{p}'', \phi \rangle + B_0[\mathbf{p}, \phi; t] \leq 0,$$

where $B_0[\mathbf{p}, \phi; t] = B[\mathbf{p}, \phi; t] - (k_0 + d_0)(\mathbf{p}, \phi)$, $k_0 = \|k\|_{L^\infty(Q)}$, $d_0 = \|d\|_{L^\infty(Q)}$, for each $\phi \in V_0 \cap L^2(Q)$. Corollary 7.2.1 gives immediately $\mathbf{r} \leq \rho$ and this completes the proof.

Theorem 7.3.1 includes several special cases of interest which we list below.

Remarks 7.3.1

(i) If $g(x, t, u) \equiv 0$ in Theorem 7.3.1, then we get the result where ρ, r are weak minimal and maximal solutions of (7.3.1).

(ii) If $f(x, t, u) \equiv 0$ in Theorem 7.3.1, we then have a result of (i) corresponding to the case when $g(x, t, u)$ is nonincreasing.

(iii) If $g(x, t, u) \equiv 0$ and $f(x, t, u)$ is not nondecreasing in u, but $\tilde{f}(x, t, u) = f(x, t, u) + d(x, t)u$ is nondecreasing in u, where $d(x, t) \geq 0$, $d \in L^\infty(Q) \times C^1[0, T]$, then we consider the IBVP

$$\begin{bmatrix} u_{tt} + Lu + d(x, t)u = \tilde{f}(x, t, u) & \text{in } Q \\ u = 0 \text{ on } \Gamma, u = 0, u_t = 0 & \text{on } \Omega \times \{t = 0\}. \end{bmatrix} \qquad (7.3.19)$$

It is easy to see that α_0, β_0 are weak lower and upper solutions of (7.3.19) and therefore Theorem 7.3.1 applied to (7.3.19) yields the same conclusion as in (i) for the IBVP (7.3.1).

(iv) Suppose that $f(x, t, u) \equiv 0$ and $g(x, t, u)$ is not nonincreasing in u but
$\tilde{g}(x, t, u) = g(x, t, u) - d(x, t)u$ is nonincreasing in u, where $d(x, t) \geq 0$
and $d \in L^\infty(Q) \times C^1[0, T]$. We then consider the IBVP

$$\left[\begin{array}{c} u_{tt} + Lu - d(x, t)u = \tilde{g}(x, t, u) \quad \text{in } Q, \\ u = 0 \text{ on } \Gamma, u = 0, u_t = 0 \text{ on } \Omega \times \{t = 0\}. \end{array} \right. \tag{7.3.20}$$

If (7.3.20) possesses α_0, β_0 as weak lower and upper solutions, then
Theorem 7.3.1 applied to (7.3.20) gives the same conclusion as in (ii)
for the IBVP (7.3.20). In addition, if $\tilde{g}(x, t, u)$ satisfies the condition
(7.3.18), then Theorem 7.3.1 applied the IBVP (7.3.20) yields the same
result for the IBVP (7.3.1).

(v) In Theorem 7.3.1, assume that $g(x, t, u)$ is nonincreasing in u, and
$f(x, t, u)$ is not nondecreasing in u but $\tilde{f}(x, t, u) = f(x, t, u) + d(x, t)u$
is nondecreasing in u, where d is a similar function as in (iii) and (iv).
Then we consider the IBVP

$$\left[\begin{array}{c} u_{tt} + Lu + d(x, t)u = \tilde{f}(x, t, u) + g(x, t, u) \quad \text{in } Q, \\ u = 0 \text{ on } \Gamma, u = 0, u_t = 0 \quad \text{on } \Omega \times \{t = 0\}. \end{array} \right. \tag{7.3.21}$$

The functions α_0, β_0 are clearly the coupled lower and upper solutions
of (7.3.21) and hence the conclusion of Theorem 7.3.1 applied to IBVP
(7.3.21) holds for the IBVP (7.3.1).

(vi) Suppose that, in Theorem 7.3.1, $f(x, t, u)$ is nondecreasing in u and
$g(x, t, u)$ is not nonincreasing in u but $\tilde{g}(x, t, u) = g(x, t, u) - d(x, t)u$
is nonincreasing in u with a similar d as in (iii). Then we consider

$$\left[\begin{array}{c} u_{tt} + Lu - du = f(x, t, u) + \tilde{g}(x, t, u) \quad \text{in } Q, \\ u = 0 \text{ on } \Gamma, u = 0, u_t = 0 \quad \text{on } \Omega \times \{t = 0\}. \end{array} \right. \tag{7.3.22}$$

If we assume that α_0, β_0 are coupled weak lower and upper solutions
of (7.3.22), then the conclusion of Theorem 7.3.1 applied to IBVP
(7.3.22) holds for the IBVP (7.3.22). Also, if the uniqueness conditions
are satisfied, then application of Theorem 7.3.1 to the IBVP (7.3.22)
gives the same conclusion for the IBVP (7.3.1).

(vii) Suppose that, in Theorem 7.3.1, $f(x, t, u)$ is not nondecreasing in u and
$g(x, t, u)$ is not nonincreasing in u, but $\tilde{f}(x, t, u) = f(x, t, u) + d_1(x, t)u$

is nondecreasing in u, and $\tilde{g}(x,t,u) = g(x,t,u) - d_2(x,t)u$ is nonincreasing in u, where $d_i \geq 0$, $d_i \in L^\infty(Q) \times C^1[0,T]$, $i = 1,2$. Then we consider the IBVP

$$
\left[
\begin{array}{c}
u_{tt} + Lu + (d_1 - d_2)u = \tilde{f}(x,t,u) + \tilde{g}(x,t,u) \quad \text{in } Q, \\
u = 0 \text{ on } \Gamma, u = 0, u_t = 0 \quad \text{on } \Omega \times \{t = 0\}.
\end{array}
\right.
$$
(7.3.23)

Assuming that α_0, β_0 are coupled lower and upper solutions of (7.3.23), we get the same conclusion of Theorem 7.3.1, which applied to IBVP (7.3.23), for the IBVP (7.3.23). Moreover, if the IBVP (7.3.23) satisfies the uniqueness conditions, then application of Theorem 7.3.1 to this problem gives the conclusion for the IBVP (7.3.1).

7.4 Generalized Quasilinearization

We shall investigate, in this section, the method of generalized quasilinearization for the second order hyperbolic IBVP (7.3.1) and prove a general result which covers several special cases of interest. We will continue to employ the same notation and necessary results of Section 7.2 to prove the following theorem. In the present case, however, we have to impose additional restrictions due to Theorem 7.2.2.

Theorem 7.4.1 *Assume that*

(A_1) *α_0, β_0 are weak lower and upper solutions of (7.3.1) such that $\alpha_0 \leq \beta_0$ and $\alpha_0, \beta_0 \in C^1[0,T]$;*

(A_2) *$f, g : Q \times R \to R$ are Caratheodory functions such that f_u, g_u, f_{uu}, g_{uu} exist and are Caratheodory functions, and $f_{uu} \geq 0$, $g_{uu} \leq 0$ for $u \in [\alpha_0(x,t), \beta_0(x,t)]$ for a.e. in Q, and in addition, f, g, f_u, g_u $\in C^1[0,T]$;*

(A_3) *There is some constant $c > 0$ such that*

$$
\|f_u(\cdot,\cdot,\eta), g_u(\cdot,\cdot,\eta), f_{uu}(\cdot,\cdot,\eta), g_{uu}(\cdot,\cdot,\eta)\|_{L^\infty(Q)} \leq c
$$

for any $\eta \in [\alpha_0, \beta_0]$.

Then there exist monotone sequences $\{\alpha_k\}$, $\{\beta_k\} \subset W_0$, $k > 1$, such that $\alpha_k \to \rho$, $\beta_k \to r$ in W_0 with $\rho = r = u$ is the unique solution of (7.3.1) satisfying $\alpha_0 \leq u \leq \beta_0$ and the convergence is quadratic.

Proof   We will prove the theorem in three steps.

(a) The iterates of generalized quasilinearization

We introduce the linearization of $f + g$ in the following form

$$G(x,t,u;\alpha,\beta) = f(x,t,\beta) + g(x,t,\beta) \qquad (7.4.1)$$
$$+ f_u(x,t,\alpha)(u-\beta) + g_u(x,t,\beta)(u-\beta),$$

and

$$F(x,t,u;\alpha,\beta) = f(x,t,\alpha) + g(x,t,\alpha) \qquad (7.4.2)$$
$$+ f_u(x,t,\alpha)(u-\alpha) + g_u(x,t,\beta)(u-\alpha),$$

and consider the related linear iteration schemes for $k = 0,1,2,\ldots$

$$\left[\begin{array}{l} \frac{\partial^2 \beta_{k+1}}{\partial t^2} + L\beta_{k+1} = G(x,t,\beta_{k+1};\alpha_k,\beta_k) \quad \text{in } Q, \\ \beta_{k+1} = 0 \text{ on } \Gamma, \beta_{k+1} = 0, \frac{\partial \beta_{k+1}}{\partial t} = 0 \quad \text{on } \Omega \times \{t=0\}, \end{array} \right. \qquad (7.4.3)$$

and

$$\left[\begin{array}{l} \frac{\partial^2 \alpha_{k+1}}{\partial t^2} + L\alpha_{k+1} = F(x,t,\alpha_{k+1};\alpha_k,\beta_k) \quad \text{in } Q, \\ \alpha_{k+1} = 0 \text{ on } \Gamma, \alpha_{k+1} = 0, \frac{\partial \alpha_{k+1}}{\partial t} = 0 \quad \text{on } \Omega \times \{t=0\}, \end{array} \right. \qquad (7.4.4)$$

whose variational forms associated with (7.4.3), (7.4.4) are given by

$$\langle \beta''_{k+1}, \phi \rangle + B[\beta_{k+1}, \phi] = \int_Q G(x,t,\beta_{k+1};\alpha_k,\beta_k)\phi\,dxdt \qquad (7.4.5)$$

and

$$\langle \alpha''_{k+1}, \phi \rangle + B[\alpha_{k+1}, \phi] = \int_Q F(x,t,\alpha_{k+1};\alpha_k,\beta_k)\phi\,dxdt \qquad (7.4.6)$$

for all $\phi \in V_0$. We wish to show that the weak solutions β_k, α_k of (7.4.3) and (7.4.4), respectively, are uniquely defined and satisfy

$$\alpha_0 \leq \alpha_1 \leq \ldots \leq \alpha_k \leq \beta_k \leq \ldots \leq \beta_1 \leq \beta_0. \qquad (7.4.7)$$

Let us consider $k = 0$. Then (7.4.1) shows that G is of the form

$$G(x,t,u;\alpha_0,\beta_0) = c_0(x,t)u + \tilde{h}(x,t),$$

where $c_0(x,t) = f_u(x,t,\alpha_0) + g_u(x,t,\beta_0)$, and

$$\tilde{h}(x,t) = f(x,t,\beta_0) + g(x,t,\beta_0) - f_u(x,t,\alpha_0)\beta_0 - g_u(x,t,\beta_0)\beta_0,$$

which implies by (A_2) that $c_0 \in L^\infty(Q) \times C^1[0,T]$ and $\tilde{h} \in L^2(Q)$. Thus by Theorem 7.2.2, there exists a unique solution $\beta_1 \in W_0$ of (7.4.3). In the same way, one can conclude that there exists a unique weak solution $\alpha_1 \in W_0$ of (7.4.4). We shall next prove that

$$\alpha_0 \leq \alpha_1 \leq \beta_1 \leq \beta_0. \tag{7.4.8}$$

Since α_0 is a lower solution of (7.3.1), we see that

$$\langle \alpha_0'', \phi \rangle + B[\alpha_0, \phi] \leq \int_Q F(x,t,\alpha_0;\alpha_0,\beta_0)\phi\,dx\,dt$$

for all $\phi \in V_0 \cap L_+^2(Q)$ and $\alpha_0(0) \leq 0$, $\alpha_0'(0) \leq 0$. By definition, the iterate α_1 is the unique weak solution of (7.4.6) with $k = 0$. Setting $\mathbf{p} = \alpha_0 - \alpha_1$, we have $\mathbf{p}(0) \leq 0$, $\mathbf{p}'(0) \leq 0$ and $\langle \mathbf{p}'', \phi \rangle + B[\mathbf{p}, \phi] -c_0(\mathbf{p}, \phi) \leq 0$ for all $\phi \in V_0 \cap L_+^2(Q)$. Since $c_0 \in L^\infty(Q) \times C^1[0,T]$, Corollary 7.2.1 implies $\mathbf{p} \leq 0$, that is $\alpha_0 \leq \alpha_1$. Similarly, we can show that $\beta_1 \leq \beta_0$. Next we show that $\alpha_1 \leq \beta_0$ and $\alpha_0 \leq \beta_1$. To prove $\alpha_1 \leq \beta_0$, we need to use the following inequalities, which are immediate consequences of $f_{uu}(x,t,u) \geq 0$ and $g_{uu}(x,t,u) \leq 0$ imposed in (A_2),

$$\begin{align}
f(x,t,u) &\geq f(x,t,v) + f_u(x,t,v)(u-v) \tag{7.4.9}\\
g(x,t,u) &\geq g(x,t,v) + g_u(x,t,u)(u-v), \tag{7.4.10}
\end{align}$$

for all $u, v \in [\alpha_0, \beta_0]$. By (7.4.9) and (7.4.10) we obtain for all $\phi \in$

$V_0 \cap L_+^2(Q)$,

$$\langle \alpha_1'', \phi \rangle + B[\alpha_1, \phi] \; = \; \int_Q F(x, t, \alpha_1; \alpha_0, \beta_0) \phi \, dx dt \qquad (7.4.11)$$

$$\leq \; \int_Q [f(x, t, \beta_0) - f_u(x, t, \alpha_0)(\beta_0 - \alpha_0)$$
$$+ g(x, t, \beta_0) - g_u(x, t, \beta_0)(\beta_0 - \alpha_0)$$
$$+ f_u(x, t, \alpha_0)(\alpha_1 - \alpha_0) + g_u(x, t, \beta_0)$$
$$\times (\alpha_1 - \alpha_0)] \phi \, dx dt$$

$$\leq \; \int_Q [f(x, t, \beta_0) + g(x, t, \beta_0) + f_u(x, t, \alpha_0)$$
$$\times (\alpha_1 - \beta_0)$$
$$+ g_u(x, t, \beta_0)(\alpha_1 - \beta_0)] \phi \, dx dt$$

$$= \; \int_Q G(x, t, \alpha_1; \alpha_0, \beta_0) \phi \, dx dt.$$

Since β_0 is a weak upper solution of (7.3.1), it satisfies $\beta_0(0) \geq 0$, $\beta_0'(0) \geq 0$ and

$$\langle \beta_0'', \phi \rangle + B[\beta_0, \phi] \geq \int_Q G(x, t, \beta_0; \alpha_0, \beta_0) \, dx dt, \qquad (7.4.12)$$

for all $\phi \in V_0 \cap L_+^2(Q)$. In view of the fact that

$$G(x, t, \alpha_1; \alpha_0, \beta_0) - G(x, t, \beta_0; \alpha_0, \beta_0)$$
$$= f_u(x, t, \alpha_0)(\alpha_1 - \beta_0) + g_u(x, t, \beta_0)(\alpha_1 - \beta_0),$$

it is easy to see that, by subtracting (7.4.12) from (7.4.11), $\mathbf{p} = \alpha_1 - \beta_0$ satisfies Corollary 7.2.1 which shows that $\alpha_1 \leq \beta_0$. Similar reasoning proves $\alpha_0 \leq \beta_1$. To prove $\alpha_1 \leq \beta_1$, we use (7.4.9) and (7.4.10) and the fact that $g_u(x, t, u)$ is nonincreasing in u and $\alpha_1 \leq \beta_0$ to obtain

$$\langle \alpha_1'', \phi \rangle + B[\alpha_1, \phi] \; = \; \int_Q F(x, t, \alpha_1; \alpha_0, \beta_0) \phi \, dx dt \qquad (7.4.13)$$

$$\leq \; \int_Q [f(x, t, \alpha_1) + g(x, t, \alpha_1) + (g_u(x, t, \beta_0)$$
$$- g_u(x, t, \alpha_1))(\alpha_1 - \alpha_0)] \phi \, dx dt$$

$$\leq \; \int_Q [f(x, t, \alpha_1) + g(x, t, \alpha_1)] \phi \, dx dt$$

$$= \; \int_Q F(x, t, \alpha_1; \alpha_1, \beta_1) \phi \, dx dt,$$

for each $\phi \in V_0 \cap L^2_+(Q)$. Similarly, for all $\phi \in V_0 \cap L^2_+(Q)$, we have

$$
\begin{aligned}
\langle \beta_1'', \phi \rangle + B[\beta_1, \phi] &= \int_Q G(x, t, \beta_1; \alpha_0, \beta_0) \phi \, dx \, dt \qquad (7.4.14) \\
&\geq \int_Q [f(x, t, \beta_1) + f_u(x, t, \beta_1)(\beta_0 - \beta_1) \\
&\quad + f_u(x, t, \alpha_0)(\beta_1 + \beta_0) + g(x, t, \beta_1)] \phi \, dx \, dt \\
&= \int_Q [f(x, t, \beta_1) + (-f_u(x, t, \beta_1) \\
&\quad + f_u(x, t, \alpha_0))(\beta_1 - \beta_0) + g(x, t, \beta_0)] \phi \, dx \, dt \\
&\geq \int_Q [f(x, t, \beta_1) + g(x, t, \beta_1)] \phi \, dx \, dt \\
&= \int_Q G(x, t, \beta_1) \phi \, dx \, dt,
\end{aligned}
$$

because of the nondecreasing nature of $f_u(x, t, u)$ in u and $\alpha_0 \leq \beta_1$. The inequalities (7.4.13) and (7.4.14) show that α_1 is a lower solution and β_1 is an upper solution of (7.3.1). Since $f(x, t, u) + g(x, t, u)$ is Lipschitzian in $u \in [\alpha_0, \beta_0]$, it follow from Theorem 7.2.1 that $\alpha_1 \leq \beta_1$ proving (7.4.8).

We shall next prove that if

$$
\alpha_{k+1} \leq \alpha_k \leq \beta_k \leq \beta_{k+1}, \qquad (7.4.15)
$$

for some $k > 1$, then it follows that

$$
\alpha_k \leq \alpha_{k+1} \leq \beta_{k+1} \leq \beta_k. \qquad (7.4.16)
$$

Since α_k satisfies

$$
\langle \alpha_k', \phi \rangle + B[\alpha_k, \phi] = \int_Q F(x, t, \alpha_k; \alpha_{k-1}, \beta_{k-1}) \phi \, dx \, dt
$$

for all $\phi \in V_0$, by following the reasoning employed to get (7.4.13) and (7.4.14), we can show, because of (7.4.15), that

$$
\langle \alpha_k'', \phi \rangle + B[\alpha_k, \phi] \leq \int_Q F(x, t, \alpha_k; \alpha_{k-1}, \beta_{k-1}) \phi \, dx \, dt, \qquad (7.4.17)
$$

and

$$
\langle \beta_k'', \phi \rangle + B[\beta_k, \phi] \geq \int_Q A(x, t, \beta_k; \alpha_k, \beta_k) \phi \, dx \, dt, \qquad (7.4.18)
$$

for all $\phi \in V_0 \cap L_+^2(Q)$. Moreover, α_{k+1}, β_{k+1} satisfy (7.4.4) and (7.4.3) respectively. We therefore obtain from Corollary 7.2.1, $\alpha_k \leq \alpha_{k+1}$ and $\beta_{k+1} \leq \beta_k$. We next show that $\alpha_{k+1} \leq \beta_k$ and $\alpha_k \leq \beta_{k+1}$. For this purpose, we find, utilizing (7.4.9) and (7.4.10), that

$$
\begin{aligned}
\langle \alpha_{k+1}'', \phi \rangle + B[\alpha_{k+1}, \phi] &= \int_Q F(x, t, \alpha_{k+1}; \alpha_k, \beta_k) \phi \, dx \, dt \\
&\leq \int_Q [f(x, t, \beta_k) - f_u(x, t, \alpha_0)(\beta_k - \alpha_k) \\
&\quad + g(x, t, \beta_k) - g_u(x, t, \beta_k)(\beta_k - \alpha_k) \\
&\quad + f_u(x, t, \alpha_k)(\alpha_{k+1} - \alpha_k) \\
&\quad + g_u(x, t, \beta_k)(\alpha_{k+1} - \alpha_k)] \phi \, dx \, dt \\
&\leq \int_Q [f(x, t, \beta_k) + g(x, t, \beta_k) + (f_u(x, t, \alpha_k) \\
&\quad + g_u(x, t, \beta_k))(\alpha_{k+1} - \beta_k)] \phi \, dx \, dt \\
&= \int_Q G(x, t, \alpha_{k+1}; \alpha_k, \beta_k) \phi \, dx \, dt,
\end{aligned}
$$

for each $\phi \in V_0 \cap L_+^2(Q)$. This inequality together with (7.4.18) yields by Corollary 7.2.1 that $\alpha_{k+1} \leq \beta_k$. In the same way, we can show that $\alpha_k \leq \beta_{k+1}$ using (7.4.17) and

$$
\begin{aligned}
\langle \beta_{k+1}'', \phi \rangle + B[\beta_{k+1}, \phi] &= \int_Q G(x, t, \beta_{k+1}; \alpha_k, \beta_k) \phi \, dx \, dt \\
&\geq \int_Q [f(x, t, \alpha_k) + f_u(x, t, \alpha_k)(\beta_k - \alpha_k) \\
&\quad + g(x, t, \alpha_k)(\beta_k - \alpha_k) \\
&\quad + f_u(x, t, \alpha_k)(\beta_{k+1} - \beta_k) \\
&\quad + g_u(x, t, \beta_k)(\beta_{k+1} - \beta_k)] \phi \, dx \, dt \\
&= \int_Q [f(x, t, \alpha_k) + g(x, t, \alpha_k) + (f_u(x, t, \alpha_k) \\
&\quad + g_u(x, t, \beta_k))(\beta_{k+1} - \alpha_k)] \phi \, dx \, dt \\
&= \int_Q F(x, t, \beta_{k+1}; \alpha_k, \beta_k) \phi \, dx \, dt,
\end{aligned}
$$

for each $\phi \in V_0 \cap L_+^2(Q)$. Finally, to show $\alpha_{k+1} \leq \beta_{k+1}$, we need to prove that α_{k+1} and β_{k+1} satisfy

$$
\langle \alpha_{k+1}'', \phi \rangle + B[\alpha_{k+1}, \phi] \leq \int_Q F(x, t, \alpha_{k+1}; \alpha_{k+1}, \beta_{k+1}) \phi \, dx \, dt,
$$

and

$$\langle \beta''_{k+1}, \phi \rangle + B[\beta_{k+1}, \phi] \geq \int_Q G(x, t, \beta_{k+1}; \alpha_{k+1}, \beta_{k+1}) \phi \, dx \, dt,$$

for each $\phi \in V_0 \cap L^2_+(Q)$. This precisely employs the same arguments as we utilized in proving (7.4.13) and (7.4.14) replacing α_0, β_0, α_1, β_1 by α_k, β_k, α_{k+1}, β_{k+1} and using (7.4.9) and (7.4.10) as well as the monotone character of f_u and g_u. Consequently, we get from Theorem 7.2.1 that $\alpha_{k+1} \leq \beta_{k+1}$, proving (7.4.16). Thus, by induction, it follows that (7.4.7) is true for each $k = 1, 2, 3, \ldots$.

(b) **Convergence of $\{\alpha_k\}$, $\{\beta_k\}$ to the unique solution of (7.3.1) in $[\alpha_0, \beta_0]$** By the monotonicity of iterates according to (7.4.7), there exist pointwise limits

$$\rho(x, t) = \lim_{k \to \infty} \alpha_k(x, t), r(x, t) = \lim_{k \to \infty} \beta_k(x, t), \text{ for } (x, t) \in Q, \text{ a.e.}$$

Furthermore, since α_k, $\beta_k \in [\alpha_0, \beta_0]$, Lebesgue's dominated convergence theorem shows that

$$\alpha_k \to \rho \text{ and } \beta_k \to \mathbf{r} \text{ in } L^2(Q). \tag{7.4.19}$$

Consider

$$\begin{aligned} G(x, t, \beta_{k+1}; \alpha_k, \beta_k) &= f(x, t, \beta_k) + g(x, t, \beta_k) \tag{7.4.20}\\ &+ (f_u(x, t, \alpha_k) + g_u(x, t, \beta_k))(\beta_{k+1} - \beta_k). \end{aligned}$$

By (A_3), we have for any $\eta \in [\alpha_0, \beta_0]$ and due to

$$f(x, t, \eta) = f(x, t, \alpha_0) + f_u(x, t, \xi)(\eta - \alpha_0)$$

where $\alpha_0 \leq \xi \leq \eta$, an estimate

$$\mid f(x, t, \eta) \mid \leq \mid f(x, t, \alpha_0) \mid + C \mid \beta_0 - \alpha_0 \mid, \eta \in [\alpha_0, \beta_0],$$

which implies that the Nemytski operator $\mathbf{f}$ related to f by $(\mathbf{fu})(x, t) = f(x, t, u(x, t))$ is continuous and bounded as a mapping $\mathbf{f} : [\alpha_0, \beta_0] \subset L^2(Q) \to L^2(Q)$. The same argument also yields that $\mathbf{g} : [\alpha_0, \beta_0] \subset L^2(Q) \to L^2(Q)$ is continuous and bounded, where $(\mathbf{gu})(x, t) =$

$g(x, t, u(x, t))$ is the Nemytski operator related to g. We therefore get by (7.4.19)

$$\mathbf{f}\beta_k \to \mathbf{fr}, \quad \mathbf{g}\beta_k \to \mathbf{gr}, \quad \text{in } L^2(Q) \text{ as } k \to \infty. \tag{7.4.21}$$

Using (7.4.21) and (A_3) we obtain for the functions $G(\cdot, \cdot, \beta_{k+1}; \alpha_k, \beta_k)$ given by (7.4.20)

$$G(\cdot, \cdot, \beta_{k+1}; \alpha_k, \beta_k) \to \mathbf{fr} + \mathbf{gr}, \quad \text{in } L^2(Q), \tag{7.4.22}$$

and similarly

$$F(\cdot, \cdot, \alpha_{k+1}; \alpha_k, \beta_k) \to \mathbf{f}\rho + \mathbf{g}\rho, \quad \text{in } L^2(Q). \tag{7.4.23}$$

Since α_k, β_k are solutions of the linear IBVP (7.4.4), (7.4.3) with homogeneous initial boundary values, we may apply the estimate (7.2.10) of Theorem 7.2.2, to get

$$\|\beta_k\|_W \leq C\|G(\cdot, \cdot, \beta_k; \alpha_{k-1}, \beta_{k-1})\|_{L^2(Q)}, \tag{7.4.24}$$

$$\|\alpha_k\|_W \leq C\|F(\cdot, \cdot, \alpha_k; \alpha_{k-1}, \beta_{k-1})\|_{L^2(Q)}. \tag{7.4.25}$$

Due to (7.4.22) and (7.4.23) the estimates (7.4.24) and (7.4.25) imply the strong convergence of the sequences $\{\alpha_k\}$, $\{\beta_k\}$ in W, whose limits must be ρ and $\mathbf{r}$ respectively, because of the compact embedding $W \subset L^2(Q)$. Since α_k, β_k, $k \geq 1$, belongs to the subset $M = [\mathbf{u} \in W : \mathbf{u}(0) = 0, \mathbf{u}'(0) = 0]$ which is closed in W, we find that the limits ρ, $\mathbf{r} \in M$, that is, the limits satisfy the homogeneous initial boundary values. The convergence result (7.4.22) and (7.4.23) together with

$$\alpha_k \to \rho, \quad \beta_k \to \mathbf{r} \text{ in } W \text{ as } k \to \infty,$$

permit us to pass to the limit in the corresponding variational forms (7.4.5) and (7.4.6) as $k \to \infty$, which yield

$$\langle \mathbf{r}'', \phi \rangle + B[\mathbf{r}, \phi] = \int_Q [f(x, t, r) + g(x, t, r)]\phi \, dx \, dt,$$

and

$$\langle \rho'', \phi \rangle + B[\rho, \phi] = \int_Q [f(x, t, \rho) + g(x, t, \rho)]\phi \, dx \, dt,$$

for all $\phi \in V_0$, showing $\mathbf{r}$, ρ are solutions of (7.3.1).

To show that $\rho = \mathbf{r} = \mathbf{u}$ is the unique weak solution of (7.3.1), we observe that, since $\rho \leq \mathbf{r}$, it is enough to show that $\mathbf{r} \leq \rho$. This follows immediately from Theorem 7.2.1, which proves that $\alpha_0 \leq \mathbf{u} \leq \beta_0$ is the unique weak solution of IBVP (7.3.1).

(c) **Quadratic convergence of** $\{\alpha_k\}$, $\{\beta_k\}$ To prove quadratic convergence of $\{\alpha_k\}$, $\{\beta_k\}$ to the unique weak solution u of (7.3.1), we get

$$\mathbf{p}_{k+1} = \mathbf{u} - \alpha_{k+1}, \qquad \mathbf{q}_{k+1} = \beta_{k+1} - \mathbf{u},$$

so that $\mathbf{p}_{k+1}(t) = 0$ on Γ, $\mathbf{p}_{k+1}(0) = 0$, $\mathbf{p}'_{k+1}(0) = 0$ on Ω and $\mathbf{q}_{k+1}(t) = 0$ on Γ, $\mathbf{q}_{k+1}(0) = 0$, $\mathbf{q}'_{k+1}(0) = 0$ on Ω. We then have, for $\phi \in V_0 \cap L^2_+(\Omega)$, and using the monotone nature of f_u and g_u,

$$\langle \mathbf{p}''_{k+1}, \phi \rangle + B[\mathbf{p}_{k+1}, \phi] \qquad\qquad (7.4.26)$$

$$= \int_Q [f(x,t,u) + g(x,t,u) - f(x,t,\alpha_k)$$
$$- f_u(x,t,\alpha_k)(\alpha_{k+1} - \alpha_k)$$
$$- g(x,t,\alpha_k) - g_u(x,t,\beta_k)$$
$$\times (\alpha_{k+1} - \alpha_k)]\phi dxdt$$

$$\leq \int_Q [f_u(x,t,u)p_k + g_u(x,t,\alpha_k)p_k$$
$$- f_u(x,t,\alpha_k)(p_k - p_{k-1})$$
$$- g_u(x,t,\beta_k)(p_k - p_{k-1})]\phi dxdt$$

$$= \int_Q [(f_u(x,t,u) - f_u(x,t,\alpha_k))p_k$$
$$- (g_u(x,t,\beta_k) - g_u(x,t,\alpha_k))p_k$$
$$+ (f_u(x,t,\alpha_k) + g_u(x,t,\beta_k))p_{k+1}]\phi dxdt$$

$$= \int_Q [f_{uu}(x,t,\xi)p_k^2 - q_{uu}(x,t,\sigma)(\beta_k - \alpha_1)p_k$$
$$+ (f_u(x,t,\alpha_k) + g_u(x,t,\beta_k))p_{k+1}]\phi dxdt,$$

where $\alpha_k \leq \xi \leq u$, $\alpha_k \leq \sigma \leq \beta_k$. The estimate (7.4.26) yields, by (A_3), the definition of p_k, q_k, and

$$-g_{uu}(x,t,\sigma)(\beta_k - \alpha_k)p_k \leq c(q_k + p_k)p_k \leq c(\frac{3}{2}p_k^2 + \frac{1}{2}q_k^2), \qquad (7.4.27)$$

$$\langle \mathbf{p}''_{k+1}, \phi \rangle + B[\mathbf{p}_{k+1}, \phi] \leq \int_Q c_0(x,t)p_{k+1}\phi dxdt + \int_Q (\frac{5}{2}cp_k^2 + \frac{1}{2}cq_k^2)\phi dxdt$$
$$(7.4.28)$$

for all $\phi \in V_0 \cap L^2_+(Q)$, where $c_0(x,t) = f_u(x,t,\alpha_k(x,t)) + g_u(x,t,\beta_k(x,t))$. The inequality (7.4.28) is equivalent to

$$\langle \mathbf{p}''_{k+1}, \phi \rangle + B_0[\mathbf{p}_{k+1}, \phi; t] \leq C(\mathbf{p}_k^2 + \mathbf{q}_k^2, \phi)$$

where $B_0[\mathbf{p}_{k+1}, \phi; t] = B[\mathbf{p}_{k+1}, \phi; t] - c_0(\mathbf{p}_{k+1}, \phi)$ and $C > 0$ is a suitable constant. Following the proof of Theorem 7.2.1, we arrive at

$$\|\mathbf{p}_{k+1}(s)\|^2_{L^2(\Omega)} + \|\mathbf{w}_{k+1}(s)\|^2_{H^1_0(\Omega)}$$

$$\leq C_1 \left[\int_0^s (\|\mathbf{p}_{k+1}\|^2_{L^2(\Omega)} + \|\mathbf{w}_{k+1}\|^2_{H^1_0(\Omega)})dt + c_1 T \|\tilde{\mathbf{h}}_k\|^2_{L^2(Q)} \right]$$

for $0 \leq s \leq T_1$, $\mathbf{w}_{k+1}(t) = \int_0^t \mathbf{p}_{k+1}(\sigma)d\sigma$, $\tilde{\mathbf{h}}_k = \mathbf{p}_k^2 + \mathbf{q}_k^2$ and $T_1 > 0$ is chosen such that $1 - 2T_1 c_1 \geq \frac{1}{2}$. Setting $\mathbf{y}(s) = \|\mathbf{p}_{k+1}(s)\|^2_{L^2(\Omega)} + \|\mathbf{w}_{k+1}(s)\|^2_{H^1_0(\Omega)}$, we have for $0 \leq s \leq T_1$

$$\mathbf{y}(s) \leq c_1 \left[\int_0^s \mathbf{y}(\sigma)d\sigma + T\|\tilde{\mathbf{h}}\|^2_{L^2(Q)} \right], \mathbf{y}(0) = 0.$$

This implies by Gronwall's inequality that

$$\mathbf{y}(s) \leq \left(\int_0^s e^{c_1(s-\sigma)}d\sigma \right) c_1 T \|\tilde{\mathbf{h}}_k\|^2_{L^2(Q)}.$$

Note that $\mathbf{y}(T_1) \leq T e^{c_1 T_1} \|\tilde{\mathbf{h}}_k\|^2_{L^2(Q)}$ and consequently, for $T_1 \leq s \leq 2T_2$, we get

$$\mathbf{y}(s) \leq \mathbf{y}(T_1)e^{c_1(s-T_1)} + \left(\int_{T_1}^s e^{c_1(s-\sigma)}d\sigma \right) c_1 T \|\tilde{\mathbf{h}}_k\|^2_{L^2(Q)}.$$

But $\mathbf{y}(2T_1) \leq k_1(t, c_1, T_1)\|\tilde{\mathbf{h}}_k\|^2_{L^2(Q)}$, where $k_1(T, c_1, T_1) = T(e^{2c_1 T_1} + e^{c_1 T_1})$, and therefore, for $2T_1 \leq s \leq 3T_2$, we get

$$\mathbf{y}(s) \leq \mathbf{y}(2T_1)e^{c_1(s-2T_1)} + \left(\int_{2T_1}^s e^{c_1(s-\sigma)}d\sigma \right) c_1 T \|\tilde{\mathbf{h}}_k\|^2_{L^2(Q)}.$$

Proceeding this way, and noting that $NT_1 \geq T$, we get finally,

$$\mathbf{y}(s) \leq k_N(T, T_1, c_1)\|\tilde{\mathbf{h}}_k\|^2_{L^2(Q)} \text{ for } (N-1)T_1 \leq s \leq NT_1 \leq T.$$

This yields, in view of the definition of $\mathbf{y}(s)$, the desired estimate

$$\|\mathbf{p}_{k+1}\|_{L^2(Q)} \leq k_N^{\frac{1}{2}}(T, T_1, c_1)[\|\mathbf{p}_k\|^2_{L^2(Q)} + \|\mathbf{q}_k\|^2_{L^2(Q)}].$$

One can get a similar estimate for $\mathbf{q}_{k+1}$. The proof is therefore complete.

The following special cases of Theorem 7.4.1 are of interest.

Remarks 7.4.1

(i) If $g(x, t, u) \equiv 0$ in (7.3.1), we get from Theorem 7.4.1 a result when $f(x, t, u)$ is convex in u.

(ii) If $f(x, t, u) \equiv 0$ in (7.3.1), Theorem 7.4.1 gives the dual result for the case when $g(x, t, u)$ is concave in u.

(iii) Consider the case when $g(x, t, u) \equiv 0$ and $f(x, t, u)$ is not convex in (7.3.1). Assume that $\tilde{f}(x, t, u) = f(x, t, u) + G(x, t, u)$ is convex, where $G(x, t, u)$ is a convex function in u. Then, it would be a special case of Theorem 7.4.1 since we can write IBVP (7.3.1) with f replaced by $\tilde{f}$ and g replaced by $\tilde{g} = -G$ and note the conditions of Theorem 7.4.1 are satisfied. Consequently we obtain the same conclusion even when f is not convex.

(iv) A dual situation of (iii) arises when $f(x, t, u) \equiv 0$ and $g(x, t, u)$ is not concave but $\tilde{g}(x, t, u) = g(x, t, u) + F(x, t, u)$ is concave with $F(x, t, u)$ being concave. We then write IBVP (7.3.1) with f and g replaced by $\tilde{f} = -F$ and $\tilde{g}$ respectively. Since all the assumptions of Theorem 7.4.1 hold, one gets the same conclusion even when $g(x, t, u)$ is not concave.

(v) Suppose that $g(x, t, u)$ is concave and $f(x, t, u)$ is not convex but $\tilde{f}(x, t, u) = f(x, t, u) + G(x, t, u)$ is convex with $G(x, t, u)$ convex. Then Theorem 7.4.1 gives the same conclusion by replacing f, g with $\tilde{f}$ and $\tilde{g} = g - G$.

(vi) A dual case of (v) is also true, that is $f(x, t, u)$ is convex but $g(x, t, u)$ is not concave and $\tilde{g} = g + F$ is concave with F concave. We then consider Theorem 7.4.1 with f and g replaced by $\tilde{f} = f - F$, $\tilde{g}$ to obtain the same conclusion of Theorem 7.4.1.

(vii) Suppose now that both f and g do not possess the required convex and concave properties in Theorem 7.4.1 for the IBVP (7.3.1), but $f + G$ is convex with G convex and $g + F$ is concave with F concave. Then we replace f, g in (7.3.1) by $\tilde{f} = f + g - F$, $\tilde{g} = g + F - G$ so that the required assumptions of Theorem 7.4.1 are satisfied and Theorem 7.4.1 yields the same result as before for the IBVP (7.3.1).

292 Chapter 7. Hyperbolic Equations

7.5   Notes and Comments

All the results of Sections 7.2, 7.3, and 7.4 related to second-order hyperbolic
IBVPs, starting with the appropriate notation, formulation, comparison results, and the methodologies of the monotone iterative technique as well as
generalized quasilinearization are taken from the work of Gnana Bhasker
et al. [7] and Gnana Bhaskar, Köksal and Lakshmikantham [8]. For the
existence and uniqueness result (Theorem 7.2.2) for linear second-order hyperbolic IBVPs, see Evans [32], Ladyzhenskaya [48], Lions [65], Lions and
Magenes [66] and Wloka [82].

Appendix A

A.1 Sobolev Spaces

Before going to Sobolev spaces, let us define simpler Hölder spaces. Assume that $\Omega \subset R^n$ is open and $0 < \gamma \leq 1$. A function $u : \Omega \to R$ satisfying

$$\mid u(x) - u(y) \mid \leq c \mid x - y \mid^\gamma, \quad x, y \in \Omega,$$

for some constant $c > 0$ is said to be Hölder continuous with exponent γ.

If $u : \Omega \to R$ is bounded and continuous, we write

$$\|u\|_{C(\bar\Omega)} = \sup_{x \in \Omega} \mid u(x) \mid .$$

The γth Hölder seminorm of $u : \Omega \to R$ is given by

$$[u]_{C^{0,\gamma}(\bar\Omega)} = \sup_{\substack{x,y \in \Omega \\ x \neq y}} \left[\frac{\mid u(x) - u(y) \mid}{\mid x - y \mid^\gamma} \right],$$

and the γth-Hölder norm is given by

$$\|u\|_{C^{0,\gamma}(\bar\Omega)} = \|u\|_{C(\bar\Omega)} + [u]_{C^{0,\gamma}(\bar\Omega)}.$$

The Hölder space $C^{k,\gamma}(\bar\Omega)$ consists of all functions $u \in C^k(\bar\Omega)$ for which the norm

$$\|u\|_{C^{k,\gamma}(\bar\Omega)} = \sum_{|\alpha| \leq k} \|D^\alpha u\|_{C(\bar\Omega)} + \sum_{|\alpha|=k} [D^\alpha u]_{C^{0,\gamma}(\bar\Omega)}$$

is finite. The space $C^{k,\gamma}(\bar\Omega)$ therefore consists of those functions u that are k-times continuously differentiable and whose kth partial derivatives are Hölder continuous with exponent γ. Such functions are well behaved and the space $C^{k,\gamma}(\bar\Omega)$ is a Banach space. The meaning of the multi-index used above is as follows: a vector $\alpha = (\alpha_1, \alpha_2, \ldots, \alpha_n)$, where each α_i is a nonnegative integer, is called a multi-index of order

$$\mid \alpha \mid = \sum_{i=1}^{n} \alpha_i.$$

Given a multi-index α, we define

$$D^{\alpha}u(x) = \frac{\partial^{|\alpha|}u(x)}{\partial x_1^{\alpha_1} \ldots \partial x_n^{\alpha_n}} = \partial_{x_1}^{\alpha_1}, \ldots, \partial_{x_n}^{\alpha_n} u.$$

If k is a nonnegative integer, $D^k u(x) = \{D^{\alpha}u(x) : \mid \alpha \mid = k\}$, the set of all partial derivatives of order k. Also, $\mid D^k u \mid = [\sum_{|\alpha|=k} \mid D^{\alpha}u \mid^2]^{\frac{1}{2}}$. If $k = 1$, we get the gradient vector $Du = (u_{x_1}, \ldots, u_{x_n})$, if $k = 2$, we find the Hessian matrix $D^2 u$, and the Laplacian of $u = \Delta = \sum_{i=1}^{n} u_{x_i x_j}$.

Since Hölder spaces are not suitable settings for the theory of elementary partial differential equations, we need spaces containing less smooth functions.

Suppose $u, v \in L^1_{\text{loc}}(\Omega)$ and α is the multi-index. Then v is said to be αth weak partial derivative of u and we write $D^{\alpha}u = v$, provided

$$\int_{\Omega} u D^{\alpha}\phi dx = (-1)^{|\alpha|} \int_{\Omega} v\phi dx$$

for all test functions $\phi \in C_c^{\infty}(\Omega)$, where $C_c^{\infty}(\Omega)$ denotes the space of infinitely differentiable functions with compact support in Ω. If $\phi \in C_c^{\infty}(\Omega)$, then ϕ is called a test function. A weak αth partial derivative of u, if it exists, is uniquely defined up to a set of measure zero.

Let $1 \leq p \leq \infty$ and let k be a nonnegative integer. The Sobolev space $W^{k,p}(\Omega)$ consists of all locally summable functions $u : \Omega \to R$ such that for each multi-index α with $\mid \alpha \mid \leq k$, $D^{\alpha}u$ exists in the weak sense and belongs to $L^p(\Omega)$. If $p = 2$, we write $H^k(\Omega) = W^{k,2}(\Omega)$, $k = 0, 1, \ldots$ and note that $H^0(\Omega) = L^2(\Omega)$.

For each $k = 1, 2, \ldots$ and $1 \leq p \leq \infty$, the Sobolev space $W^{k,p}(\Omega)$ is a Banach space and $H^k(\Omega)$ is a Hilbert space.

If $u \in W^{k,p}$, we define its norm to be

$$\|u\|_{W^{k,p}(\Omega)} = \begin{bmatrix} (\sum_{|\alpha|\leq k} \int_{\Omega} \mid D^{\alpha}u \mid^p)^{\frac{1}{p}}, & 1 \leq p < \infty, \\ \sum_{|\alpha|\leq k} \text{ess sup}_{\Omega} \mid D^{\alpha}u \mid, & p = \infty. \end{bmatrix}$$

We denote by $W_0^{k,p}(\Omega)$, the closure of $C_c^{\infty}(\Omega)$ into $W^{k,p}(\Omega)$. We interpret $W_0^{k,p}(\Omega)$ as comprising those functions $u \in W^{k,p}(\Omega)$ such that

$$D^{\alpha}u = 0 \text{ on } \partial\Omega \text{ for all } \mid \alpha \mid \leq k - 1.$$

Let $1 \leq p < \infty$ and assume that Ω is bounded and $\partial\Omega$ is C^1. Then there exists a bounded linear operator $T : W^{1,p}(\Omega) \to L^p(\partial\Omega)$ satisfying

(i) $Tu = u \mid_{\partial\Omega}$ if $u \in W^{1,p}(\Omega) \cap C(\bar{\Omega})$, and

(ii) $\|Tu\|_{L^p(\partial\Omega)} \leq C\|u\|_{W^{1,p}(\Omega)}$,

for each $u \in W^{1,p}(\Omega)$, where C depends on p and Ω. We call Tu the trace of u on $\partial\Omega$. Also,

$$u \in W_0^{1,p}(\Omega) \text{ if and only if } Tu = 0 \text{ on } \partial\Omega.$$

We usually set $H_0^k(\Omega) = W_0^{k,2}(\Omega)$.

The crucial analytic tools we need are Sobolev-type inequalities, which establish the estimates of arbitrary functions in the relevant Sobolev spaces. Let $1 \leq p < n$. Then the Sobolev conjugate of p is $p^* = \frac{np}{n-p}$. We note that $\frac{1}{p^*} = \frac{1}{p} - \frac{1}{n}$, $p^* > p$. Then we can write the Gagliardo–Nirenberg–Sobolev inequality, which says that there exists a constant c, depending only on p and n, such that

$$\|u\|_{L^{p^*}(R^n)} \leq C\|Du\|_{L^p(R^n)},$$

for all $u \in C_c^1(R^n)$.

Assume that Ω is a bounded open set in R^n and $u \in W_0^{1,p}(\Omega)$ for some $1 \leq p < n$. Then we have Poincaré's inequality

$$\|u\|_{L^q(\Omega)} \leq C\|Du\|_{L^p(\Omega)}$$

for each $q \in [1, p^*]$, C being a constant depending only on p, q, n and Ω.

The Friedrich inequality claims that for $u \in C_0^\infty(\Omega)$, the estimate

$$\int_\Omega |u(x)|^p \, dx \leq C \int_\Omega |Du(x)|^p dx$$

holds with $c > 0$ independent of u but depending on Ω. For Ω bounded, the expression

$$\|u\|_{W^{k,p}(\Omega)} = \left(\sum_{|\alpha|=k} \int_\Omega |D^\alpha u|^p \right)^{\frac{1}{p}}$$

is a norm on $W_0^{k,p}(\Omega)$ equivalent to $\|\cdot\|_{W^{k,p}(\Omega)}$, which is a consequence of Friedrich's inequality.

Let X, Y be Banach spaces such that $X \subset Y$. Then we say X is compactly embedded in Y, written as

$$X \subset\subset Y,$$

provided

(i) $\|x\|_Y \le C\|x\|_X$, $x \in X$ for some constant C, and

(ii) each bounded sequence in X is precompact in Y.

We now state the Rellich–Konrachov compactness theorem.

Theorem A.1.1 *Assume $\Omega \subset R^n$ is a bounded open set and $\partial\Omega \in C^1$. Suppose that $1 \le p < n$. Then $W^{1,p}(\Omega) \subset\subset L^q(\Omega)$ for each $1 \le q < p^*$.*

We denote by $H^{-1}(\Omega)$ the dual space of $H_0^1(\Omega)$, that is, f belongs to $H^{-1}(\Omega)$ provided f is a bounded linear functional on $H_0^1(\Omega)$. We write $\langle\ ,\ \rangle$ to denote the pairing between $H^{-1}(\Omega)$ and $H_0^1(\Omega)$. Also, if $f \in H^{-1}(\Omega)$, we define the norm

$$\|f\|_{H^{-1}(\Omega)} = \sup[\langle f, u \rangle : u \in H_0^1(\Omega), \|u\|_{H_0^1(\Omega)} \le 1].$$

We next consider the spaces involving time. The space $L^p(0, T; X)$ consists of all measurable functions $u : [0, T] \to X$ with

(i) $\|u\|_{L^p(0,T;X)} = \left[\int_0^T \|u(t)\|^p dt\right]^{\frac{1}{p}} < \infty$ for $1 \le p < \infty$, and

(ii) $\|u\|_{L^\infty(0,T;X)} = \operatorname{ess\,sup}_{0 \le t \le T} \|u(t)\| < \infty$.

If $u \in L^2(0, T; H_0^1(\Omega))$ with $u' \in L^2(0, T; H^{-1}(\Omega))$, then

(i) $u \in C([0, T]; L^2(\Omega))$;

(ii) the mapping $t \to \|u(t)\|_{L^2(\Omega)}^2$ is absolutely continuous with $\frac{d}{dt}\|u(t)\|_{L^2(\Omega)}^2$
$= 2\langle u'(t), u(t)\rangle$ for $0 \le t \le T$, a.e.

(iii) $\max_{0 \le t \le T} \|u(t)\|_{L^2(\Omega)} \le C\left[\|u\|_{L^2(0,T;H_0^1(\Omega))} + \|u'\|_{L^2(0,T;H^{-1}(\Omega))}\right]$, the constant C depending only on T.

Next we consider weighted Sobolev spaces. Let k be a nonnegative integer and $1 \le p < \infty$. Let w be a given family of weight functions w_α, $|\alpha| \le k$: $w = [w_\alpha(x) : x \in \Omega; |\alpha| \le k]$. We denote by $W^{k,p}(\Omega, w)$ the set of all functions $u \in L^p(\Omega, w_\theta)$, where θ is the zero multi-index $\theta = (0, 0, \ldots, 0)$ for which the weak derivatives $D^\alpha u$ with $|\alpha| \le k$ belong to $L^p(\Omega, w_\alpha)$. $W^{k,p}(\Omega, w)$ is a normed linear space if it is equipped with the norm

$$\|u\|_{W^{k,p}(\Omega,w)} = \left[\sum_{|\alpha|=k} \int_\Omega |D^\alpha u|_{w_\alpha}^p\, dx\right]^{\frac{1}{p}}.$$

If $1 < p < \infty$ and the weight functions satisfy $w_\alpha^{\frac{-1}{-1+p}} \in L_{loc}^1(\Omega)$, $\mid \alpha \mid \leq k$, then $W^{k,p}(\Omega, w)$ is a uniformly convex Banach space i.e. a reflexive Banach space. If, in addition, $w_\alpha \in L'_{loc}(\Omega)$, $\mid \alpha \mid \leq k$, then $C_0^\infty(\Omega)$ is a subset of $W^{k,p}(\Omega, w)$, and one can introduce the space $W_0^{k,p}(\Omega, w)$ as the closure of $C_0^\infty(\Omega)$ with respect to the norm $\| \cdot \|_{W^{k,p}(\Omega,w)}$. In the above discussion, the weighted Lebesgue space $L^p(\Omega, w)$, $1 \leq p < \infty$, is introduced, where $w = w(x)$ is a weight function, i.e. a function measurable and positive a.e. in Ω, that is, $L^p(\Omega, w) = [u = u(x) : uw^{\frac{1}{p}} \in L^p(\Omega)]$. It is a Banach space with the norm $\|u\|_{L^p(\Omega,w)} = [\int_\Omega \mid u(x) \mid^p w(x)dx]^{\frac{1}{p}}$.

Let us consider as an example, $W^{1,p}(\Omega, w)$ with a special choice of the family w, namely,

$$w_0(x) = 1, w_1(x) = w_2(x) = \ldots = w_n(x) \equiv w(x).$$

In this case, $\|u\|_{W^{1,p}(\Omega,w)} = [\int_\Omega \mid u(x) \mid^p dx + \int_\Omega \mid Du \mid^p w(x)dx]^{\frac{1}{p}}$. As before, we set

$$H^1(\Omega, w) = W^{1,2}(\Omega, w) \text{ and } H_0^1(\Omega, w) = W_0^{1,2}(\Omega, w),$$

which are weighted Hilbert spaces.

For more details on Sobolev spaces including weighted Sobolev spaces, refer to Adams [1], Evans [32], DiBenedetto [25], Kufner [45], Drábek, Kufner and Nicolosi [27], Murthy and Stampachia [71], and Zeidler [83].

A.2 Elliptic Equations

Consider the linear second-order elliptic BVP

$$\left[\begin{array}{ll} Lu = F(x), & x \in \Omega, \\ (Bu)(x) = \phi(x), & x \in \partial\Omega. \end{array} \right. \tag{A.2.1}$$

where L and B are as defined in (1.2.1) and (1.2.3) respectively. Concerning the BVP (A.3.1), we have the following classical existence and uniqueness result whose proof may be found in [65] and [50].

Theorem A.2.1 *Assume that*

(i) *a_{ij}, b_i, c and $F \in C^\alpha[\bar{\Omega}, R]$, $c(x) \geq 0$ and L is a strictly uniformly elliptic operator in Ω;*

(ii) *$p, q \in C^{1,\alpha}[\partial\Omega, R]$, p and q are nonnegative functions which do not vanish simultaneously;*

(iii) $\partial\Omega$ belongs to class $C^{2,\alpha}$;

(iv) $\phi \in C^{1,\alpha}[\partial\Omega, R]$.

Then the linear elliptic BVP (A.2.1) has a unique solution u such that $u \in C^{2,\alpha}[\bar\Omega, R]$.

Definition A.2.1 Let Ω be an m-dimensional domain with boundary $\partial\Omega$. We say that $\partial\Omega$ belongs to class $C^{2,\alpha}$, if for every $x \in \partial\Omega$, there exists a neighborhood U of x such that $\partial\Omega \cap U$ can be represented in the form

$$x_i = h(x_1, x_2, \ldots, x_{i-1}, x_{i+1}, \ldots, x_m),$$

for some i, $1 \le i \le m$, where $h \in C^{2,\alpha}[\partial\Omega, R]$.

The next result provides the global *a priori* Schauder estimate for classical solutions of (A.2.1).

Theorem A.2.2 Let the hypotheses of Theorem A.2.1 be satisfied. Then for any $u \in C^{2,\alpha}(\bar\Omega, R]$, there exists a constant

$$C = C(\Omega, \alpha, K, \|p\|_\alpha^{\partial\Omega}, \|v\|_1^{\partial\Omega}, \max_{\substack{1\le i\le m \\ 1\le j\le m}} \|a_{ij}\|_\alpha^{\bar\Omega}, m)$$

such that

$$\|u\|_{2+\alpha}^{\bar\Omega} \le C(\|Bu\|_{1+\alpha}^{\partial\Omega} + \|Lu\|_\alpha^{\bar\Omega}). \tag{A.2.2}$$

Moreover, if u is a classical solution of (A.2.1), then (A.2.2) reduces to

$$\|u\|_{2+\alpha}^{\bar\Omega} \le C(\|\phi\|_{1+\alpha}^{\partial\Omega} + \|F\|_\alpha^{\bar\Omega}). \tag{A.2.3}$$

The following result gives the global *a priori* Agmon–Douglas–Nirenberg estimates for generalized solutions of (A.2.1).

Theorem A.2.3 Assume that

(i) a_{ij}, b_i and $c \in C[\bar\Omega, R]$, $c(x) \ge 0$ and L is a strictly uniformly elliptic operator in Ω;

(ii) $p, q \in C^1[\bar\Omega, R]$, p and q are nonnegative functions which do not vanish simultaneously;

(iii) $\partial\Omega$ belongs to class C^2;

(iv) $F \in L^q(\bar\Omega, R]$;

(v) $\phi \in C^1[\bar{\Omega}, R]$.

Then for any u in $W^{2,q}[\bar{\Omega}, R]$, there exists a constant $C = C(m, K, q, \partial\Omega)$, the modulus of continuity of a_{ij} and norms b_i and C such that

$$\|u\|_{W^{2,q}[\bar{\Omega}, R]} \leq C(\|Lu\|_{L^q[\bar{\Omega}, R]} + \|Bu\|_{W^{1,q}[\partial\Omega, R]}). \tag{A.2.4}$$

Moreover, if u is a generalized solution belonging to $W^{2,q}[\bar{\Omega}, R]$, then

$$\|u\|_{W^{2,q}[\bar{\Omega}, R]} = (\|u\|_{L^q[\bar{\Omega}, R]} + \|\phi\|_{W^{1,q}[\bar{\Omega}, R]}), \tag{A.2.5}$$

where

$$\|u\|_{W^{2,q}[\bar{\Omega}, R]} = (\|u\|_{L^q[\bar{\Omega}, R]} + \|Du\|_{L^q[\bar{\Omega}, R]} + \|D^2u\|_{L^q[\bar{\Omega}, R]}),$$

and

$$\|D^iu\|_{L^q[\bar{\Omega}, R]} = \left(\sum_{j=1} \int_\Omega |D^ju|^q \, dx \right)^{\frac{1}{q}}.$$

Definition A.2.2 *Let X and Y be normed spaces. We say that the normed space X is embedded in the normed space Y iff*

(i) *X is a vector subspace of Y, and*

(ii) *the identity operator e defined on X into Y by $ex = x$ for all $x \in X$ is continuous.*

We shall next state the following embedding theorem which is essential to our discussion.

Theorem A.2.4 (Embedding Theorem). *Let $\Omega \subseteq R^m$ and let $\partial\Omega$ be of class C^2.*

(i) *Suppose $2q > m > q$. Then*

$$W^{2,q}[\bar{\Omega}, R] \quad \text{is embedded in} \quad C^{1,\mu}[\bar{\Omega},], \tag{A.2.6}$$

where $0 < \mu \leq \frac{1-2}{q}$.

(ii) *Suppose $m = q$. Then*

$$W^{2,q}[\bar{\Omega}, R] \quad \text{is embedded in} \quad C^{1,\mu}[\bar{\Omega}, R], \tag{A.2.7}$$

where $0 < \mu < 1$; moreover, if $m = q = 1$, then (A.2.6) holds for $\mu = 1$.

For the elliptic operator L, a general eigenvalue problem is given in the form

$$(L + \lambda r)u = 0 \text{ in } \Omega, \; Bu = 0 \text{ on } \partial\Omega, \qquad (A.2.8)$$

where $r = r(x) > 0$ is a continuous function in Ω. The following result concerning the eigenvalue problem (A.2.8) is well known.

Theorem A.2.5 *The principal eigenvalue $\lambda(r)$ of (A.2.8) is real, nonnegative and the corresponding eigenfunction ϕ is positive in Ω. Furthermore, $\lambda(r) > 0$ when c and p are not both identically zero, and $\phi > 0$ in Ω, when $q > 0$.*

The principal eigenvalue $\lambda(r)$ in Theorem A.2.5 is in the sense that no other eigenvalue has a smaller real part. When $\lambda(r)$ is real, it is the smaller eigenvalue (A.2.8). In the special case $c = 0$, the principal eigenvalue is denoted by λ_0. Usually, the eigenfunction ϕ is normalized so that $\max_\Omega \phi(x) = 1$. Of special interest in applications is the eigenvalue problem

$$(L + \lambda)u = 0 \text{ in } \Omega, \; Bu = 0 \text{ on } \partial\Omega, \qquad (A.2.9)$$

with $c = 0$ and $r = 1$. When L and B are self-adjoint operators, all the eigenvalues λ_i of (A.2.8) are real and the eigenfunctions ϕ_i corresponding to distinct eigenvalues are orthogonal relative to the weight function $r(x)$, that is $\int_\Omega r(x)\phi_i(x) \, \phi_j(x)dx = 0$, when $\lambda_i \neq \lambda_j$. This implies that the principal eigenfunction ϕ is the only positive eigenfunction on Ω.

If in (A.2.9), $L = \nabla^2$ is the Laplacian operator, there exist a countable number of real eigenvalues $0 \leq \lambda_0 \leq \lambda_1 \leq \ldots \leq \lambda_n \leq \ldots$, and the corresponding eigenfunctions $\phi_0, \phi_1, \phi_2, \ldots, \phi_n, \ldots$, satisfy the orthogonality condition $\int_\Omega \phi_i(x) \, \phi_j(x)dx = 0$ when $\lambda_i \neq \lambda_j$. The eigenfunction ϕ_0 possesses the sign in Ω and the one that is positive in Ω is normalized as before. In particular, $\phi(x) = 1$ and $\lambda_0 = 0$ when $p = 0$.

For more details, see Adams [1], Gilberg and Trudinger [34], Ladyzhenskaya and Uralćeva [50], Evans [32], DiBenedetto [25], Ladde, et al. [47], and Pao [74].

A.3   Parabolic Equations

Let $\mathcal{L}$ be a second-order differential operator defined by

$$\mathcal{L} = \frac{\partial}{\partial t} + L,$$

where

$$L = - \sum_{i,j=1}^{n} a_{ij}(t,x) \frac{\partial^2}{\partial x_i \partial x_j} + \sum_{i=1}^{m} b_i(t,x) \frac{\partial}{\partial x_i} + c(t,x).$$

Consider the linear second-order parabolic initial boundary value problem (IBVP)

$$\left[\begin{array}{ll} \mathcal{L}u = F(t,x), & (t,x) \in Q_T \\ (Bu)(t,x) = \phi(t,x), & (t,x) \in \Gamma_T, \\ u(0,x) = \phi_0(x), & x \in \bar{\Omega}, \end{array} \right. \qquad (A.3.1)$$

where B is defined by $Bu = p(t,x)u + q(t,x)\frac{du}{dv}$ and $\frac{du}{dv}$ stands for the normal derivative of u; Ω is a bounded domain; $Q_T = (0,T) \times \Omega$; $\Gamma_T = (0,T) \times \partial\Omega$; and $D_x u = (\frac{\partial u}{\partial x_1}, \dots, \frac{\partial u}{\partial x_m})$.

Let us now state the classical existence and uniqueness theorem.

Theorem A.3.1 *Assume that*

(a_1) a_{ij}, b_i, c *and* $F \in C^{\frac{\alpha}{2},\alpha}[\bar{\Omega}_T, R]$, $c(t,x) \geq 0$ *and* $\mathcal{L}$ *is strictly uniformly parabolic in* Q_T;

(a_2) $p, q \in C^{\frac{(1+\alpha)}{2},1+\alpha}[\bar{\Gamma}_T, R]$, p *and* q *are nonnegative functions which do not vanish simultaneously;*

(a_3) $\partial\Omega$ *belongs to class* $C^{2,\alpha}$;

(a_4) $\phi \in C^{\frac{(1+\alpha)}{2},1+\alpha}[\bar{\Gamma}_T, R]$ *and* $\phi_0 \in C^{2+\alpha}[\bar{\Omega}, R]$;

(a_5) *the IBVP (A.3.1) satisfies the compatibility condition of order* $[\frac{1+\alpha}{2}]$.

Then the linear parabolic IBVP (A.3.1) has a unique solution u such that $u \in C^{\frac{1+\alpha}{2},2+\alpha}[\bar{Q}_T, R]$.

Definition A.3.1 *We say that the compatibility conditions of order $k \geq 0$ are fulfilled for IBVP (A.3.1) if*

$$\sum_{i=0}^{j} \binom{j}{i} \left(p^{(j-i)}(x)u^{(i)}(x) + q^{(j-i)}(x)\frac{du^{(i)}(x)}{d\gamma} \right) = \phi^{(j)}(x) \quad on \quad \partial\Omega,$$

for $j = 0, 1, \dots, k$, *where* $u^{(j)}(x) = u^{(j)}(0,x) = \frac{\partial^j u(t,x)}{\partial t^j}|_{t=0}$,

$$\phi^{(j)}(x) = \phi^{(j)}(0,x) = \frac{\partial^j \phi(t,x)}{\partial t^j}|_{t=0},$$

and

$$\sum_{i=0}^{j} \binom{j}{i} \left(p^{(j-1)}(0,x)u^{(j)}(0,x) + q^{(j-i)}(0,x)\frac{du^{(j)}(0,x)}{d\gamma} \right)$$

$$= \frac{\partial^j}{\partial t^j} \left(p(t,x)u(t,x) + q(t,x)\frac{du(t,x)}{d\gamma} \right)|_{t=0}.$$

The following result provides global *a priori* Schauder-type estimates for classical solutions of (A.3.1).

Theorem A.3.2 *Assume that the hypotheses of Theorem A.3.1 hold. Then for any* $u \in C^{\frac{1+\alpha}{2},2+\alpha}[\bar{Q}_T, R]$, *there exists a positive constant* C *which is independent of* u *such that*

$$\|u\|_{2+\alpha}^{\bar{Q}_T} \le C(\|Bu\|_{1+\alpha}^{\bar{\Gamma}} + \|\mathcal{L}u\|_{\alpha}^{\bar{Q}_T} + \|\phi_0\|_{2+\alpha}^{\bar{\Omega}}). \qquad (A.3.2)$$

Moreover, if u *is the classical solution of (A.3.1), then (A.3.2) reduces to*

$$\|u\|_{2+\alpha}^{\bar{Q}_T} \le C(\|F\|_{\alpha}^{\bar{Q}_T} + \|\phi\|_{1+\alpha}^{\bar{\Gamma}_T} + \|\phi_0\|_{2+\alpha}^{\bar{\Omega}}). \qquad (A.3.3)$$

We need the following definitions and notation which will be used subsequently. $L^{r,q}[\bar{Q}_T, R]$ is the Banach space consisting of all equivalence classes of Lebesgue measurable functions u defined on $\bar{Q}_T$ into R with a finite norm

$$\|u\|_{L^{r,q}[\bar{Q}_T,R]} = \left(\int_0^T \left[\int_{\bar{\Omega}} |u(t,x)|^1 \, dx \right]^{\frac{r}{q}} \right)^{\frac{1}{r}},$$

where $q \ge$ and $r \ge 1$. If $r = q$, then $L^{q,q}[\bar{Q}_T, R]$ is denoted by $L^q[\bar{Q}_T, R]$ and the norm $\|u\|_{L^{q,q}[\bar{Q}_T,R]}$ by $\|u\|_{L^q[\bar{Q}_T,R]}$. For a nonnegative integer l, $W_q^{l,2l}[\bar{Q}_T, R]$ is the Banach space consisting of the elements $L^q[\bar{Q}_T, R]$ having generalized (weak or distributional) derivatives of the form $D_t^r D_x^s$ with any r and s satisfying the inequality $2r + s \le 2l$. The norm in it is defined by

$$\|u\|_{W^{l,2l}[\bar{Q}_T,R]} = \sum_{j=0}^{2l} \sum_{2r+s=j} \|D_t^r D_x^s u\|_{L^q(\bar{Q}_T,R]}, \qquad (A.3.4)$$

where the summation $\sum_{2r+s=j}$ is taken over all nonnegative integers r and s satisfying the condition $2r + s \le 2l$. For nonintegral l, $W_q^l[\bar{\Omega}, R]$ is a Banach space consisting of the elements u of $W_q^{[l]}[\bar{\Omega}, R]$ with finite norm

$$\|u\|_{W_q^l[\bar{\Omega},R]} = \|u\|_{W_q^{[l]}[\bar{\Omega},R]} + \|u\|_{L_q^{l-[l]}[\bar{\Omega},R]} \qquad (A.3.5)$$

where

$$\|u\|_{W_q^{[l]}[\bar{\Omega},R]} = \sum_{j=0}^{[l]} \sum_{j} \|D_x^j u\|_{L^q[\bar{\Omega},R]};$$

$$\|u\|_{L_q^{l-[l]}[\bar{\Omega},R]} = \sum_{j=[l]} \left(\int_{\bar{\Omega}} dx \int_{\bar{\Omega}} | D_x^j u(x) - D_y^j u(y) |^q \frac{dy}{\|x - y\|^{n+q(l-[l])}} \right)^{\frac{1}{q}},$$

$\sum_j$ denotes summation over all possible derivatives of u of order j satisfying the condition $j \le [l]$; $\sum_{j=[l]}$ denote summation over all possible derivatives of u of order $j = [l]$. For nonintegral l, the Banach space $W_q^{\frac{1}{2},l}[\bar{\Gamma}_T,R]$ is defined analogously. These spaces are needed to obtain the existence result on the solvability of boundary value problems for linear parabolic equations in spaces $W_q^{1,2}[\bar{Q}_T,R]$. This is related to the fact that the differential properties of the bounded values of functions from classes $W_q^{1,2}[\bar{Q}_T,R]$ and of their derivatives can be exactly described in terms of the spaces $W_q^{\frac{l}{2},l}[\bar{\Gamma}_T,R]$ with nonintegral $l : l = k - \frac{1}{q}$, where k is an integer. This relationship is given by the following result.

Lemma A.3.1 *If $u \in W_q^{1,2}[\bar{Q}_T,R]$, then for all nonnegative integers r and s, $2r + s < 2 - \frac{2}{q}$, $D_t^r D_x^s u \mid_{t=0} \in W_q^{2-2r-s-\frac{2}{q}}[\bar{\Omega},R]$ and*

$$\|u\|_{W^{2-2r-s-\frac{2}{q}}[\bar{\Omega},R]} \le C\|u\|_{W_q^{1,2}[\bar{Q}_T,R]}.$$

Moreover, for $2r + s < 2 - \frac{1}{q}$

$$D_t^r D_x^s u \mid_{\bar{\gamma}_T} \in W^{1-r-\frac{s}{2}-\frac{1}{2}q,2-2r-s-\frac{1}{q}}[\bar{\Gamma}_T,R]$$

and

$$\|u\|_{W^{1-r-\frac{s}{2}-\frac{1}{2q},2-2r-s-\frac{1}{q}}[\bar{\Gamma}_T,R]} \le C\|u\|_{W_q^{1,2}[\bar{Q}_T,R]}.$$

The next result provides the global *a priori* Agmon–Douglas–Nirenberg type of estimates for generalized (weak) solutions of (A.3.1).

Theorem A.3.3 *Assume that*

(i) *a_{ij}, b_i, and $c \in C^{\frac{\alpha}{2},\alpha}[\bar{Q}_T,R]$, $c(t,x) \ge 0$ and $\mathcal{L}$ is strictly uniformly parabolic in Q_T;*

(ii) *$p, q \in C^{\frac{(1+\alpha)}{2},1+\alpha}[\bar{\Gamma}_T,R]$, p and q are nonnegative functions which do not vanish simultaneously;*

(iii) $\partial\Omega$ *belongs to class* $C^{2+\alpha}$;

(iv) $F \in L^q[\bar{Q}_T, R]$ *for* $q > 1$;

(v) $\phi \in W^{\frac{1}{2}-\frac{1}{2q},1-\frac{1}{q}}[\bar{\Gamma}_T, R]$ *and* $\phi_0 \in W_q^{2-\frac{2}{q}}[\bar{\Omega}, R]$;

(vi) *the IBVP* (A.3.1) *satisfies the compatibility condition of order* $[\frac{(q-3)}{2q}]$.

Then for any $u \in W_q^{1,2}(\bar{Q}_T, R]$, *there exists a constant* C *which is independent of* u *such that*

$$\|u\|_{W_q^{1,2}[\bar{Q}_T,R]} \;\leq\; C(\|\mathcal{L}u\|_{L^q[\bar{Q}_T,R]} + \|Bu\|_{W_q^{\frac{1}{2}-\frac{1}{2q},1-\frac{1}{q}}[\bar{\Gamma}_T,R]} \qquad (A.3.6)$$
$$+ \|\phi_0\|_{W^{2-\frac{2}{q}}[\bar{\Omega},R]}).$$

Moreover, if u *is a generalized solution belonging to* $W_q^{1,2}[\bar{Q}_T, R]$, *then*

$$\|u\|_{W_q^{1,2}[\bar{Q}_T,R]} \;\leq\; C(\|F\|_{L^q[\bar{Q}_T,R]} + \|\phi\|_{W^{\frac{1}{2}-\frac{1}{2q},1-\frac{1}{q}}[\bar{\Gamma}_T,R]} \qquad (A.3.7)$$
$$+ \|\phi_0\|_{W^{2-\frac{2}{q}}[\bar{\Omega},R]}).$$

Theorem A.3.4 (Embedding Theorem.) *Let* $\partial\Omega$ *be of class* $C^{2+\alpha}$. *Suppose that* $q = \frac{(m+2)}{(1-\alpha)}$ *for* $0 < \alpha < 1$. *Then* $W_q^{1,2}[\bar{Q}_T, R]$ *is embedded in* $C^{\frac{(1+\alpha)}{2},1+\alpha}[\bar{Q}_T, R]$.

The standard result concerning the asymptotic behavior of solutions of the linear IBVP (A.3.1) is as follows.

Theorem A.3.5 *Let* $p(t,x) > 0$, $c(t,x) \geq 0$ *and either* p *or* c *be strictly positive. Suppose further that* $F(t,x)$, $\phi(t,x)$ *converge to zero uniformly in* $\bar{\Omega}$ *and on* $\partial\Omega$, *respectively as* $x \to \infty$. *Then for any initial function* $\phi_0(x)$, *the solution* $u(t,x)$ *of (A.3.1) converges uniformly in* $\bar{\Omega}$ *to zero as* $t \to \infty$. *The convergence of* u *to zero is in* $L^2(\Omega)$, *if* $F(t,\cdot)$, $\phi(t,\cdot)$ *converge to zero in* $L^2(\Omega)$ *and* $L^2(\partial\Omega)$ *respectively, as* $t \to \infty$.

For more details, see Ladyzhenskaya et al. [49], Friedman [33], DiBenedetto [25], Ladde et al. [47], and Pao [74].

A.4   Impulsive Differential Equations

In Section 3.2, we get the bounds for the solutions of impulsive parabolic IBVPs by means of the maximal solution of ordinary impulsive differential equations. We give a comparison result for impulsive differential equations.

Theorem A.4.1 *Assume that*

(A_0) *the sequence $\{t_k\}$ satisfies $0 \leq t_0 < t_1 < t_2 < \cdots$ with $t_k = \infty$ as $k \to \infty$;*

(A_1) *$m \in PC^1[R_+, R]$ and $m(t)$ is left continuous at t_k, $k = 1, 2, \ldots$;*

(A_2) *$g \in C[R_+ \times R, R]$, $\psi_k : R \to R$, $\psi_k(u)$ is nondecreasing in u and for each $k = 1, 2, \ldots$*

$$D_- m(t) \leq g(t, m(t)), t \neq t_k, m(t_0) \leq u_0,$$

$$m(t_k^+) \leq \psi_k(m(t_k));$$

(A_3) *$r(t)$ is the maximal solution of*

$$u' = g(t, u), \ t \neq t_k, \ u(t_0) = u_0,$$

$$u(t_k^+) = \psi_k(u(t_k)), \ t_k > t_0 \geq 0,$$

existing on $[t_0, \infty)$.

Then $m(t) \leq r(t)$, $t \geq t_0$.

As an example, let $g(t, u) = p(t)u + q(t)$ $t \neq t_k$, $\psi_k(u) = d_k u + b_k$ where $d_k \geq 0$ and b_k are constants. Then we have

$$
\begin{aligned}
m(t) \ \leq \ & m(t_0) \prod_{t_0 < t_k < t} d_k \exp\left(\int_{t_0}^t p(s)ds\right) \\
& + \sum_{t_0 < t_k < t}\left(\prod_{t_k < t_j < t} d_j \exp\left(\int_{t_k}^t p(s)ds\right) b_k\right. \\
& + \int_{t_0}^t \prod_{s < t_k < t} d_k \exp\left(\int_s^t p(\sigma)d\sigma\right) q(s)ds, \ t \\
\geq \ & t_0.
\end{aligned}
$$

See Lakshmikantham, Bainov and Simeonov [52] for more information on impulsive differential equations.

A.5   Hyperbolic Equations

For $a, b \in R$, $a > 0$, $b > 0$, let I_a, I_b denote the intervals $[0, a]$ and $[0, b]$ respectively, and I_{ab} the rectangle $I_a \times I_b$. By $v \in C^{1,2}[I_{ab}, R]$, we mean that v is a continuous function on I_{ab} and its partial derivatives v_x, v_y and v_{xy} exist and are continuous on I_{ab}.

For two functions $v, w \in C^{1,2}[I_{ab}, R]$, the 'sector' is defined by

$$\langle v, w \rangle = \{ z \in C^{1,2}[I_{ab}, R] : v \leq z \leq w, (x, y) \in I_{ab} \}.$$

The notation $(v, v_x, v_y) \leq (w, w_x, w_y)$ means that $v(x, y) \leq w(x, y)$, $v_x(x, y) \leq w_x(x, y)$ and $v_y(x, y) \leq w_y(x, y)$.

Consider the IBVP

$$\left[\begin{array}{ll} u_{xy} = f(x, y, u, u_x, u_y), & (x, y) \in I_{ab}, \\ u(x, 0) = \sigma(x), & x \in I_a \\ u(0, y) = \tau(y), & y \in I_b \\ \sigma(0) = \tau(0) = u_0, & \end{array} \right. \tag{A.5.1}$$

where $f \in C[I_{ab} \times R^3, R], \sigma \in C^1[I_a, R]$ and $\tau \in C^1[I_b, R]$.

We have immediately the following result.

Theorem A.5.1 *Suppose that f in (A.5.1) is bounded everywhere. Then there exists a solution $u(x, y)$ for the IVP (A.5.1) on I_{ab}.*

Definition A.5.1 *A function $z \in C^{1,2}[I_{ab}, R]$ is said to be an upper solution of (A.5.1) on I_{ab} if*

$$\left[\begin{array}{ll} z_{xy} \geq f(x, y, z, z_x, z_y), & (x, y) \in I_{ab} \\ z_x(x, 0) \geq \sigma'(x), & x \in I_a \\ z_y(0, y) \geq \tau'(y), & y \in I_b \\ z(0, 0) \geq u_0, & \end{array} \right.$$

and, a lower solution of (A.5.1) on I_{ab} if the reversed inequalities hold.

We shall now state the following fundamental result on hyperbolic differential inequalities.

Theorem A.5.2 *Assume that*

(i) $v, w \in C^{1,2}[I_{ab}, R]$, *and*

$$v_{xy} \leq f(x, y, v, v_x, v_y),$$

$$w_{xy} \geq f(x, y, w, w_x, w_y),$$

$$v(0, 0) \leq w(0, 0), v_x(x, 0) \leq w_x(x, 0), v_y(0, y) \leq w_y(0, y);$$

(ii) $f(x, y, z, p, q)$ is monotone nondecreasing in (z, p, q);

(iii) $f(x, y, z, p, q) - f(x, y, \tilde{z}, \tilde{p}, \tilde{q}) \leq L[(z - \tilde{z}) + (p - \tilde{p}) + (q - \tilde{q})]$, whenever $(z, p, q) \geq (\tilde{z}, \tilde{p}, \tilde{q})$ on I_{ab}, $L > 0$, a constant.

Then we have

$$(v, v_x, v_y) \leq (w, w_x, w_y) \text{ on } I_{ab}. \tag{A.5.2}$$

Utilizing the ideas of Theorem A.5.2, we can prove that any solution u of (A.5.1) lies in the sector $\langle v, w \rangle$. We can dispense with the assumption of monotonicity of f, if we strengthen the notion of lower and upper solutions. This is precisely the content of the next result.

Theorem A.5.3 *Assume that*

(i) $v, w \in C^{1,2}[I_{ab}, R]$ *are such that* $(v, v_x, v_y) \leq (w, w_x, w_y)$ *on* I_{ab};

(ii) $v_{xy} \leq f(x, y, z, p, q)$, $w_{xy} \geq f(x, y, z, p, q)$ *whenever* $(v, v_x, v_y) \leq (z, p, q) \leq (w, w_x, w_y)$ *on* I_{ab};

(iii) $v(0, 0) \leq u_0 \leq w(0, 0)$

$$v_x(x, 0) \leq \sigma'(x) \leq w_x(x, 0), x \in I_a$$

$$v_y(0, y) \leq \tau'(y) \leq w_y(0, y), y \in I_b;$$

(iv) f *satisfies*

$$| f(x, y, z, p, q) - f(x, y, \tilde{z}, \tilde{p}, \tilde{q}) | \leq L(| z - \tilde{z} | + | p - \tilde{p} | + | q - \tilde{q} |), \tag{A.5.3}$$

for $(x, y) \in I_{ab}$ *and* $z, \tilde{z}, p, \tilde{p}, q, \tilde{q} \in R$, *where* $L \geq 0$ *is a constant.*

Then, for any solution u of (A.5.1), we have

$$(v, v_x, v_y) \leq (u, u_x, u_y) \leq (w, w_x, w_y) \text{ on } I_{ab}. \tag{A.5.4}$$

The next result asserts the existence of a solution u of the IBVP (A.5.1), such that $(v, v_x, v_y) \leq (u, u_x, u_y) \leq (w, w_x, w_y)$.

Let D be the domain

$$D = \{(x, y, z, p, q) : (x, y) \in I_{ab}, (z, p, q) \in (\langle v, w \rangle, \langle v_x, w_x \rangle, \langle v_y, w_y \rangle)\}.$$

Theorem A.5.4 *Let* $f \in C[D, R]$. *Suppose that*

(i) $v_{xy} \leq f(x, y, z, p, q)$, $w_{xy} \geq f(x, y, z, p, q)$ *whenever* $(x, y, z, p, q) \in D$;

(ii) $v(0,0) \le u_0 \le w(0,0)$,

$$v_x(x,0) \le \sigma'(x) \le w_x(x,0), \quad x \in I_a$$

$$v_y(0,y) \le \tau'(y) \le w_y(0,y), \quad y \in I_b.$$

Then, there exists a solution u of (A.5.1) *satisfying*

$$(v, v_x, v_y) \le (u, u_x, u_y) \le (w, w_x, w_y) \quad \text{on } I_{ab}.$$

Corollary A.5.1 *Conclusion* (A.5.4) *in Theorems A.5.2 and A.5.3 holds if we assume that v and w are respectively the lower and upper solutions of* (A.5.1) *and $f(x,y,u,p,q)$ is monotone nondecreasing in (u,p,q).*

Remark A.5.1 If we assume in Theorem A.5.4 that f also satisfies a Lipschitz condition (A.5.3), then u is the unique solution of the IVP (A.5.1) satisfying (A.5.4).

For more details, see Ladde, et al. [47].

Bibliography

[1] R.A. Adams. *Sobolev Spaces*. Academic Press, New York, 1975.

[2] K. Ako. On the Dirichlet problem for quasilinear elliptic differential equations of the second order. *J. Math. Soc. Japan*, 13:45–62, 1961.

[3] H. Amman. On the existence of positive solutions of nonlinear elliptic boundary value problems. *Indian Univ. Math. J.*, 21:125–146, 1971.

[4] J.W. Bebernes and K. Schmitt. On the existence of maximal and minimal solutions for partial differential equations. *Proc. of Amer. Math. Soc.*, 73:211–218, 1979.

[5] R. Bellman. *Methods of Nonlinear Analysis*, volume II. Academic Press, New York, 1973.

[6] R. Bellman and R. Kalaba. *Quasilinearization and Nonlinear Boundary Value Problems*. American Elsevier, New York, 1965.

[7] T. Gnana Bhaskar, S. Köksal, and V. Lakshmikantham. Generalized quasilinearization for semilinear hyperbolic PDEs via variational method. *Nonlinear Studies*. to appear.

[8] T. Gnana Bhaskar, S. Köksal, and V. Lakshmikantham. Monotone iterative techniques for semilinear hyperbolic PDEs via variational method. to appear.

[9] R.D. Blakley and S.G. Pandit. On a sharp linear comparison result and an application to the nonlinear Cauchy problem. *Dyn. Sys. and Appl.*, 3:135–140, 1994.

[10] H. Brezis and F. Browder. Some properties of higher order Sobolev spaces. *J. Math. Pures Appl.*, 61:245–259, 1992.

[11] L. Byszewski. Monotone iterative method for a system of nonlocal initial-boundary parabolic problems. *J. of Math. Anal. and Appl.*, 177:445–458, 1993.

[12] L. Byszewski and V. Lakshmikantham. Monotone iterative technique for nonlocal hyperbolic differential problem. *J. Math. Phy. Sci.*, 26(4):345–359, 1992.

[13] S. Carl. A monotone iterative scheme for nonlinear reaction–diffusion systems having nonmonotone reaction terms. *J. of Math. Anal. and Appl.*, 134:81–93, 1988.

[14] S. Carl and C. Grossman. Monotone enclosure for elliptic and parabolic systems with monotone nonlinearities. *J. of Math. Anal. and Appl.*, 151:190–202, 1990.

[15] S. Carl and S. Heikkilä. *Nonlinear Differential Equations in Ordered Spaces*. Chapman and Hall/CRC, Boca Raton, FL, 2000.

[16] S. Carl and V. Lakshmikantham. Generalized quasilinearization and semilinear parabolic problems. *J. of Nonlinear Analysis*. to appear.

[17] S. Carl and V. Lakshmikantham. Generalized quasilinearization for quasilinear parabolic equations with nonlinearities of d.c. type. *J. Optim. Theory Appl.*, 109(1):27–50, 2001.

[18] S. Carl and V. Lakshmikantham. Generalized quasilinearization method for reaction–diffusion equations under nonlinear and nonlocal flux conditions. *J. of Math. Anal. and Appl.*, 271:182–205, 2002.

[19] J. Chabrowski. On nonlocal problems for parabolic equations. *Nagoya Math. J.*, 93:109–131, 1984.

[20] J. Chandra, F.G. Dressel, and D. Norman. A monotone method for a system of nonlinear parabolic differential equation. *Proc. Royal Soc. of Edinburgh*, 87(A):209–217, 1981.

[21] S.A. Chaplygin. *Collected Papers on Mechanics and Mathematics*. Nauka, Moscow, 1954.

[22] E.N. Dancer and G.S. Sweers. On the existence of a maximal weak solution for a semilinear elliptic equation. *Diff. Integral Eqns.*, 2:533–540, 1989.

[23] S.G. Deo and S.G. Pandit. Method of generalized quasilinearization for hyperbolic initial-boundary value problems. *Nonlinear World*, 3:267–275, 1996.

[24] J.I. Diaz. *Nonlinear Partial Differential Equations and Free Boundaries.* Pitman, Boston, 1985.

[25] E. DiBenedetto. *Partial Differential Equations.* Birkhauser, Dordrecht, 1995.

[26] P. Drábek and J. Hernandez. Existence and uniqueness of positive solutions for some quasilinear elliptic problems. *J. of Nonlinear Analysis*, 44:189–204, 2001.

[27] P. Drábek, A. Kufner, and F. Nicolosi. *Quasilinear Elliptic Equations with Degenerations and Singularities.* Walter de Gruyter, Berlin, 1997.

[28] Z. Drici, N. Kouhestani, and A.S. Vatsala. An extension of quasilinearization to reaction diffusion equations with fixed moments of impulses. *Dyn. of Cont., Disc. and Impulsive Sys.*, 7:33–50, 2000.

[29] Z. Drici, A.S. Vatsala, and N. Kouhestani. Generalized quasilinearization for impulsive reaction–diffusion equations with fixed moments of impulses, dynamic systems and applications. *Dyn. Sys. and Appl.*, 7:245–257, 1998.

[30] J. Duel and P. Hess. Criterion for the existence of solutions of nonlinear elliptic boundary value problems. *Proc. Roy. Soc. Edinburgh*, 74(A):49–54, 1975.

[31] L.H. Erbe, H.I. Freedman, X.Z. Liu, and J.H. Wu. Comparison principles for impulsive parabolic equations with applications to models of single species growth. *J. Austral. Math. Soc.*, 32(Ser. B):382–400, 1991.

[32] L.C. Evans. *Partial Differential Equations.* American Math. Society, Providence, RI, 1998.

[33] A. Friedman. *Partial Differential Equations of Parabolic Type.* Prentice-Hall, Englewood Cliffs, NJ, 1964.

[34] D. Gilbarg and N.S. Trudinger. *Elliptic Partial Differential Equations of Second Order.* Springer-Verlag, Berlin, 1977.

[35] J.L. Gouze and K.P. Hadeler. Monotone flows and ordered intervals. *Nonlinear World*, 1:23–34, 1994.

[36] F. Guglielmino and F. Nicolosi. Sulle w. soluzioni dei problemi di contorno per operatori ellitici degneri. *Riceche di Matematica*, 35:59–72, 1987.

[37] R. Kalaba. On nonlinear differential equations, the maximum operation, and monotone convergences. *J. Math. Mech.*, 8:519–574, 1959.

[38] H.B. Keller. Elliptic boundary value problems suggested by nonlinear diffusion processes. *Arch. Rat. Mech. Anal.*, 35:363–381, 1969.

[39] H.B. Keller and D.S. Cohen. Some positone problems suggested by nonlinear heat generation. *J. Math. Mech.*, 16:1361–1376, 1967.

[40] M. Kirane and Yu. V. Rogovenchko. Comparison results for systems of impulsive parabolic equations with application to population dynamics. *Annali Math. Pura Appl.* to appear.

[41] S. Köksal and V. Lakshmikantham. Monotone iterative technique for degenerate elliptic boundary value problems via the variational method. *Dyn. Sys. and Appl.* to appear.

[42] S. Köksal and V. Lakshmikantham. Unified approach to the monotone iterative technique for semilinear parabolic problems. *Dyn. of Cont., Disc. and Impulsive Sys.* to appear.

[43] S. Köksal and V. Lakshmikantham. A unified approach to the monotone iterative technique for quasilinear elliptic problems. *J. of Applicable Analysis*, 79:391–403, 2001.

[44] S. Köksal and V. Lakshmikantham. Unified approach to the monotone iterative technique for semilinear elliptic problems. *J. of Nonlinear Analysis*, 51:567–586, 2002.

[45] A. Kufner. *Weighted Sobolev Spaces*, volume 2. Wiley, Chichester, 1985.

[46] T. Kura. The weak supersolution–subsolution method for second order quasilinear elliptic equations. *Hiroshima Math. J.*, 19:1–36, 1989.

[47] G.S. Ladde, V. Lakshmikantham, and A.S. Vatsala. *Monotone Iterative Techniques for Nonlinear Differential Equations*. Pitman, Boston, 1985.

[48] O.A. Ladyzhenskaya. *The Boundary Value Problems of Mathematical Physics*. Springer-Verlag, New York, 1995.

[49] O.A. Ladyzhenskaya, V.A. Solonikov, and N.N. Uralćeva. *Linear and Quasilinear Equations of Parabolic Type.* Amer. Math. Soc. Transl., Providence, RI, 1968.

[50] O.A. Ladyzhenskaya and N.N. Uralćera. *Linear and Quasilinear Elliptic Equations.* Academic Press, New York, 1968.

[51] V. Lakshmikantham. Comparison results for reaction–diffusion equations in Banach space. In *Lecture Notes of Survey of Theoretical and Numerical Trends in Nonlinear Analysis*, G. Laterza, editor. Bari, Italy, 1979.

[52] V. Lakshmikantham, D.D. Bainov, and S. Simeonov. *Theory of Impulsive Differential Equations.* World Scientific Publishers, Singapore, 1989.

[53] V. Lakshmikantham and Z. Drici. Positivity and boundedness of solutions of impulsive reaction–diffusion equations. *J. of Computational and Appl. Math.*, 88:175–184, 1998.

[54] V. Lakshmikantham and S. Leela. Generalized quasilinearization and semilinear degenerate elliptic problems. *Disc. and Cont. Dyn. Sys.*, 7:801–808.

[55] V. Lakshmikantham and S. Leela. *Differential and Integral Inequalities*, volume I. Academic Press, New York, 1968.

[56] V. Lakshmikantham and S. Leela. Reaction–diffusion systems and vector lyapunov functions. *Diff. and Integral Eqns.*, I(1):41–47, 1988.

[57] V. Lakshmikantham and S. Leela. Generalized quasilinearization and extremal solutions of semilinear elliptic problems. *Nonlinear Studies*, 6:181–189, 1999.

[58] V. Lakshmikantham and S. Leela. Generalized quasilinearization and quasilinear elliptic problem. *J. of Nonlinear Analysis*, 46:1101–1110, 2001.

[59] V. Lakshmikantham and A.S. Vatsala. Stability results for solutions of reaction–diffusion equations by the method of quasi-solutions. *Applicable Analysis*, 12:229–235, 1981.

[60] V. Lakshmikantham and A.S. Vatsala. *Generalized Quasilinearization for Nonlinear Problems.* Kluwer Academic Publishers, Dordrecht, 1995.

[61] V. Lakshmikantham and A.S. Vatsala. Monotone flows and mixed monotone reaction diffusion systems. *Ana. Sti. Ale Univ. Al I Cuz, Iasi,* XLIV:263–269, 1998.

[62] V. Lakshmikantham and A.S. Vatsala. Generalized quasilinearization and semilinear elliptic boundary value problems. *J. of Math. Anal. and Appl.,* 249:199–220, 2000.

[63] V.K. Le. On some equivalent properties of sub-supersolutions in second order quasilinear elliptic equations. *Hiroshima Math. J.,* 28(2):373–380, 1998.

[64] V.K. Le and K. Schmitt. On boundary value problems for degenerate quasilinear elliptic equations and inequalities. *J. Diff. Eqns.,* 144:170–218, 1998.

[65] J.-L. Lions. *Equations differentielles operationelles et problémes aux limites.* Springer-Verlag, Berlin and Göttingen and Heidelberg, 1961.

[66] J.-L. Lions and E. Magenes. *Problémes aux limites nonhomogénes et applications,* volume 1. Dunod, Paris, 1968.

[67] R.G. Marshall and S.G. Pandit. Numerical analysis of monotone methods for nonlinear hyperbolic partial differential equations. *Proc. of UCF Conference,* 1999.

[68] F.A. McRae and V. Lakshmikantham. Generalized quasilinearization for semilinear parabolic partial differential equations. To appear.

[69] F.A. McRae and V. Lakshmikantham. Monotone iterative technique for semilinear parabolic partial differential equations. To appear.

[70] M. Müller. Uber das fundamental theorem in der theorie der gewohnlichen differential gleichungen. *Math. Z.,* 26:619–645, 1926.

[71] M.V.K. Murthy and G. Stampacchia. Boundary value problems for some degenerate elliptic operators. *Annali di Matematica,* 80:1–122, 1968.

[72] M. Nagumo. Uber das randwertproblem der nichtlinearen gewohnlichen differentiagleichung zweiter ordnung. *Proc. Phys. Math. Soc. Japan,* 24:845–851, 1942.

[73] S.G. Pandit. Variation of parameters formulas and maximum principles for linear hyperbolic problems in two independent variables. *Dyn. of Cont., Disc. and Impulsive Sys.*, 4:295–312, 1998.

[74] C.V. Pao. *Nonlinear Parabolic and Elliptic Equations*. Plenum, New York, 1992.

[75] C.V. Pao. Nonlinear elliptic systems in unbounded domains. *J. of Nonlinear Analysis*, 22(11):1391–1407, 1994.

[76] C.V. Pao. Quasisolutions and global attractor of reaction–diffusion systems. *J. of Nonlinear Analysis*, 26:1889–1903, 1996.

[77] C.V. Pao. Parabolic systems in unbounded domains. II equations with time delays. *J. of Math. Anal. and Appl.*, 225:557–586, 1998.

[78] D.H. Sattinger. Monotone methods in nonlinear elliptic and parabolic boundary value problems. *Indian Univ. Math. J.*, 21:979–1000, 1972.

[79] A.S. Vatsala and J. Wang. The generalized quasilinearization method for reaction–diffusion equations on an unbounded domain. *J. of Math. Anal. and Appl.*, 237:644–656, 1999.

[80] J. Wang. Monotone method for diffusion equations with nonlinear diffusion coefficients. *J. of Nonlinear Analysis*, 34:113–142, 1998.

[81] P. Wildenauer. Existence of a minimal solution and a maximal solution of nonlinear elliptic boundary value problems. *Indiana Univ. Math. J.*, 29:455–462, 1980.

[82] J. Wloka. *Partial Differential Equations*. Cambridge University Press, UK, 1987.

[83] E. Zeidler. *Nonlinear Functional Analysis and Its Applications*, volume IIA/IIB. Springer-Verlag, Berlin, 1990.

[84] L. Zhenhai. On mixed quasimonotone parabolic systems. *Annales Univ. Sci. Budapest*, 37:229–241, 1994.

[85] K. Zygourakis and R. Aris. Weakly coupled systems of nonlinear elliptic boundary value problems. *J. of Nonlinear Analysis*, 6:555–569, 1982.

Index

9 780367 395407